Oral Delivery of Macromolecular Drugs

Andreas Bernkop-Schnürch
Editor

Oral Delivery of Macromolecular Drugs

Barriers, Strategies and Future Trends

Springer

Editor
Andreas Bernkop-Schnürch
Universität Innsbruck
Fakultät für Chemie und Pharmazie
Inst. Pharmazie
Innrain 52, 6020 Innsbruck, Austria
andreas.bernkop@uibk.ac.at

ISBN 978-1-4419-0199-6 e-ISBN 978-1-4419-0200-9
DOI 10.1007/978-1-4419-0200-9
Springer Dordrecht Heidelberg London New York

Library of Congress Control Number: 2009929310

Printed on acid-free paper

Springer is part of Springer Science+Business Media (www.springer.com)

Preface

Due to a rapid development in biotechnology more and more macromolecular drugs such as therapeutic peptides, oligosaccharides and nucleic acids are entering the pharmaceutical arena representing unprecedented challenges from the drug delivery point of view. One of the likely greatest challenges is their oral administration presenting a series of attractive advantages. These advantages are in particular of high relevance for the treatment of pediatric patients and include the avoidance of additional risks, pain and discomfort associated with injections. Furthermore, oral formulations are less expensive to produce, as they do not need to be manufactured under sterile conditions. The oral administration is by far the most favored one. The majority (84%) of 50 most-sold pharmaceutical products in US and Europe markets are given orally. Although there have been major advances in delivering macromolecular drugs in humans by other non-invasive routes, including the pulmonary delivery of insulin, we did so far not succeed in the development of oral delivery systems for these therapeutic agents. Apart from a few exceptions such as ciclosporin, desmopressin, chondroitin sulphate and bromelain macromolecular drugs cannot be administered orally. The development of oral formulations for macromolecular drugs is therefore highly on demand. Due to the scientific progress having been made within the 1990s and this decade numerous novel oral delivery systems for macromolecular drugs are meanwhile subject of clinical trials. This book addresses the most critical issues for a successful oral delivery of macromolecular drugs by a detailed characterisation of the 'enemy's strength'. Furthermore, an overview on the likely most promising strategies to overcome barriers encountered with the gastrointestinal (GI) tract is provided. These barriers are mainly the enzymatic barrier (Chapter 1), the mucus gel layer barrier (Chapter 2) and the absorption barrier (Chapter 3).

The enzymatic barrier is based on various classes of enzymes including proteases/peptidases, nucleases, glycosidases and lipases. Taking the most important macromolecular drugs into consideration, which are likely candidates for oral administration, the enzymatic barrier is primarily represented by proteases/ peptidases and nucleases. Proteases/peptidases are on the one hand based on luminally secreted proteases including pepsin, trypsin, chymotrypsin, elastase and carboxypeptidases A and B and on the other hand based on membrane-bound peptidases including various endo- as well as amino- and carboxypeptidases. In the colon numerous additional enzymes originating from the local microflora have to be taken into consideration. In terms of nucleases the enzymatic barrier is much less characterized.

The mucus gel layer barrier is based on mucus glycoproteins being crosslinked via disulphide bonds. Macromolecular drugs have to diffuse through this 50–200 μm thick three-dimensional network in order to reach the absorption membrane. In addition, due to its negative net charge being based on sialic acid and sulphonic acid substructures therapeutic macromolecules exhibiting a positive charge can be immobilized on the mucus gel barrier because of ionic interactions.

As the GI mucosa is highly vascularized, macromolecular drugs have to 'merely' permeate the epithelial cell layer in order to reach the systemic circulation. More lipophilic and relatively small drugs are primarily absorbed via the transcellular route, whereas more hydrophilic and relatively bigger drugs enter the systemic circulation via the paracellular route. In addition, efflux pumps can significantly further reduce the absorption of macromolecular drugs such as the case for ciclosporin. Because of this absorption barrier the size of orally administered macromolecular drugs is more or less limited to up to 10 kDa in maximum. Macromolecules greater than that are still absorbed to a certain extent; however, gained oral bioavailabilities are in most cases not anymore of therapeutic and/or commercial relevance.

Strategies to overcome the enzymatic barrier (Chapter 4) include the design of macromolecular drugs remaining more stable in the GI environment. From the drug delivery point of view, a protective effect towards enzymatic degradation can be achieved by using auxiliary agents such as enzyme inhibitors and/or polymers displaying enzyme inhibitory properties. In particular in combination with appropriate dosage forms shielding towards enzymatic attack such as micro- and nanoparticulate delivery systems and patch systems sufficient protection towards this barrier can be achieved.

Strategies to overcome the absorption barrier focus on the other hand on low molecular mass permeation enhancers (Chapter 5) such as medium chain fatty acids, which can still be regarded as a kind of gold standard. As low molecular mass permeation enhancers are per se rapidly uptaken from the gastrointestinal mucosa, however, the macromolecular drug is to a considerable high extent left alone behind in the gastrointestinal tract. In addition, local and systemic toxic side effects of low molecular mass permeation enhancers cannot be excluded. In contrast, polymeric permeation enhancers (Chapter 6) are simply too big to be absorbed from the GI tract. Consequently, systemic toxic side effects can be excluded. More recently various excipients could be identified as potent efflux pump inhibitors which can be subdivided into low molecular mass efflux pump inhibitors and polymeric efflux pump inhibitors (Chapter 7). Certain polymeric excipients exhibit various favourable properties for oral macromolecular delivery such as high mucoadhesive, enzyme inhibitory, permeation enhancing and efflux pump inhibitory properties. Among them thiolated polymers – designated thiomers – showed most encouraging results (Chapter 8). From the drug delivery point of view certain formulations could be identified to show beneficial properties in order to improve the oral uptake of macromolecular drugs. Matrix tablets, patch systems, micro- and nanoparticulate delivery systems and liposomes seem to be most promising. Micro- and nanoparticles (Chapter 9) offer the advantage to penetrate into the mucus gel layer. Consequently, their gastrointestinal residence time is strongly prolonged resulting in a prolonged period of time for drug absorption. Furthermore, a presystemic degradation, for instance, of therapeutic peptides or oligonucleotides by luminally secreted enzymes can be avoided. The likely best protection towards enzymatic degradation in the GI tract is provided by liposomal formulations (Chapter 10), which can also guarantee an intimate contact with the mucosa when being coated with a mucoadhesive polymer.

Special and to some extent different approaches are needed for oral immunization utilizing various types of antigens and immunostimulating auxiliary agents (Chapter 11). Particulate delivery systems such as nanoparticles and liposomes accumulating in the region of Peyer's patches seem to be highly beneficial. Oral nucleic acid delivery systems are designed for local and systemic treatment (Chapter 12). The systemic delivery of nucleic acids via the oral route is likely the most challenging aim in oral macromolecular delivery.

The combination of suitable and comparatively more stable macromolecular drugs (**I**), highly efficient, multifunctional and non-toxic excipients (**II**) and appropriate formulations (**III**) is certainly the key for success in oral macromolecular delivery. On the one hand macromolecular drugs can be produced more and more effectively making also low oral bioavailabilities in the range of 0.5–5% commercially interesting. The oral bioavailability of desmopressin tablets, which are for almost 20 years on the market, for instance, is just 0.5%. On the other hand, oral macromolecular drug formulations are becoming more and more efficient. Taking these developments into consideration, the number of oral macromolecular delivery systems entering the market will increase considerably over the years. 'Invasive-to-oral-conversions' promise great rewards for those investing in this market. This book should encourage and motivate scientists in academia and industry to move on or intensify their activities in this challenging research field of great future.

Finally, I wish to thank all the contributing authors for their excellent chapters, which were certainly not easy to write in such a complex and challenging field. Moreover, I wish to thank the Editorial Director Andrea Macaluso from Springer Science and Business Media for inviting me to edit this book.

Innsbruck, Austria Andreas Bernkop-Schnürch

Contents

Contributors

Andreas Bernkop-Schnürch Institute of Pharmacy, University of Innsbruck, Innsbruck, Austria, andreas.bernkop@uibk.ac.at

Gerrit Borchard Laboratory of Pharmaceutics and Biopharmaceutics, School of Pharmaceutical Sciences, University of Geneva, Geneva, Switzerland, gerrit.borchard@unige.ch; gerrit.borchard@pharm.unige.ch

Florian Föger Diabetes Research Unit, Oral Formulation Research, Novo Nordisk A/S, Målϕv, Denmark, faf@novonordisk.com

Hans E. Junginger Department of Pharmaceutics, Faculty of Pharmaceutical Sciences, Naresuan University, Phitsanulok, Thailand, hejunginger@yahoo.com

Ronny Martien Department of Pharmaceutics, Faculty of Pharmacy, Gadjah Mada University, Sekip Utara, Yogyakarta, Indonesia, ronnymartien@ugm.ac.id

Gioconda Millotti Department of Pharmaceutical Technology, Leopold-Franzens-University Innsbruck, Innsbruck, Austria, gioconda.millotti@uibk.ac.at

Peter Moyle School of Chemistry & Molecular Biosciences, School of Pharmacy, University of Queensland, Brisbane, QLD, Australia, pmoyle@rockefeller.edu

Colleen Olive School of Chemistry & Molecular Biosciences, School of Pharmacy, University of Queensland, Brisbane, QLD, Australia, colleen.olive@qimr.edu.au

Juan Perez-Vilar Cystic Fibrosis/Pulmonary Research and Treatment Center, University of North Carolina at Chapel Hill, Chapel Hill, NC, USA, jp_vilar@yahoo.com

Pavla Simerska School of Chemistry & Molecular Biosciences, School of Pharmacy, University of Queensland, Brisbane, QLD, Australia, p.simerska@uq.edu.au

Pornsak Sriamornsak Department of Pharmaceutical Technology, Faculty of Pharmacy, Silpakorn University, Nakhon Pathom, Thailand, pornsak@email.pharm.su.ac.th

Hirofumi Takeuchi Laboratory of Pharmaceutical Engineering, Department of Drug Delivery Technology and Science, Gifu Pharmaceutical University, Gifu, Japan, takeuchi@gifu-pu.ac.jp

Jringjai Thongborisute Laboratory of Pharmaceutical Engineering, Gifu Pharmaceutical University, Gifu, Japan, jringjai@gmail.com

Istvan Toth School of Chemistry & Molecular Biosciences, School of Pharmacy, University of Queensland, Brisbane, QLD, Australia, i.toth@uq.edu.au

Claudia Vigl Thiomatrix Forschungs Beratungs GmbH, Innsbruck, Austria, c.vigl@thiomatrix.com

Martin Werle Laboratory of Pharmaceutical Engineering, Department of Drug Delivery Technology and Science, Gifu Pharmaceutical University, Gifu, Japan, martin.werle@uibk.ac.at

John Woodley Biopharmtech, Toulouse, France, woodley.john@orange.fr

About the Editor

Andreas Bernkop-Schnürch is the Dean of the Faculty of Chemistry and Pharmacy and Professor of Pharmaceutical Technology at the University of Innsbruck, Austria. His main interests focus on non-invasive drug delivery systems, multifunctional polymers such as thiomers, mucoadhesion and permeation enhancement. He is author of more than 200 original research and review articles and received the Research-Award of the City of Vienna (Vienna), EURAND-Award (Boston), Best of Biotech Award (Vienna), MBPW-Award (Munich), PHÖNIX Award (Mannheim), Houska-Award (Vienna) and Austrian-Nanoaward (Vienna). In addition, Andreas Bernkop-Schnürch is founder and owner of the following companies: ThioMatrix GmbH, Dr. K. Schnürch KG and Green River Polymers GmbH.

Chapter 1
Enzymatic Barriers

John Woodley

Contents

Abstract The major enzymatic barrier to the absorption of macromolecules, particularly therapeutic peptides, is the pancreatic enzymes: the peptidases, nucleases, lipases and esterases that are secreted in considerable quantities into the intestinal lumen and rapidly hydrolyse macromolecules and lipids. In the case of the peptidases, they work in a co-ordinated fashion, whereby the action of the pancreatic enzymes is augmented by those in the brush borders of the intestinal cells. The sloughing-off of mucosal cells into the lumen also furnishes a mixture of enzymes that are a threat to macromolecules. As the specificity and activity of the enzymes are not always predictable, during pharmaceutical development it is important to test the stability of therapeutic macromolecules, and novel macromolecular-containing or lipid-containing formulations, in the presence of mixtures of pancreatic enzymes and bile salts, or in animal intestinal washouts or ideally, aspirates of human intestinal contents.

J. Woodley (✉)
Biopharmtech, Toulouse, France
e-mail: woodley.john@orange.fr

A. Bernkop-Schnürch (ed.), *Oral Delivery of Macromolecular Drugs*,
DOI 10.1007/978-1-4419-0200-9_1,

1.1 Introduction

Due to advances in biotechnology the last decades have seen an explosion in 'biopharmaceutics': the production of biological molecules with pharmacological activity. Most of these are active peptides of varying sizes, and many are now on the market to treat a range of diseases. The FDA has approved over 30 different proteins produced by recombinant techniques and there are probably many more in the pipeline (George and Abraham 2006). Not all are new: the best known and most widely administered is insulin, which has been used to treat diabetes for nearly a century. Calcitonin is a peptide hormone that has long been used for the treatment of osteoporosis in post-menopausal women. At the present time, none of these active peptides are available in a form that patients can take orally. Much effort by researchers and pharmaceutical companies has been devoted to trying to find ways of making these molecules bioavailable and active via the oral route, which, as is well known, is the most convenient and acceptable way for drugs to be administered. Insulin has probably been the most studied, and with the forecasts of the prevalence of type 2 diabetes escalating, particularly in rich Western societies, there is both a clinical need and market opportunity for an oral formulation of insulin. While no such product has yet reached the market place, various preparations are in clinical trials.

While the emphasis in recent years has been on peptides, and to a lesser extent, proteins, there are new players in the field: nucleic acids and their derivatives, as described in further detail in Chapter 12. This is due to the rapid advances and perceived potential of gene therapy whereby specific genes, in the form of DNA and/or its analogues, are introduced into cells to make them synthesise molecules with therapeutic potential or to change cell function. A promising therapeutic development with wide possible applications is the ability to 'switch off' or silence specific gene expression using small, usually 15–25 bp, oligonucleotide analogues (antisense oligonucleotides, ASOs) that bind to mRNA by Watson–Crick base pairing. The analogues are chemically modified to make them more stable, and recently it has been shown that a methoxyethyl-modified antisense oligonucleotide could be absorbed from the GI tract of human volunteers in the presence of an absorption enhancer (Tillman et al. 2008). A more recent development for silencing genes that is seen by some researchers as having considerable therapeutic potential is RNA interference, whereby translation within cells is modified by the delivery to cells of small fragments of double-stranded RNA (small interfering RNA, siRNA) (Felekkis and Deltas 2006). To date, such therapeutic applications are still mainly at the research stage and require the modification of cells ex vivo, with their return to the body following modification, or the administration of the oligonucleotides and their analogues by parenteral routes. This is usually carried out using viruses, polymeric constructions or liposomes to protect the therapeutic oligonucleotide and enhance its entry into the target cells.

Of particular interest for the oral administration of macromolecules, be they peptides or nucleic acids, is the potential of oral vaccines that could revolutionise the health care of millions of people, particularly in developing countries.

Other macromolecules that may eventually be taken orally by patients are also making appearances on the scene in the formulation of drugs for the oral route, notably polymers, from both natural (such as chitosan) and synthetic (such as poly(D,L-lactide-*co*-glycolide) PLGA) sources. Chitosan is a β1,4-linked copolymer of *N*-acetyl-D-glucosamine and D-glucosamine prepared by the deacylation of chitin, which is the second (after cellulose) most abundant polysaccharide in nature: the polymeric base of the shells of crustacea and exoskeleton of insects. Chitosan has been extensively used in experimental oral formulations, notably microparticles and nanoparticles, for the delivery of peptides such as insulin and calcitonin as well as oligonucleotides (Hejazi and Amiji 2003). Derivatives of chitosan have also been prepared with enhanced bioadhesive properties for oral delivery (see Chapter 8). Similarly, PLGA has been extensively researched and used as a material to make nanoparticles and microparticles for the oral delivery of peptides and other drugs and there are many reports in the literature of such studies. Other natural macromolecules such as carrageen and plant gums are actively being studied as coatings for pharmaceutical formulations to deliver drugs to the colon (see Section 1.7).

Modified and natural lipids and polymeric surfactants are increasingly being used in pharmaceutical formulations to increase the bioavailability of 'difficult' drugs: 'difficult' usually because they have very poor solubility in aqueous environments. While these molecules may not themselves be biologically active, what happens to them in the gastrointestinal (GI) tract may influence their intended function, especially if they are acting to protect the active principle, for example, in the case of therapeutic peptides.

The first major hurdle that macromolecular drugs have to overcome in the gastrointestinal tract is that of the digestive enzymes. The intestinal mucosa is a formidable barrier separating the organism from its external environment, and as such only small low molecular weight molecules can cross the barrier in any appreciable quantities, hence the evolution of the highly sophisticated and efficient process of digestion that breaks down most naturally occurring macromolecules into their component low molecular weight sub-units. This process of digestion is the co-ordinated action of the digestive enzymes found at various sites in the gastrointestinal tract.

The purpose of this chapter is to document the nature of the enzyme barrier that macromolecular drugs will encounter during their passage down the human gastrointestinal tract. I will discuss the peptidases that digest peptides and proteins and the nucleases that will hydrolyse nucleic acids and also briefly consider other enzymes that may affect the behaviour of the newer generation of pharmaceutical formulations. It is very important to consider both the *qualitative* aspects of the problem, that is, the specificity of the digestive enzymes for particular chemical bonds and particular macromolecular substrates, and the *quantitative* aspects, that is, how much enzyme activity will be present at any particular site within the digestive tract.

At the end I will discuss how the knowledge of the enzymology of the GI tract can be put to good use in devising the essential in vitro tests that are required to screen the stability of potential therapeutic peptides and oligonucleotides and thus help in the development of formulation strategies to protect them.

For a very detailed consideration of the problem with respect to peptide and protein delivery, the reader is referred to a previous review article published in 1994 (Woodley 1994).

1.2 The Peptidases

Peptidases are enzymes that hydrolyse the bond between two amino acids in a peptide or protein chain. The term 'peptidase' is used to define any enzyme that hydrolyses such a peptide bond, regardless of the size of the substrate. In the author's view, the terms 'proteases' and 'proteinases' may lead to confusion by implying that such enzymes may attack proteins but not small peptides. This is generally not the case: most peptidases will hydrolyse peptide bonds in proteins and small peptides, albeit often at different rates. What is important is the specificity of the bonds being hydrolysed, and that is determined by the amino acids adjacent to the bond. Peptidases can be subdivided into two groups on the basis of their mode of action: endopeptidases and exopeptidases. Endopeptidases hydrolyse peptide bonds at the interior of the peptide chain, whereas exopeptidases hydrolyse terminal peptide bonds, that is, they remove amino acids from the ends of the chains. Carboxypeptidases remove amino acids from the carboxy terminus and aminopeptidases from the amino terminus of the chain. A further class of exopeptidases remove dipeptides (or occasionally tripeptides) from the ends of protein and peptide chains: termed peptidyl dipeptidases if from the C terminus and dipeptidyl peptidases if from the N terminus. Peptidase specificity is usually described in terms of the amino acids on the amino side of the target peptide bond for endopeptidases and by the amino acids removed from the chain ends by exopeptidases. These specificities can be broad or narrow in terms of the number of sites of actions that any particular peptidase will display. Thus the pancreatic enzyme trypsin has a narrow specificity in that it hydrolyses the peptide bonds on the carboxyl side of the amino acids, namely, lysine and arginine. On the other hand, another pancreatic peptidase, elastase, hydrolyses peptide chains at a greater number of sites, all involving aliphatic amino acids (see Table 1.1). A knowledge of the specificity of the enzymes that a therapeutic peptide will encounter is very important in assessing the vulnerability of the particular peptide. It must also be borne in mind that for many therapeutic peptides the biological activity may be lost by just one or two cleavages of the molecule. For example, cleavage of one cystinyl–phenylalanyl bond in the biologically active peptide atrial natriuretic peptide (ANP) causes loss of biological activity (Seymour et al. 1987). Thus if a biologically active peptide contained an arginine or lysine residue at a key site, it would be attacked by the pancreatic enzyme trypsin, which is the most abundant enzyme in the upper part of the small intestine, and the biological activity would be rapidly destroyed. If, on the other hand the amino acid at a key site was proline this would not be attacked by any of the pancreatic enzymes and they would therefore pose less of a threat. The wheat protein gliadin, responsible for the malabsorption syndrome celiac disease, contains high levels of the amino acid proline and as a consequence after digestion by pancreatic enzymes, the resulting digest contains relatively large peptides, whereas other 'normal' proteins will be degraded to much smaller peptides and amino acids.

Table 1.1 Bond specificity of pancreatic peptidases

Enzyme	**Bond hydrolysed**
Trypsin	—○—●↓○—○— Arginine, lysine
Chymotrypsin	—○—●↓○—○— Phenylalanine, tyrosine (leucine, methionine, aspartic, glutamine, tryptophan)
Elastase	—○—●↓○—○— Alanine, glycine, leucine, valine, isoleucine
Carboxypeptidase A	○—○—○—○↓● Phenylalanine, tyrosine, isoleucine (threonine, glutamic, histidine, alanine)
Carboxypeptidase B	○—○—○—○↓● Lysine, arginine, hydroxylysine, ornithine

1.3 The Nucleases

The nucleases are enzymes that hydrolyse nucleic acids, either deoxyribonucleases (DNases) that have DNA as the substrate or ribonucleases (RNases) that have ribonucleic acids as the substrate. The DNases hydrolyse the phosphodiester linkages between the deoxyribose molecules of DNA, and similarly, the RNases attack the equivalent bonds in RNA. There are many nucleases found in mammalian tissues, and as in the case of the peptidases, they can be divided into the categories endo and exo based on whether they attack bonds in the interior of the nucleic acid molecule or remove nucleosides from the end termini of the chains. They may also have broad specificity whereby they hydrolyse the oligonucleotide chain at multiple sites or they may be highly specific for certain base sequences. The latter is not usually the case when the nucleases are functioning in a digestive role, but more so when they are part of complex interactions taking place during cell division, transcription and translation. The most important nucleases in the digestive processes are DNase I and RNase A.

1.4 Other Enzymes

Lipases are enzymes that hydrolyse triglycerides in fats and phospholipases, as the name indicates, hydrolyse phospholipids. Lipases remove long-chain fatty acids from triglycerides, and they are also frequently described as having esterase activity. There are also specific esterases described in the GI tract, for example, carboxylesterase that is secreted by the pancreas. These enzymes are included in the discussion because their activity may be relevant to the use of macromolecular materials in novel formulations, particularly for oral peptide and nucleic acid delivery.

A major group of macromolecules that will be present in the GI tract through ingestion in the diet are the polysaccharides. The digestion/hydrolysis of polysaccharides in the upper gastrointestinal tract is an under-researched and under-documented subject. Apart from the digestion of the simple polysaccharides, starch and glycogen that are polyglucose with 1-4α and 1-6α glycosidic links between the glucose monomers, there seems to be no information in the literature concerning the digestion of more complex polysaccharides in the upper GI tract. This seems to be very strange given that probably the majority of proteins ingested in the human diet are in fact glycoproteins, that is, they have groups of polysaccharides attached to the protein backbone. In eukaryotes the polysaccharide arrays attached to many proteins, particularly cell-surface proteins, represent a highly variable molecular recognition system: the simplest and best-known example being the human blood groups. In addition, there are many complex and variable polysaccharides in the diet from plant sources. Thus apart from salivary and pancreatic amylases, which are doted with limited specificity, i.e. 1-4α bonds between glucoses, and some α-glucosidase activity and limited disaccharidase activity on the brush border there are no other enzymes documented to be present in the upper small intestine able to hydrolyse the myriad of sugar–sugar bonds present in complex carbohydrates and glycoproteins. There will be some polysaccharidase activity present in the lumen emanating from the lysosomes of sloughed-off cells (see below), but the contribution to the digestion of complex carbohydrates is unknown. It seems that these complex polysaccharides are mostly hydrolysed by enzymes produced by the bacteria that inhabit the lower intestine in vast numbers (see Section 1.7).

1.5 Where Are the Enzymes in the GI Tract?

Figure 1.1 shows the major sites of enzyme activity in the GI tract, and we will consider each of these in turn. While most of the enzymes that hydrolyse macromolecules enter the gut in the pancreatic fluid and hence are found in the lumen of the gut, there is significant peptidase activity located on the membranes of the intestinal cells, the so-called brush border. Consideration should also be given to the enzymes that are located inside the cells of the intestinal mucosa, namely, the epithelial cells or enterocytes. This is for two reasons: first, the intestinal mucosa has a turnover of 3–6 days in humans and this means that the enterocytes are constantly being sloughed-off into the lumen of the gut. Thus intracellular enzymes and brush border enzymes will be found in the lumen of the gut, though the precise quantity is difficult to assess (see later in Section 1.6).

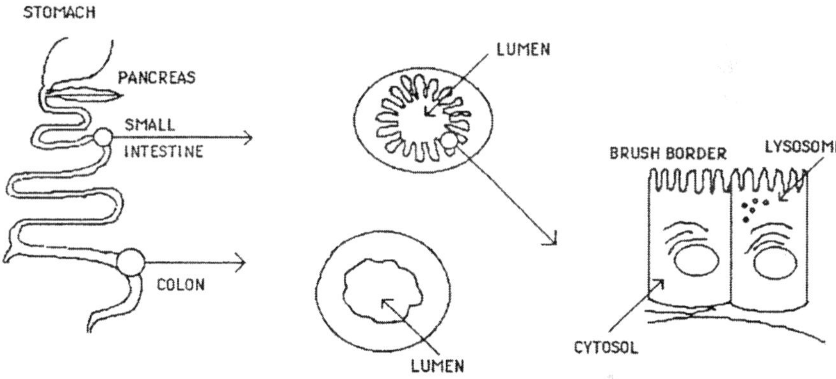

Fig. 1.1 Location of enzymatic activity in the gastrointestinal tract.

The second reason to consider the intracellular enzymes is because of the absorption mechanisms by which macromolecules may cross the intestinal mucosa. There are two possible mechanisms: for relatively small macromolecules such as therapeutic peptides and oligonucleotides, they *may* be able to pass via the paracellular route between the cells, particularly if some absorption enhancers are present. For example, Tsutsumi et al. (2008) have shown in vitro that in the presence of chenodeoxycholate as an absorption enhancer modified oligonucleotides with molecular weights of nearly 3,700 and 7,400 Da could cross rat intestine via the paracellular route. In the case of the paracellular route the macromolecules will not be exposed to the intracellular enzymes and thus they will not be subject to intracellular hydrolysis. However, macromolecules, especially larger ones, will cross the epithelium via the transcellular mechanism of endocytosis. In this case they will be taken into the lysosomes that contain a formidable array of digestive enzymes (see later in Section 1.5.2).

1.5.1 *The Stomach*

The stomach is the site of the initiation of protein digestion due to the presence of the peptidase enzyme pepsin. Pepsin was the first peptidase to be described and it is an unusual enzyme in that it is active at a very low pH, with the optimum being around pH 2.0 for most proteins. It is a broadly specific endopeptidase with a specificity directed towards cleaving bonds at phenylalanine, methionine, leucine, tryptophan and other hydrophobic residues. It must be noted, however, that much of the literature describing peptidase specificities is very old and is often based on the study of a limited number of peptide – or protein–enzyme interactions. It is highly likely that the sites of action of broadly specific peptidases will vary according to the substrate, for example, due to the amino acids adjacent to the amino acids at the primary site of attack, and the information on bond specificity may need updating in the literature. This has been elegantly demonstrated recently in a study by Werle et al. who showed that while the 34-amino acid peptide teriparatide contained no 'classic' pepsin sites, experimentally it was rapidly degraded by pepsin (Werle et al. 2006).

In terms of oral systems for peptide delivery, modern formulation technology ensures that the activity of pepsin can be avoided by using polymeric enteric coatings to protect the target pharmaceutical peptide. These coatings are resistant to the acid environment of the stomach but dissolve when the pH rises in the small intestine. It is *absolutely essential* that any formulation being considered for the oral delivery of therapeutic peptides be coated with such materials to ensure that the peptide is protected from the action of pepsin.

The stomach also contains an active lipase that can hydrolyse triglycerides and so may start acting against the components of the lipidic formulations that are becoming more common. While this enzyme may be less abundant and less active than pancreatic lipase, there is sufficient activity to substantially initiate lipid hydrolysis (Carriere et al. 2000).

1.5.2 *The Lumen of the Small Intestine*

The lumen of the intestine, especially that of the jejunum, contains the highest concentration and diversity of enzymes that are a threat to macromolecular drugs and macromolecular pharmaceutical formulations. The main sources of these enzymes will be in the pancreatic juice and the sloughed-off epithelial cells.

1.5.2.1 Pancreatic Enzymes

The major group of enzymes produced by the pancreas and secreted into the duodenum as an aqueous bicarbonate solution is the peptidases. There are three endopeptidases, trypsin, chymotrypsin and elastase, and two exopeptidases, carboxypeptidases A and B. These enzymes are secreted as non-active precursor molecules (zymogens). On the membrane of the small intestinal cells is located the highly specific peptidase enzyme, enteropeptidase, that activates the precursor of trypsin (trypsinogen) by cleaving a small fragment off the molecule. The activated trypsin then activates the other pancreatic peptidases as well as being autoactivating. Note, therefore, that for the secreted peptidases of the pancreas, they are only fully active *after* contact with the surface of the intestinal epithelium.

The generally accepted and preferred specificities of the pancreatic peptidases are shown in Table 1.1. While chymotrypsin has the preferred sites of phenylalanine it has also been reported to cleave at leucine, methionine, asparagine, glutamine, and tryptophan (Berezin and Martinek 1970). This has been clearly demonstrated by Werle et al. (2006), who showed that while the peptide teriparatide contained neither of the preferred cleavage sites, the presence of several leucines, asparagines and methionines meant that it was rapidly hydrolysed by pure chymotrypsin. It is very important to note that the enzymes work together in a *co-ordinated* way: the endopeptidases trypsin and chymotrypsin have relatively narrow specificities but after they split bonds in the peptide chain, they leave fragments that are then substrates for the carboxypeptidases A and B. The removal of these particular amino acids will then allow further attack by other carboxypeptidases, for example, those on the surface of the epithelial cells in the brush border membrane. Meanwhile, the amino termini of the peptides generated by the endopeptidase attack will also be sites for hydrolysis by the many aminopeptidases present in the brush border. In addition, the attack of one endopeptidase on a folded protein or peptide may then open up sites to access by another endopeptidase. This idea of co-ordinated attack has a very important bearing on the design of in vitro tests of macromolecular stability as discussed at the end of the chapter. The endopeptidase elastase is considered to have a broader specificity than trypsin and chymotrypsin, but as noted earlier, the published information on peptidase site specificity may need updating.

The other major enzymes in pancreatic juice of relevance to our discussion are the nucleases. Human pancreatic DNase I has a molecular weight of around 30 kDa, shows maximal activity in the pH range 7.2–7.6 and requires metal ions such as Mn^+, Mg^+ or Co^+ in the presence of Ca^+. It is an endonuclease that hydrolyses the phosphodiester linkages in both single- and double-stranded DNA at multiple sites (Funakoshi et al. 1977; Takeshita et al. 2000).

The main RNase is RNase A, one of the smallest enzymes known (molecular weight: 15 kDa) and remarkably stable, able to survive boiling. It attacks single-stranded RNA, cleaving the 3′-end of unpaired cytosine and uracil residues.

The pancreas produces several lipases, the 'classical' lipase (pancreatic lipase, PL) and the pancreatic lipase-related proteins 1 (PLRP1) and 2 (PLRP2). In addition there is a protein named colipase, which is necessary for the activity of PL. PLRP1 seems to be devoid of lipase activity, while PLRP2 hydrolyses retinyl palmitate and galactosyl lipids (De Caro et al. 2004; Reboul et al. 2006). In humans, the expression in pancreas of PLRP2 is much lower than PL (Giller et al. 1992). In addition to the lipases there is also esterase activity in the form of an enzyme called carboxylesterase or carboxy ester lipase (CEL). As discussed later these enzymes may be relevant to the stability of macromolecular pharmaceutical formulations.

1.5.2.2 Cellular Enzymes

The brush border membrane (BBM): The surface area of the absorbing cells (enterocytes) of the small intestine is considerably increased by the surface membrane being folded into microvilli. This membrane structure is known as the microvillus membrane (MVM) or brush border membrane (BBM) and it is rich in hydrolytic (digestive) enzymes of which the majority are peptidases. These are at least 12 in number, and comprise endopeptidases and both amino- and carboxy-exopeptidases. The specificity of these enzymes is shown in Table 1.2. It can be seen that between them these enzymes can remove a wide range of amino acids from the termini of peptide chains, particularly aminopeptidase N, which has a broad specificity and is the most abundant of the brush border peptidases. The substrates will be small peptides that have been produced as a result of the action of the gastric and pancreatic enzymes on dietary proteins. The physiological function of the brush border peptidases is to reduce any peptides down to single amino acids, dipeptides and possibly tripeptides. These are the maximum-sized units that are absorbed across the cell membranes and subsequently into the cells by specialised transport proteins. Any therapeutic peptides that have the target amino acids at either their carboxy or amino termini will also be hydrolysed. In addition, the presence of endopeptidase activity, particularly the widely distributed neutral endopeptidase 24.11, ensures that larger peptide and protein fragments are capable of being hydrolysed at the level of the brush border as well as in the lumen of the intestine. Thus the brush border membrane constitutes the second major enzyme barrier to therapeutic peptides. As in the case of the pancreatic peptidases, again it is important to reiterate that the brush border membrane presents a panel of enzymes that work in a co-ordinated manner: the action of one enzyme may create substrates for its neighbour.

The lysosomal enzymes: The lysosomes are membrane vesicles ubiquitous to mammalian cells and contain a panoply of hydrolytic enzymes, estimated to be over 60 in number, that function to 'digest' practically any biological macromolecule. They are important to the discussion of oral macromolecular drug delivery for two reasons. First, any macromolecules that escape digestion by the pancreatic and brush border enzymes are likely to be taken up into the epithelial cells by the process of endocytosis. In this process, the apical membrane invaginates and the target molecules enter endocytic vesicles that then fuse with the lysosomes and are subjected to intracellular hydrolysis by the lysosomal enzymes. Second, the sloughing-off of the epithelial cells means that the lysosomal enzymes will be released into the lumen of the intestine. They *may* be

Table 1.2 Bond specificity of brush border peptidases

Enzyme	**Bond hydrolysed**
Endopeptidases	
Endopeptidase 24.11	—○—○—○↓●—○—○— Hydrophobic amino acids
Endopeptidase 24.18	—○—○—○↓●↓○—○— Aromatic amino acids
Exopeptidases: amino terminus	
Aminopeptidase N	●↓○—○—○—○— Many different amino acids
Aminopeptidase A	●↓○—○—○—○— Aspartic, glutamic
Aminopeptidase P	●↓○—○—○—○— Proline
Aminopeptidase W	●↓○—○—○—○— Tryptophan, tyrosine, phenylalanine
γ-Glutamyl transpeptidase	●↓○—○—○—○— γ-Glutamic acid
Dipeptidyl peptidase IV	○—●↓○—○—○—○— Proline, alanine
Exopeptidases: carboxy terminus	
Carboxypeptidase P	—○—○—○—○↓● Proline, glycine, alanine
Carboxypeptidase M	—○—○—○—○↓● Lysine, arginine
Peptidyl dipeptidase A	—○—○—○—○↓●—● Many, but particularly histidine-leucine
γ-Glutamyl carboxypeptidase	—○—○—○—○↓● (γ-Glutamic acid)$_n$

less active or less stable than in the vesicles inside the cells as they will be diluted and the pH in the intestinal lumen may be above their optimum as they are generally most active in the pH range 5.0–5.5.

The lysosomal enzymes most relevant to our discussion are the peptidases and the nucleases. The peptidases, also referred to as the cathepsins, comprise at least eight exopeptidases and nine endopeptidases, which between them have a broad range of specificities that enable them to reduce any proteins or peptides to their constituent amino acids.

The other principal lysosomal enzyme of relevance is DNase II. This is a ubiquitous DNase found in the lysosomes of all tissues that functions at an acidic pH in the absence of metal ions. DNase II is considered to be a barrier to the transfection of cells by potentially therapeutic gene sequences (Howell et al. 2003): any oligonucleotides delivered to cells for therapeutic reasons are highly likely to be endocytosed and end up in lysosomes and exposed to DNase. Given its acid pH optimum, whether DNase II would be very active in the lumen of the intestine following the sloughing-off of enterocytes is not certain.

1.6 Quantitative Aspects of Intestinal Enzymes: How Much and How Active?

The consequences for macromolecules exposed to enzymes at different sites in the GI tract will depend not only on the specificity of the enzymes present, but also on the quantities and activities at any particular site. This consideration has been considerably overlooked by many researchers. In the case of therapeutic peptides this is particularly important, as it is the peptidase enzymes that not only present a considerable range of specificities between them, as discussed earlier and shown in Tables 1.1 and 1.2, but that are present in the gut in the greatest quantities. The human GI tract has evolved to be a highly efficient organ for the digestion of proteins and consequently contains sufficient peptidase activity to completely digest *gram* quantities of protein in the minutes and hours following a meal, while for therapeutic peptides, the quantities likely to be administered are in the *milligram* range. The greatest threat comes from the pancreatic peptidases: estimates of the quantity of enzymes produced by the pancreas vary from a few grams up to 40 or 50 g a day (Kukral et al. 1965; Magee and Dalley 1986). The quantities will show considerable inter-subject variation, and considerable fluctuations, with increased output following meals. What is important to note is that the quantities are *gram* quantities, that is, very large amounts in catalytic terms. However, as pointed out by Bernkop-Schnürch (1998), the amounts in protein terms do not tell us about the *activity* of the enzymes, which will depend on the prevailing pH, as well as factors such as the concentration of ions and bile salts. From published data he calculated the activity of secreted pancreatic enzymes, which is useful for designing in vitro stability tests, but the activities are based on the hydrolysis of synthetic enzyme substrates. There is a serious need for information on the activity of enzymes with physiological substrates actually present under physiological conditions, but such information is hard to obtain. While there are many studies on the constituents of pancreatic fluid, any enzyme activity measure will not be a reflection of the intraduodenal or jejunal activity as the pancreatic peptidases only reach full activity after interaction with enteropeptidase, and that is located in the apical membrane of the mucosal cells. The most useful would be to measure enzyme activity in the fluid recovered from human jejunum. Lindahl and colleagues (1997) intubated the jejunum in fasted human volunteers. They blocked-off the intestine and held a tube in position with inflatable balloons and then collected the fluid in the upper jejunum for two and a half hours. They measured the ionic composition but unfortunately not the enzyme activity. Such experiments, with a measure of all the enzyme activities, would be most helpful, but are obviously ethically and technically complicated.

The activities of some of the peptidases in the brush border membranes of rats have been measured in attempts to identify which of the many enzymes present are the most active and hence represent the greatest threat to therapeutic peptides (Woodley 1994). The aminopeptidases N, A, and P appear to be the most active, but one has to be cautious in interpreting the data as in the two examples cited different substrates were used and in all cases the substrates were synthetic substrates designed to yield fluorimetric or spectrophotometric moieties on hydrolysis and *not* natural peptides.

The other group of enzymes of major concern to our discussion are the nucleases secreted by the pancreas, which are very much in the minority as they constitute less than 1% of the pancreatic juice protein (Scheele et al. 1981). Again this is in quantitative protein terms, and I can find no information concerning their actual activity in the lumen of the intestine.

As discussed earlier, the other source of multiple enzyme activity in the intestinal lumen will be from the sloughed-off cells of the intestinal epithelium, and this will mean that the enzymes of the brush border membrane and intracellular activities such as the lysosomal enzymes will be present. Again it is very difficult to assess how much hydrolytic activity this might represent. Glaeser et al. have used the intestinal lumen as a source of epithelial cells to study metabolism (Glaeser et al. 2002). They intubated the intestine in humans and isolated a section of 20 cm with inflatable balloons. Perfusion of the segment yielded 56 million epithelial cells, clearly demonstrating that in man the sloughing-off of cells into the lumen is a real phenomenon, but how much hydrolytic enzyme activity this may represent remains to be determined. In earlier studies, Andersen et al. had estimated in the rat that 1.6% of the total mucosal content of the brush border aminopeptidase N was released into the lumen per hour (Andersen et al. 1988).

1.7 The Colon

The colon has frequently been described in the pharmaceutical literature as a site with low enzymatic activity, particularly of peptidases, and therefore potentially suitable for therapeutic peptide delivery. The word ‘low’ here is very relative: while the activity of the peptidases may be lower than in the small intestine, there is still *considerable* enzyme activity. Pancreatic peptidases can survive transit through the GI tract, indeed, they have been measured in faeces for diagnostic purposes, and so will be present in the colon. More importantly, the colon is home to vast numbers of bacteria: up to 10^{11}–10^{13} per gram of faecal dry weight, and they produce a wide array of enzymes (including peptidases) capable of hydrolysing more or less any macromolecule. In addition, the slow transit time in the colon, and viscous contents, means that any orally administered macromolecule will be exposed to the enzymes for a considerable time, compared within the upper jejunum, for example. Therefore as a site for the delivery and absorption of therapeutic peptides or oligonucleotides, the colon is clearly *not* suitable. For a more detailed discussion of the problem with respect to therapeutic peptides, including peptide degradation by colon contents, the reader is referred to an earlier review (Woodley 1994). However, while the colon may be a poor site for the delivery of peptides or oligonucleotides for absorption into the parenteral system, there is considerable interest in delivering drugs to the colonic tissues. Diseases of the colonic tissue, notably inflammatory diseases, constitute a major health problem in the world today as they include ulcerative colitis, Crohn's disease, irritable bowel syndrome and colorectal cancer.

As mentioned earlier, there is little or no degradation of complex carbohydrates in the upper GI tract. On the other hand the gut flora produce a vast range of enzymes, including saccharidases with specificities for a wide range of different glycosidic bonds: mannosidases, xylosidases, galactosidases, galacturonidases, glucuronidases, fucosidases, etc., and with specificities for β-linkages as well as α-linkages, and thus the colon is likely to be the site for the degradation of the complex carbohydrate side chains of glycoproteins and of polysaccharides. For this reason, drug delivery systems are being investigated and developed whereby polysaccharide macromolecules are used as coating materials for drug formulations, where the need is to release the drug for action specifically into the colon to treat colonic diseases. The macromolecular coating material should remain intact during the passage through the stomach and small intestine and then be hydrolysed by the bacterial enzymes to release the drug in the colon. Candidate macromolecules include plant gums, chitosans, pectins, chondroitin sulphates, dextrans, and alginates (Jain et al. 2007). The enzyme activities within the colon will vary considerably between individuals, according to the populations of the microflora, which in turn will vary considerably depending on the diet and other physiological factors. However, given the very high numbers of bacteria present it is hard to imagine any formulation 'saturating' the enzymatic capacity.

Precise information on the enzyme specificities, and particularly the quantity of enzyme activity present in the normal human colon is not easy to find, and underlines the principal that in the development of oral delivery systems either for therapeutic macromolecules or using macromolecules as part of pharmaceutical formulations, then in vitro testing is an essential and valuable tool as described in the next section.

1.8 The Importance of In Vitro Testing

Given the uncertainties of the quantities and the precise specificities of many digestive enzymes, notably the endopeptidases, and the surprises that biological systems sometimes throw up (who would have expected chitosan to be hydrolysed by pepsin, pancreatin, a lipase or α-amylase as has been reported? (Muzzarelli 1997; Muzzarelli et al. 1995; Yalpani and Pantaleone 1994)) any therapeutic macromolecule or formulation destined for oral use should be tested in vitro at an early stage of development. Such testing can be invaluable and can save much time and money. A brief personal anecdote illustrates my point. Some years ago I was asked by a pharmaceutical company to explain some pharmacokinetic data they had obtained. A formulation section of the company had produced a controlled release formulation for oral use, whereby a drug was to be encapsulated and thus slowly released in the GI tract. They did classic release experiments as described by the Pharmacopeia, that is, stirring the formulation in a 'paddle' apparatus in phosphate buffer (*note*: the intestinal tract does not contain phosphate buffer, but bicarbonate). Sure enough the drug was steadily released over several hours and so confident that they had a good controlled release formulation, the company then tested it in humans. To their incomprehension, the pharmacokinetics were *exactly* the same as an unformulated dose of the drug in solution. One look at the chemical structure of the components of the formulation was enough for me to explain the result. The formulation contained several components that were substrates for pancreatic lipase and esterase, especially in the presence of bile salts, and therefore the drug was immediately released as the formulation was rapidly hydrolysed in the upper intestine. Had the company tested the drug release in vitro with pancreatic enzymes and bile salts, instead of just phosphate buffer, they would have saved considerable expense.

1.8.1 Strategies for In Vitro Testing of the Stability of Therapeutic Macromolecules and Macromolecular Formulations

Given the physiological conditions in the intestinal lumen and the co-ordinated action of the peptidase, for example, it is important that the test macromolecule or formulation be incubated in the presence of a *mixture* of enzymes, not just with individual ones. The simplest type of experiment is to incubate the macromolecules with a mixture of enzymes designed to simulate the conditions in the lumen of the upper intestine. One could use either a mixture of the individual purified pancreatic enzymes or a pancreatic extract such as 'pancreatin', which should contain all the enzymes, and is the preparation suggested in the US Pharmacopeia for simulated intestinal fluid (SIF), although whether the enzymes are fully active is another question. It is important also that bile salts be added as these can influence the enzyme activity and the solubility of the test molecules. In addition, the enzymes of the intestinal brush border membrane should also be present. It is difficult to simulate conditions in the human, but could be achieved by adding a preparation of such membranes obtained from laboratory animals by standard methods using differential centrifugation (Kessler et al. 1978). The incubation buffer used should be bicarbonate and not phosphate. Alternatively in the case of studies using animals (e.g. rats), the enzymes, bile salts and test materials could be incubated in the medium surrounding an everted gut sac. This would be a relatively realistic simulation of the actual in vivo situation, as both the pancreatic enzymes and the surface enzymes of the enterocytes would be present. The difficulty of such experiments is deciding on the concentrations of enzymes to use: the calculations of Bernkop-Schnürch are perhaps a good starting point, but with the enzymes as mixtures not as individual enzymes (Bernkop-Schnürch 1998). The best strategy would be to use a range of different concentrations, which in vivo would anyway vary considerably between individuals and between the fed and fasted states. The composition and concentrations of the bile salts in the intestine are more readily available and have formed the basis of dissolution test media pioneered by Dressman and her colleagues (Vertzoni et al. 2004). While these dissolution media contain bile salts, they do not, however, contain enzymes, an absolute prerequisite for testing macromolecular stability. While it may be possible to simulate the presence of pancreatic, brush border enzymes and bile salts, these artificially created media would not contain the enzymes emanating from the sloughed-off cells of the epithelium as discussed earlier.

Therefore, another possibility in animal studies is to wash out the upper jejunum with a small volume of bicarbonate buffer, and use this fluid as a medium to determine the stability of a test macromolecule or formulation. A similar approach can be used to study enzyme activities in the colon in animals. It should be noted that the preparations should not be centrifuged, but used in their entirety as enzymes can bind to particulate matter or are on the surfaces of bacteria in the case of colon contents (Woodley 1991). Fluid thus obtained from the upper intestine should contain all the pancreatic enzymes, bile salts and sloughed-off cells, but again getting the concentrations right is not obvious.

In many ways, the ideal test medium for pharmaceutical development and testing the stability of putative therapeutic macromolecules would be intestinal aspirates from the upper GI tract of human subjects, and under fed and fasted conditions. Obtaining such aspirates is now technically quite feasible and a number of groups have published data on the ionic and bile salt components of the fluids thus obtained (Kalantzi et al. 2006; Lindahl et al. 1997; Perez de la Cruz Moreno et al. 2006).

It is thus highly recommended that for any studies on the development of formulations to orally deliver therapeutic macromolecules or develop novel formulations composed of or containing macromolecules (including lipids and surfactants), the first step should be a thorough investigation of the macromolecular stability using one or all of the above strategies. In the view of the author such studies are *essential* and can save much development time and expense by enabling unstable therapeutics or formulations to be identified or more importantly, the reverse, that is, enabling the identification of the most stable macromolecules. For example, in the case of therapeutic oligonucleotides, the chemical analogues that are the most stable in the GI tract would be identified.

1.9 Conclusions

The human intestine has evolved as a highly efficient organ to digest (i.e. hydrolyse) practically all the macromolecules in the human diet (albeit with the help of a few trillion bacteria!) with the exception of some plant fibres. To do this it possesses a formidable array of enzymes. This is particularly true for the digestion of proteins and peptides where peptidases are found in the stomach, are secreted by the pancreas in considerable quantities and are found on the surface of and inside intestinal epithelial cells. These enzymes work in a co-ordinated fashion to rapidly hydrolyse proteins. They present the major difficulty for designing oral delivery systems for therapeutic peptides, which may explain why 86 years after the first attempt to orally administer insulin (Bliss 1982), there is still not an oral insulin product available for diabetics.

While there are enzymes in the GI tract that hydrolyse oligonucleotides, it would appear that these enzymes are produced in lower amounts, but more precise qualitative and quantitative data are badly needed. The same is true for other macromolecules such as chitosan, used in experimental formulations, where there are some reports of its susceptibility to enzymes found in the GI tract, but no direct data. Many researchers have either ignored or are not aware of the considerable enzymatic potential of the human intestine, particularly with respect to peptides. It is essential that research in this field be multi-disciplinary involving pharmaceutical scientists, chemists, and biologists who understand the processes of digestion and absorption. It is also essential that recognised and established in vitro testing strategies be developed to test the stability of therapeutic macromolecules and macromolecular formulations in the early stages of pre-clinical development.

References

Andersen V, Hansen GH, Sjöström H and Norén O (1988) The occurrence of aminopeptidase N and desquamated cell proteins in intestinal perfusate of rat. Biochim Biophys Acta 967:43–48

Berezin IV and Martinek K (1970) Specificity of alpha-chymotrypsin. FEBS Lett 8(5):257–260

Bernkop-Schnürch A (1998) The use of inhibitory agents to overcome the enzymatic barrier to perorally administered therapeutic peptides and proteins. J Cont Rel 52(1–2):1–16

Bliss M. (1982) The discovery of insulin. Chicago University Press, Chicago

Carriere F, Renou C, Lopez V, De Caro J, Ferrato F, Lengsfeld H, De Caro A, Laugier R and Verger R (2000) The specific activities of human digestive lipases measured from the in vivo and in vitro lipolysis of test meals. Gastroenterology 119(4):949–960

De Caro J, Sias B, Grandval P, Ferrato F, Halimi H, Carrière F and De Caro A (2004) Characterisation of pancreatic lipase-related protein 2 isolated from human pancreatic juice. Biochim Biophys Acta 1701:89–99

Felekkis K and Deltas C (2006) RNA interference: A powerful laboratory tool and its therapeutic implications. Hippok 10(3):112–115

Funakoshi A, Tsubota Y, Wakasugi H, Ibayashi H and Takagi Y (1977) Purification and properties of human pancreatic deoxyribonuclease I. J Biochem 82(6):1771–1777

George M and Abraham TE (2006) Polyionic hydrocolloids for the intestinal delivery of protein drugs: alginate and chitosan – a review. J Cont Rel 114(1):1–14

Giller T, Buchwald P, Blum-Kaelin D and Hunziker W (1992) Two novel human pancreatic lipase related proteins, hPLRP1 and hPLRP2. J Biol Chem 267(23):16509–16516

Glaeser H, Drescher S, van der Kuip H, Behrens C, Geick A, Burk O, Dent J, Somogyi A, Von Richter O, Griese EU, Eichelbaum M and Fromm MF (2002) Shed human enterocytes as a tool for the study of expression and function of intestinal drug-metabolizing enzymes and transporters. Clin Pharmacol Ther 71(3):131–140

Hejazi R and Amiji M (2003) Chitosan-based gastrointestinal delivery systems. J Cont Rel 89(2):151–165

Howell DP, Krieser RJ, Eastman A and Barry MA (2003) Deoxyribonuclease II is a lysosomal barrier to transfection. Mol Ther 8(6):957–963

Jain A, Gupta Y and Jain SK (2007) Perspectives of biodegradable natural polysaccharides for site-specific delivery to the colon. J Pharm Pharmaceut Sci 10(1):86–128

Kalantzi L, Goumas K, Kalioras V, Abrahamsson B, Dressman JB and Reppas C (2006) Characterization of the human upper gastrointestinal contents under conditions simulating bioavailability/bioequivalence studies. Pharm Res 23(1):165–176

Kessler M, Acuto O, Storelli C, Murer H, Müller M and Semenza G (1978) A modified procedure for the rapid preparation of efficiently transporting vesicles from small intestinal brush border membranes. Their use in investigating some properties of D-glucose and choline transport systems. Biochim Biophys Acta 506(1):136–154

Kukral J, Adams A and Preston F (1965) Protein producing capacity of the human exocrine pancreas. Surg 162:67–71

Lindahl A, Ungell AL, Knutson L and Lennernas H (1997) Characterization of fluids from the stomach and proximal jejunum in men and women. Pharm Res 14(4):497–502

Magee DF and Dalley AF. (1986) Digestion and structure and function of the gut. Karger, Basel

Muzzarelli RAA (1997) Human enzymatic activities related to the therapeutic administration of chitin derivatives. Cell Mol Life Sci 53:131–140

Muzzarelli RAA, Xia W, Tomasetti M and Ilari P (1995) Depolymerization of chitosan and substituted chitosans with the aid of a wheat germ lipase preparation. Enz Microbial Tech 17:541–545

Perez de la Cruz Moreno M, Oth M, Deferme S, Lammert F, Tack J, Dressman J and Augustijns P (2006) Characterization of fasted-state human intestinal fluids collected from duodenum and jejunum. J Pharm Pharmacol 58(8):1079–1089

Reboul E, Berton A, Moussa M, Kreuzer C, Crenon I and Borel P (2006) Pancreatic lipase and pancreatic lipase-related protein 2, but not pancreatic lipase-related protein 1, hydrolyze retinyl palmitate in physiological conditions. Biochim Biophys Acta 1761(1):4–10

Scheele G, Bartelt D and Biegzr W (1981) Characterisation of human exocrine pancreatic proteins. Gastroent 80(3):461–473

Seymour AA, Swerdel JN, Delaney NG, Rom M and Cushman DW (1987) Effects of ring opened antrial natriuretic peptide (ANP) on mean arterial pressure MAP and renal excretion in SHR. Fed Proc 46:1296

Takeshita H, Mogi K, Yasuda T, Nakajima T, Nakashima Y, Mori S, Hoshino T and Kishi K (2000) Mammalian deoxyribonucleases I are classified into three types: pancreas, parotid, and pancreas-parotid (mixed), based on differences in their tissue concentrations. Biochim Biophys Acta 269(2):481–484

Tillman LG, Geray RS and Hardee GE (2008) Oral delivery of antisense oligonucleotides in man. J Pharm Sci 97(1):225–236

Tsutsumi K, Kevin Li S, Hymas RV, Teng C-L, Tillman LG, Hardee GE, Higuchi WI and Ho NFH (2008) Systematic studies on the paracellular permeation of model permeants and oligonucleotides in the rat small intestine with chenodeoxycholate as enhancer. J Pharm Sci 97(1):350–367

Vertzoni M, Fotaki N, Kostewicz E, Stippler E, Leuner C, Nicolaides E, Dressman J and Reppas C (2004) Dissolution media simulating the intralumenal composition of the small intestine: physiological issues and practical aspects. J Pharm Pharmacol 56(4):453–462

Werle M, Samhaber A and Bernkop-Schnurch A (2006) Degradation of teriparatide by gastro-intestinal proteolytic enzymes. J Drug Target 14(3):109–115

Woodley JF (1994) Enzymatic barriers for GI peptide and protein delivery. Crit Rev Ther Drug Car Syst 11(2 & 3):61–95

Woodley JF (1991) Peptidase activity in the G.I. tract: distribution between luminal contents and mucosal tissue. Proc Int Symp Contr Rel Bioact Mater 18:337

Yalpani M and Pantaleone D (1994) An examination of the unusual susceptibilities of aminoglycans to enzymatic hydrolysis. Carbohydr Res 256(1):159–175

Chapter 2
Gastrointestinal Mucus Gel Barrier

Juan Perez-Vilar

Contents

Juan Perez-Vilar (✉)
Cystic Fibrosis/Pulmonary Research and Treatment Center, University of North Carolina, Chapel Hill, NC 27599, USA
e-mail: jp_vilar@yahoo.com

A. Bernkop-Schnürch (ed.), *Oral Delivery of Macromolecular Drugs*,
DOI 10.1007/978-1-4419-0200-9_2, © Springer Science+Business Media, LLC 2009

Abstract A family of glycoproteins, known as gel-forming mucins, endow gastrointestinal mucus with its characteristic viscoelastic and biological properties. In the mucus, these large oligomeric glycoproteins are organized into entangled networks that occasionally can be stabilized by non-covalent interactions as in the stomach lumen. This network is a formidable chemical and physical barrier that not only protects the underlying epithelia but also limits the usefulness of orally administered drugs. In this chapter, I review the molecular and cellular properties of gel-forming mucins and how these macromolecules are organized into a tri-dimensional network to form the gastrointestinal mucus gel barrier.

2.1 Introduction

Between the gastrointestinal lumen and the epithelial cells there is a viscous solution, known as mucus, which lubricates the epithelia, helping in the passage of substances and particles through the digestive tract, and forms a protective layer against noxious chemicals, microbial infections, dehydration, and changing luminal conditions. Moreover, the gastrointestinal mucus harbors antimicrobial peptides and proteins involved in the epithelia innate and adaptative immunity systems and hundreds of different microorganisms inhabiting the different areas of the gastrointestinal tract. Mucus appears to be a source of energy for the enteric microbiota, allowing intestinal colonization by the adhering microorganisms. Simultaneously, mucus permits the selective diffusion of nutrients and other compounds from the lumen toward the epithelial layer and accordingly is a major barrier against orally administered drugs. Although mucus is a complex aqueous solution of proteins, lipids, ions, carbohydrates, etc., a family of glycoproteins, the gel-forming mucins, is largely responsible for mucus viscoelastic and adhesive properties. All five gel-forming mucins identified in humans are found in the gastrointestinal system, although they are differentially distributed along the tract. MUC5B and MUC19 are expressed in the oral mucosa while MUC5AC and MUC6 are the major mucins expressed by gastric goblet and submucosal gland mucous cells, respectively. Finally, MUC2 is the main intestinal mucin under non-pathological conditions.

Gel-forming mucins are among the largest proteins known in nature with thousands of amino acid residues per monomer and molecular weights in the millions. Considering the key functions of the mucus barrier, it is not surprising that mucin gene expression, biosynthesis, and secretion are highly regulated. This chapter is focused on the molecular and cellular properties of gel-forming mucins and how these macromolecules are organized into a tri-dimensional network to form the mucus gel barrier. Excellent reviews on physiological and pathological aspects of the gastrointestinal mucus barrier can be found in the literature (e.g., Allen and Flemstrom 2005; Bi and Kaunitz 2003; Lievin-Le Moal and Servin 2006; Nochi and Kiyono 2006; Ham and Kaunitz 2007).

2.2 Mucin Structural Properties

2.2.1 Multi-domain Organization

Mucin polypeptides comprise thousands of amino acid residues organized into different protein domains with specific structural and/or functional properties (Fig. 2.1) (Perez-Vilar and Hill 1999; Dekker et al. 2002).

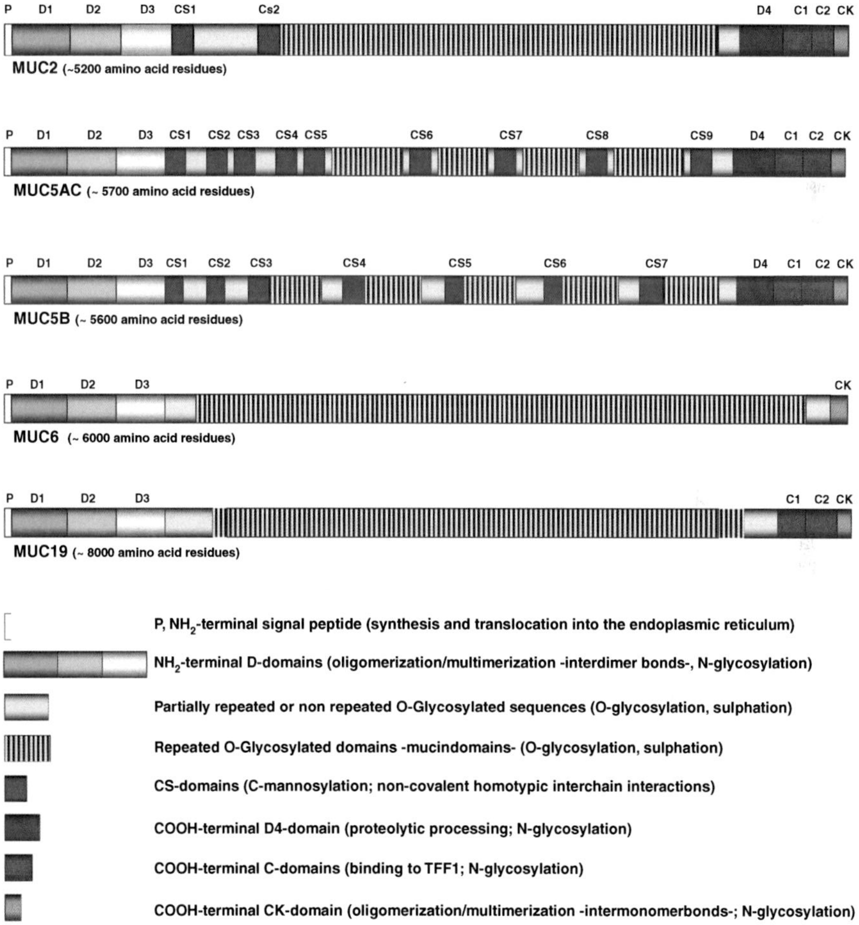

Fig. 2.1 Multi-domain structural organization of gastrointestinal gel-forming mucins. Schematic drawings (not at scale) of the different domains found in gastrointestinal mucins and their roles. Note that while the cysteine-rich domains (D-, CS-, C-, and CK-domains) have a similar length among mucin alleles, the O-glycosylated regions, except in the case of MUC5B, differ in size, a feature not shown in the drawings

2.2.1.1 Mucin (O-Glycosylated) Domains

The structure of mucins is dominated by the presence of a prominent, centrally located domain formed by threonine and/or serine-rich repeated sequences in their polypeptide chains. O-Linked oligosaccharides are covalently bound to these residues, resulting in the formation of highly glycosylated domains, known as the O-glycosylated domains, mucin domains, or the tandem repeat domains (TRDs). The sequence, length, and number of the serine/threonine-rich repeats vary among mucins. While in MUC2, MUC5AC, and MUC5B the central O-glycosylated domains are separated by unglycosylated domains, the structure of the mucin domain of MUC6 and likely MUC19 is less complex as it comprises a single, continuous O-glycosylated region (Chen et al. 2004; Rousseau et al. 2004) (Fig. 2.1). It has been long assumed that the lack of secondary structures in the mucin domains is an indication that these domains function as a scaffold for O-linked oligosaccharides (Eckhardt et al. 1987). However, it can be argued that the lack of stable secondary structure plus the high density of *O*-glycans in the mucin domains are the two critical factors that make mucins highly flexible, random-coil macromolecules suited for forming mucus gels (see Sections 2.2.2.1 and 2.5.1).

2.2.1.2 CS-Domains

Besides the prominent mucin domains, gel-forming mucins have other domains (i.e., CS-, D-, C-, and CK-domains as shown in Fig. 2.1) (Perez-Vilar and Hill 1999; Dekker et al. 2002) characterized by being (a) under-glycosylated, although most have one or more N-linked oligosaccharide chains (see Section 2.2.2.2); (b) rich in cysteine residues; (c) conserved among different mucins and other families of extracellular proteins; and (d) globular structures with α-helices and pleated sheets, as predicted by their primary sequences, and few or no free thiols (e.g., Perez-Vilar et al. 1998). In MUC2, MUC5AC, and MUC5B, but not MUC6 and likely MUC19, the mucin domain is interrupted by several copies of an $\sim$100 amino acid-long Cys-rich domain, known as the Cys-subdomain or CS-domain, which is likely C-mannosylated (see Section 2.2.2.3) and could be involved in weak, non-covalent mucin-mucin interactions (Perez-Vilar et al. 2004). These rather hydrophilic domains are highly homologous with amino acid identities that range between 80 and 100% (Dessein et al. 1997; Escande et al. 2001).

2.2.1.3 D-Domains

Three NH_2-terminal D-domains, designated D1, D2, and D3, are found in all gel-forming mucins whereas a fourth D-domain, named D4, is at the COOH terminus of MUC2, MUC5AC, and MUC5B but not MUC6 and MUC19 (Fig. 2.1) (Perez-Vilar and Hill 1999; Dekker et al. 2002; Chen et al. 2004; Rousseau et al. 2004). A partial D-domain, D', is between D2 and D3 in all of them. Each

D-domain, which contains up to 30 cysteines and up to ~400 amino acid residues, shows significant sequence identity with the other D-domains, especially the cysteine, glycine, and proline residues. Further analysis of the D-domain sequences has defined a potential domain, known as the trypsin inhibitor-like cysteine-rich domain (TIL) within each D-domain sequence (Lang et al. 2004). The NH_2-terminal D-domains are involved in formation of mucin disulfide-linked oligomers/multimers (Perez-Vilar and Hill 1999; Perez-Vilar and Mabolo 2007; see Section 2.3.2).

2.2.1.4 C- and CK-Domains

The COOH-terminal of 240–350 residues in MUC2, MUC5AC, MUC5B, and MUC19 comprises two different Cys-rich domains (Fig. 2.1). The first 140–250 residues in this domain have sequence identity with the C-domains of von Willebrand factor whereas the remaining residues from the COOH terminus have sequence identities with the CK-domain at the COOH terminus of many extracellular proteins and the "cystine-knot" super family of proteins (Sun and Davies 1995; Sadler 1998; Perez-Vilar and Hill 1999). In contrast to the other gel-forming mucins, the CK-domain is the only Cys-rich domain at the COOH terminus of MUC6 (Rousseau et al. 2004). The CK-domain provides the Cys residues that form interchain disulfide in mucin dimers, which later are assembled into mucin oligomers/multimers (see Section 2.3.1).

2.2.2 Glycosylation

2.2.2.1 O-Glycosylation

Mucin-type O-linked oligosaccharides are covalently bound to the hydroxyl groups of Ser and Thr residues in the mucin domains. In these glycan chains, the first monosaccharide residue is always *N*-acetyl-galactosamine, which is followed by variable numbers of galactose, fucose, *N*-acetyl-galactosamine, sialic acid, and other monosaccharides arranged in linear or branched structures (Spiro 2002). Sialic acid and fucose residues are usually sulfated (Brockhausen 2003) whereas the former rests can be O-acetylated (Klein and Roussel 1998). Since *O*-glycans comprise up to 90% of the weight of the native mucins, it is not surprising that they are largely responsible for the biochemical, biophysical, and functional properties of these glycoproteins. First, *O*-glycans are very diverse and, accordingly, mucins can interact with many different compounds and microorganisms. Second, the high density of *O*-glycan chains makes difficult the rotation of the peptide bonds in the mucin domains, which result in extended, though flexible, polypeptide chains with large hydrodynamic sizes (Jentoft 1991; Gerken 1993; Hong et al. 2005). Since mucins occupy large volumes in solution, mucin chains overlap and form entanglements at very low concentrations (see Section 2.5.1). Third, sialic and sulfate residues in *O*-glycans are responsible for the polyanionic nature of mucins. High negative charge density contributes to the stiffness of mucin polypeptides (Jentoft 1991; Gerken 1993) and also permits mucins to absorb large amounts of water molecules. Moreover, polyanionic mucins likely are necessary for the accumulation of mucin inside mucin granules and for the expansion of the intragranular mucins upon exocytosis (see Sections 2.4.1 and 2.4.2.2). Besides these roles, it has been recently reported that *O*-glycans in the human gastric mucins likely function as a natural antibiotic against *Helicobacter pylori* (Kawakubo et al. 2004).

2.2.2.2 N-Glycosylation

The D-, C-, and CK-domains in mucins are modified with one or more N-linked oligosaccharide chains. This kind of glycan differs from the mucin-type O-linked oligosaccharides in that they are added to asparagine residues in N-glycosylation Asn-X-Ser/Thr acceptor motifs, have mannose residues, a monosaccharide absent in *O*-glycans, and are less diverse (Spiro 2002). Similar to *O*-glycan chains, *N*-glycans can be sialylated and sulfated. The role of protein N-glycosylation in mucins is still somewhat controversial but appears to be related to the folding of nascent polypeptides in the endoplasmic reticulum (Strous and Dekker 1990; Perez-Vilar et al. 1996; Bell et al. 2003). Both N- and O-linked oligosaccharide chains display cell-type differences in their composition and length that likely reflect the particular repertoire of glycosyltransferases available.

2.2.2.3 C-Mannosylation

The major structural characteristic of mucin CS-domain primary sequences is the presence of one C-mannosylation Trp-X-X-Trp acceptor motif in the NH_2-terminal side of the domain. Studies with recombinant mucin CS-domains suggest that these acceptor motifs are C-mannosylated (Perez-Vilar et al. 2004), although detection of C-mannoses in native mucins has not yet been reported. C-Mannosylation differs from O- and N-glycosylation in that it involves a covalent attachment of a single α-mannose residue to the indole C2 carbon atom of the first tryptophan residue in Trp-X-X-Trp peptide motifs (Spiro 2002). The function of protein C-mannosylation is unknown at present, although the available evidence suggests that it contributes to the folding of the mucin polypeptide or some aspect of the secretion process (Perez-Vilar et al. 2004; Mabolo and Perez-Vilar, unpublished observations). The weak homotypic interactions among mucin Cys subdomains appear to be independent of C-mannosylation (Perez-Vilar et al. 2004).

2.2.3 Oligomerization/Multimerization

In the mucus, gel-forming mucins are found as disulfide-linked oligomers with variable degree of oligomerization (Fig. 2.2) (Perez-Vilar and Hill 1999; Perez-Vilar and Mabolo 2007). Only mucin oligomers with 16 or less units have been reported, although the mechanism of mucin covalent assembly (see Section 2.3) should make possible the formation of larger oligomers and perhaps even

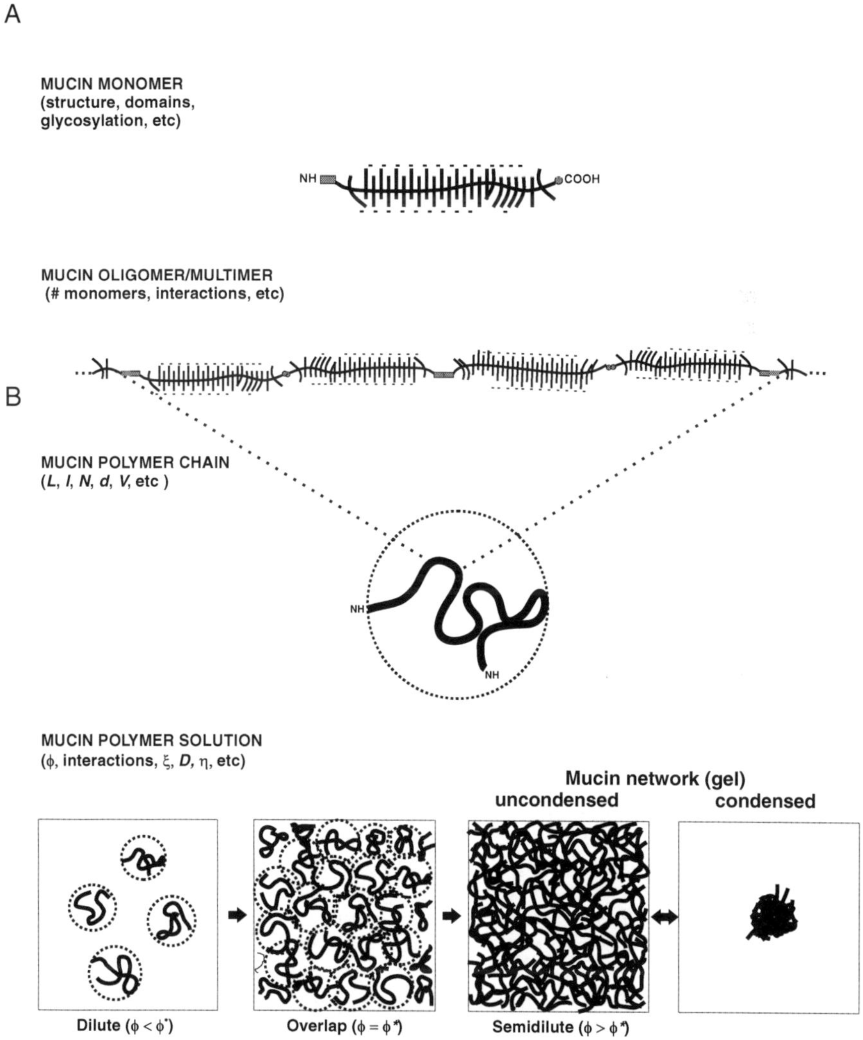

Fig. 2.2 Different levels of structural organization in gel-forming mucins. From the biochemical point of view (**a**), mucin polypeptides have very complex multi-domain structures and glycosylation patterns and thousands of amino acids per monomer. Moreover, the monomers are assembled into disulfide-linked oligomers/multimers that have contour sizes of several microns.

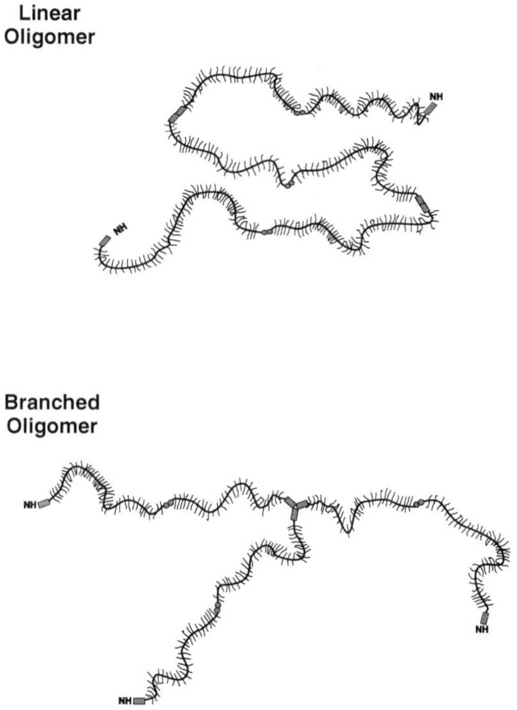

Fig. 2.3 Organization of mucin disulfide-linked oligomers/multimers. Schematic representation of the two types of disulfide-linked oligomers/multimers in mucins. Biophysical and electron microscopy studies are consistent with mucin forming linear oligomers/multimers such as in the case of MUC5B, MUC5AC, and MUC6. However, biochemical and microscopy studies with recombinant mucins have provided strong evidence that pMUC19, the porcine counterpart of MUC19, and MUC2 form branched oligomers/multimers. In all cases, the monomeric subunits are bound by linkages linking their respective NH- and COOH-terminal regions. *Rectangles* indicate the NH-terminal D-domains while *circles* and the *bars* represent the COOH-terminal CK-domains and the *O*-glycan chains, respectively

Fig. 2.2 (continued) Because the O-glycosylated regions, which lack stable secondary structures, are predominant in size, mucin oligomers/multimers are basically very long, flexible random coils. Hence, from the physical point of view (**b**), mucin oligomers/multimers can be defined as polymers of *N* freely jointed Kuhn segments (see Section 2.5.1). At this level, the biochemical features of mucins are secondary and, indeed, a number of properties can be described in terms of *N*. Mucin chains occupy large spherical volumes (*V*) and continually change from one conformation to another. As mucin concentration increases, the chains start to overlap and eventually form entangled networks or gels, likely stabilized by other interactions (see Section 2.5.2). Due to mucin high intracellular concentration, cellular mucins are presumably forming gels (see Section 2.4.1). Interestingly, volume of polymeric gels can be modulated by phase transitions, which shrink or swell the gel upon changes on environmental parameters (pH, ions, temperature, etc.). Phase transitions are likely exploited during mucin granule biogenesis and also exocytosis (see Section 2.4). *Rectangles* indicate the NH-terminal D-domains while *circles* and the *bars* represent the COOH-terminal CK-domains and the *O*-glycan chains, respectively multimers (Perez-Vilar and Hill 1999; Perez-Vilar and Mabolo 2007). In mucin oligomers/multimers, the NH_2-terminal D-domains and COOH-terminal CK-domain of each monomer form interchain disulfide bonds with the corresponding domains in different monomers, i.e., D-to-D and CK-to-CK (Fig. 2.2). Although it was long presumed that all gel-forming mucin oligomeric/multimeric chains would be linear such as in the case of MUC5AC and MUC5B, other studies support the notion that some mucins (e.g., MUC2 and likely MUC19) are assembled into branched oligomers/multimers (Fig. 2.3) (Perez-Vilar et al. 1998; Perez-Vilar and Hill 1999; Gold et al. 2002). Such disparate mechanisms might originate from the potential capability of the mucin amino-terminal regions (containing the D1-, D2-, and D3-domains) to form disulfide-linked homodimers, in the case of linear chains, or alternatively disulfide-linked homotrimers in branched mucins, whereas the corresponding CK-domains can only form disulfide-linked homodimers (see Section 2.3). The predicted tertiary structure of dimeric CK-domains (e.g., Meitinger et al. 1993) suggests that the two monomers are oriented in an antiparallel configuration, which projects the respective mucin polypeptides toward opposite directions. This is important to maintain the extended structure of the mucin disulfide-linked oligomeric chains.

2.2.4 Molecular Polydispersity

Native or purified gel-forming mucin solutions are molecularly polydisperse, i.e., contain a mixture of chains with different masses. Polydispersity originates by four major mechanisms. First, mucin genes have different alleles (e.g.,Vinall et al. 2000). Mucin genes consist of large, central exons, encoding the entire mucin domains, while genomic sequences encoding other domains are comprised of short exons interrupted by large introns. Mucin alleles differ in the number of tandem repeats encoded by their respective central exon, a feature known by the acronym VNTR (variable number of tandem repeats). The biological and pathological significances of having a larger or a shorter allele are not yet fully understood. However, the fact that, for instance, *MUC2* has more than six recognized alleles that range from 3 to 9 kb (Vinall et al. 2000) clearly points to important differences in the viscoelastic property of mucus with different alleles combinations as many properties of random-coil macromolecules such as mucins depend on its length (see Section 2.5.1).

Second, the type of N- and O-linked oligosaccharides present in mucins can change within the organ, the tissue, the differentiation state, or during pathological conditions (e.g., Veerman et al. 2003). This variability creates glycoforms of otherwise identical mucins that contribute to the differential location of carbohydrates along the gastrointestinal tract revealed in histological preparations using carbohydrate- and/or sulfate-staining solutions (e.g., neutral vs. acidic mucins) and lectins or sugar-specific antibodies (see Deplancke and Gaskins 2001 and references therein). For instance, acidic mucins are predominant in the large intestine, although it must be noted that all these carbohydrate-specific probes will stain, besides gel-forming mucins, membrane/tethered mucins such as MUC1, MUC3, and MUC4, proteoglycans, and other glycoproteins.

Third, Proteolysis is a well known contributor to mucin molecular polydispersity. Though some of the proteolytic cleavages can be artifactual (e.g., Eckhardt et al. 1991), there are a wide range of mucinases that contribute to mucin degradation and some mucins can be cleaved intracellularly during their biosynthesis (see Section 2.3.2). Finally, the mechanism of mucin covalent oligomerization/multimerization likely generates a mixture of chains that differ in the degree of oligomerization with a minimal number of two monomers per chain (see Section 2.3) (Perez-Vilar and Mabolo 2007).

2.3 Mucin Biosynthesis and Intracellular Processing

Mucins are predominantly synthesized in and secreted from superficial goblet and, when present, submucosal glandular mucus cells by a complex process that involves the endoplasmic reticulum and the subcompartments of the Golgi complex (Fig. 2.4) (reviewed by Perez-Vilar and Hill 1999; Perez-Vilar and Mabolo 2007). Although the major biochemical steps and key cellular aspects occurring during mucin biosynthesis have been uncovered, virtually no significant data on the intracellular trafficking of mucin precursors, except for the early microscopic studies (e.g., Neutra et al. 1984; Roth et al. 1994), are currently available (Perez-Vilar 2007; Perez-Vilar 2008).

2.3.1 Endoplasmic Reticulum

Nascent mucin polypeptides are translocated into the endoplasmic reticulum lumen, where they are folded, N-glycosylated, and C-mannosylated (McCool et al. 1999; Perez-Vilar and Hill 1999; Perez-Vilar et al. 2004). The coordination and timing of these modifications as well as the specific roles of resident chaperones and foldases are still unclear. Mucin covalent assembly begins in the endoplasmic reticulum lumen by formation of disulfide-linked homodimers via the COOH-terminal CK-domains (Fig. 2.4). Of the 11 cysteine residues in the CK-domain only 3 cysteine residues appear to be critical for interchain disulfide bond formation in mucins (Perez-Vilar et al. 1998a). Specifically, the Cys-X-Cys-Cys peptide motif is likely to form an interchain disulfide in all the proteins with COOH-terminal CK-domains, including among others von Willebrand factor (Sadler 1998) and norrin (Perez-Vilar and Hill 1997). Studies with de-glycosylated mucins (e.g., Rose et al. 1984, Shogren and Gerken 1989) support the notion that in the endoplasmic reticulum mucin polypeptides are largely forming rather globular structures.

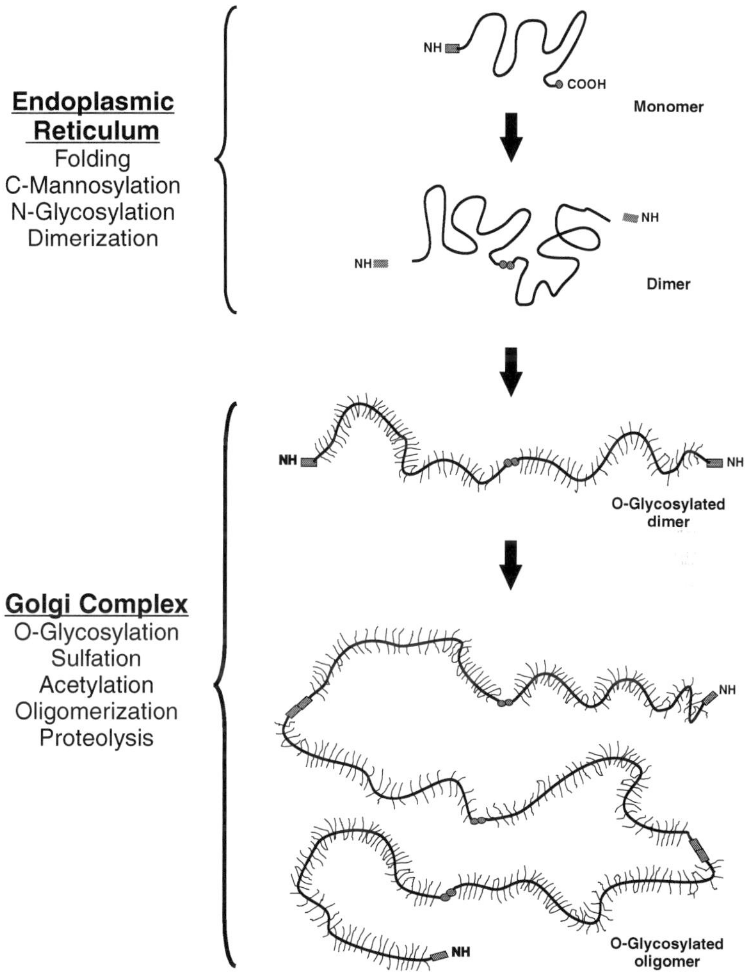

Fig. 2.4 Biosynthesis of gel-forming mucins. Schematic representation of the major steps occurring during the biosynthesis of mucins. Mucin biosynthesis is a sequential process that starts in the endoplasmic reticulum and ends in the *trans*-compartments of the Golgi complex. Formation of disulfide-linked oligomers/multimers involves two steps: dimerization in the endoplasmic reticulum and interdimeric disulfide bonding in the Golgi complex. O-Glycosylation of dimeric precursors results in more extended but still flexible chains. See text for further details. *Rectangles* indicate the NH-terminal D-domains while *circles* and the *bars* represent the COOH-terminal CK-domains and the *O*-glycan chains, respectively

2.3.2 Golgi Complex

Once the mucin dimeric precursors reach the Golgi complex, they are O-glycosylated and N-glycosylation is completed. In vitro studies with purified, de-glycosylated gel-forming mucins (Rose et al. 1984, Shogren et al. 1989;

Gerken 1993; Hong et al. 2005) suggest that initiation of O-glycosylation, i.e., the addition of *N*-acetyl-galactosamine to the hydroxyl groups in Ser and Thr residues, results in a drastic structural change. As mentioned above, the high numbers of consecutive *O*-*N*-acetyl-galactosamine residues in the mucin domains prevent free rotation around the anomeric carbon, which results in the formation of random coil dimeric molecules with extended, rather than globular, flexible conformations (Gerken 1993). Mucin O-glycosylation and sulfation continues in the *medial*- and *trans*-Golgi compartments where the requisite glycosyltransferases for elongation and termination of the oligosaccharides and sulfotransferases are located. Repulsion among sulfate and sialic acid residues likely contribute to the stiffness of the mucin polypeptides (Jentoft 1991, Gerken 1993).

Mucin covalent oligomerization/multimerization is also completed in the late compartments of the Golgi complex by formation of disulfide bonds connecting the NH_2-terminal D-domains in different mucin dimeric precursors (Fig. 2.4) (see reviews by Perez-Vilar and Hill 1999 and Perez-Vilar and Mabolo 2007 and refs. therein). The first cysteine in the conserved Cys-X-Trp-X-Tyr-X-Pro-Cys-Gly peptide motif in the D3-domain is forming an interchain disulfide bond in mucin multimers whereas the Cys-Gly-Leu-Cys-Gly peptide motifs at the D1- and D3-domains, respectively, are critical for the oligomerization/multimerization mechanism (Perez-Vilar and Hill 1998b). Because similar motifs are found in the catalytic center of cysteine/cystine oxido/reductases (Noiva 1994), it was proposed that mucin multimerization is a self-catalyzed, pH-dependent process in which the Cys-Gly-Leu-Cys-Gly in the D1-domain would be required for disulfide bond formation in the acidic *trans*-Golgi compartments (Perez-Vilar et al. 1998; Perez-Vilar and Hill 1999). Consistent with this view, purified polypeptides comprising the entire mucin NH_2-terminal D1-, D2-, and D3-domains are able to form interchain disulfide bond species when incubated in acidic buffers in the absence of any cellular component (Perez-Vilar, unpublished observations). A self-catalytic mechanism raises the intriguing possibility of mucin extracellular oligomerization/multimerization if certain conditions are met (Perez-Vilar and Boucher 2004).

Some proteolytic processing of the mucin precursors likely takes place while MUC2 and MUC5AC are in the endoplasmic reticulum and/or the acidic *trans*-Golgi complex compartments (e.g., Lidell et al. 2003; Lidell and Hansson 2006). The cleavage involves the peptide bond between Asp and Pro in the Gly-Asp-Pro-His peptide motif, which is located in the COOH-terminal D4-domain. Since the resulting fragments can only be detected when disulfide bonds are reduced, it is unclear what the biological significance of this proteolytic processing is, although they may contribute to further cross-linking of the mucin chains by the reactive group in the new COOH terminus (Lidell and Hansson 2006). Proteolytic processing of native gel-forming mucin amino-terminal D-domains has been previously reported (e.g., Wickstrom and Carlstedt 2001).

2.4 Mucin Intracellular Storage and Secretion

Gel-forming mucins are secreted from the apical domain of goblet/mucus cells by constitutive (baseline) secretion via the continuous (low rate) discharge of mucin granules and by regulated secretion, which involves the rapid discharge of mucin granules in response to specific agonists (Forstner 1995; McCool et al. 1995). Whether mucins are also constitutively secreted by small vesicles/tubules departing from the *trans*-Golgi network, as in other secretory cell types, is an interesting possibility that awaits experimental evidence.

2.4.1 Mucin Granule

The biogenesis of the 0.5–2.0-μm mucin granule is virtually an unexplored field and subject of speculations mostly based on data obtained from other cell types (reviewed in Perez-Vilar 2008). However, the generation of HT29-18N2 colon adenocarcinoma cells expressing a fluorescent protein that is sorted and accumulated in the mucin granule has made possible to study the granule lumen by FRAP (fluorescence recovery after photobleaching) under different experimental conditions (Perez-Vilar et al. 2005a, 2006). FRAP provides a conceptually simple and straightforward method to characterize organelles in live cells (Snapp et al. 2003). In a typical FRAP experiment, an area of interest is bleached at high laser power and the recovery of the fluorescence, i.e., the diffusion of the non-bleached molecules inside the bleached area followed over time. Important information can be derived from the fluorescence recovery curves that is pertinent to the environment where the fluorescent tracer is confined. The results from these FRAP studies are consistent with the following conclusions (reviewed in Perez-Vilar 2007, 2008): (a) the mucin granule lumen is organized into a pH-dependent immobile, condensed (mucin) matrix meshwork, which has an average pore size of 5–10 nm, embedded in a fluid phase; (b) proteins and ions in the fluid phase can diffuse through the meshwork pores and interact with their components; (c) charge density (i.e., the degree of mucin/glycoprotein sialylation and sulfation) of the mucin matrix and other intragranular glycoproteins, and also the length of their *O*-glycans, determine the mobility of secretory proteins through the matrix pores; (d) mucin *O*-glycans affect the accessibility of luminal proteins to the matrix protein-rich regions; (e) intraluminal acidic pH and high [Ca^{2+}] (Kuver et al. 2000; Chin et al. 2002) maintain the condensed meshwork by stabilizing the entanglements among the mucin chains (see Section 2.5.1); and (f) protein–protein interactions involving mucin disulfide-rich domains (i.e., D-, CS-, C- and/or CK-domains) are not required to maintain the matrix organization but they may be important during the early stages of granule biogenesis when the pH is less acidic. As discussed in Section 2.4.2.2, once in the extracellular medium the intragranular matrix will be swollen, i.e., the average pore of the meshwork will increase, but the mucin meshwork and its biochemical properties are likely the same.

2.4.2 *Mucin Granule Exocytosis*

2.4.2.1 Regulatory Aspects

Mucin granule exocytosis is triggered by a wide variety of stimuli, including inflammatory and neuroendocrine mediators, nucleotides (ATP and UTP), nutritional factors, bacterial toxins and exoproducts, hormones, reactive oxygen species, and apical membrane disruption. Although much work is still needed to understand the regulatory networks involved, three tentative conclusions can be proposed from the published data. First, most stimuli increase both mucin gene expression and secretion (e.g., Hong et al. 1997, 1999; Plaisancie et al. 2006; Zoghi et al. 2006). It appears, therefore, that mucin-regulated discharge is coupled with the corresponding increase of *MUC* gene expression, which may just reflect the need to replenish the intracellular mucin pool of goblet /mucous cells.

Second, irrespective of the stimuli, a protein kinase C (PKC)-dependent pathway is part of the mucin regulatory network (e.g., Hong et al. 1997, 1999; El Homsi et al. 2007). Specifically, studies with mucin-producing cell lines suggest the participation of the nPKCϵ isoform (Hong et al. 1999). A similar PKC-dependent signal pathway could function in mucin-producing cells of the middle ear (Lin et al. 2000), intrahepatic biliary tract (Zen et al. 2002), and respiratory tract (e.g., Ehre et al. 2007; Park et al. 2007), although conflicting results have been reported regarding the specific PKC isoform that mediates the response in respiratory cells. MARCKS (myristolated alanine-rich C-kinase substrate) is likely one of the proteins phosphorylated by PKC in the airway goblet cells during mucin-regulated secretion (Park et al. 2007). Thus, upon phosphorylation by PKC, MARCKS would bind to the mucin granule and somehow mediate interactions with microfilaments necessary for granule sorting and attachment to the plasma membrane (Li et al. 2001). Besides the PKC-dependent pathway, other pathways can mediate mucin secretion in gastrointestinal goblet cells, including protein kinase A- (e.g., Bradbury 2000) and phosphatidylinositol 3-kinase-mediated. (Plaisancie 2006) pathways.

Third, increase of cytosolic [Ca^{2+}] is a crucial event during the exocytotic mechanism of mucin granule as for any regulated secretory granule (Burgoyne and Morgan 2003). Indeed, it is well established that Ca^{2+} ionophores (e.g., ionomycin) are robust mucin secretagogues whereas Ca^{2+} chelators (e.g., BAPTA) inhibit secretagogue-triggered mucin secretion. Inositol-tri-phosphate (IP_3)-dependent Ca^{2+} release from the endoplasmic reticulum, which in living goblet cells appears to engulf the granules (Perez-Vilar et al. 2005b), the mucin granule lumen (Chin et al. 2002), or the extracellular medium (e.g., Miyake et al. 2006), has been proposed as likely sources for calcium. Extracellular Ca^{2+} is the main source when mucin secretion is triggered by plasma membrane disruption (Miyake et al. 2006). In any case, an exchange of intragranular Ca^{2+} by cytosolic monovalent ions (K^+ and/or Na^+) might trigger the Ca^{2+}-dependent fusion of the mucin granule and the plasma membrane (Verdugo 1990; Nguyen et al. 1998). In this respect, our FRAP studies suggest that a change in the intragranular organization is sufficient to trigger granule partial discharge in the absence of secretagogue activation (Perez-Vilar et al. 2006).

2.4.2.2 Biochemical/Biophysical Aspects

Once the intragranular condensed mucin network (see Section 2.5.1) makes contact with the extracellular medium, it diffuses into the latter compartment and gradually swells. The swelling of the mucin granule matrix, as determined by video-enhanced microscopy (Verdugo 1990; Verdugo 1991), follows the same kinetics observed during the swelling of spherical synthetic gels (Li and Tanaka 1992), which is characterized by the following relation:

$$\tau \equiv \frac{R^2}{D} \tag{2.1}$$

where τ is the characteristic time of swelling of the gel, i.e., the time taken by the gel to reach a radius half the value of its final radius (R), i.e., once the gel is fully swollen, and D is the collective diffusion of the gel into the aqueous medium (Li and Tanaka 1992). Since the swelling of these synthetic gels can be explained by a volume phase transition, a universal physical phenomena that occurs when a small change in an environment parameter (pH, temperature, ionic composition, etc.) results in a drastic change in one or more of the forces operating in a gel (i.e., chain–chain interactions, osmotic pressure, and elasticity; Li and Tanaka 1992) and ultimately the collapse (or swelling) of the gel, it was proposed that the same mechanism could explain mucin gel expansion during granule discharge (Fig. 2.2) (Verdugo 1990, 1991). The displacement of the intragranular Ca^{2+}, which cross-links and shields the polyanionic mucin chains in the granule lumen, by extracellular monovalent ions is thought to allow the repulsion among mucin chains, i.e., alters the chain–chain interactions, eventually resulting in gel swelling. Consistent with this notion, intragranular matrix expansion can be prevented by increased levels of extracellular Ca^{2+} and acidic pH (Verdugo 1991). Because phase transitions are reversible, a volume phase transition could in part explain the condensation of the mucin matrix during granule formation (Perez-Vilar 2007, 2008).

The "freshly" discharged and swelled mucus is now free to flow and anneal with the extracellular mucus gel. Since the same tracer protein diffuses at least 150-fold faster in the extracellular matrix than in the intragranular mucin matrix (Perez-Vilar et al. 2005a; Perez-Vilar, unpublished observations), which has an average pore size of 5–10 nm, the mucus network mesh size might be in the micron range (see Section 2.7).

2.5 Mucin Extracellular Organization

The rheological properties of mucus and purified mucin solutions range from those of a gel (e.g., gastric mucus), i.e., with viscoelastic properties, to a viscous aqueous solution (e.g., saliva) depending on the mucus source, water content (i.e., mucin concentration), and the presence of other components, including ions, proteins, DNA, and lipids. However, in its simple state, mucus can be considered an aqueous mucin solution and, accordingly, its properties are governed by the properties of mucin macromolecules in solution. Because mucin polypeptides are dominated by the central, O-glycosylated mucin domain, which essentially does not have secondary structure (see Sections 2.2.1.1 and 2.2.2.1), it is not surprising that mucin macromolecules in solution behave like random coil polymers, meaning that the mucin chains are highly flexible and continually change from one conformation to another (Fig. 2.2). Accordingly, the same physical principles that describe the properties of random coils of less complex, synthetic polymers are valid to fully appreciate the importance of mucin macroscopic organization.

2.5.1 Mucin Entanglements

The formation of mucin entangled networks can be reasonably understood when experimental parameters obtained with mucin solutions are considered under basic principles of polymer physics (Doi and See 1995; see also Perez-Vilar 2007 and references herein). For instance, an MUC5AC linear 16-mer oligomer has a contour length (L), i.e., the end-to-end length, of ~8 μm. Since gel-forming mucin chains in solution behave like random coils and have Kuhn or effective segments (l), i.e., the minimal polypeptide segment that can be considered rigid, ranging from 30 to 60 nm and a diameter d between 5 and 10 nm (Verdugo 1990; Bansil et al. 1995; Bansil and Turner 2006), mucin chains in diluted solutions can be considered to be physical polymers comprising $N \sim l^{-1}L$ freely jointed cylindrical subunits and a spherical solution volume (V) (Fig. 2.2) that can be approximated by the following expression:

$$V \sim 4 \times \frac{\pi}{3} \times \left(\frac{l^{4/5} \times d^{1/5} \times N^{0.588}}{2.45}\right)^3 \qquad (2.2)$$

Hence, for a 16-mer mucin chain, N and V would range from ~134 to ~268 and from ~0.076 to 012 μm^3, respectively. It can be concluded that due to their long sizes and flexibility mucin chains tend to occupy large volumes of the solution in which they are diluted.

Above certain concentration (the threshold concentration or c^*), the polymer chains begin to overlap and form entanglements and eventually a heavily entangled network or gel (Fig. 2.2) (Doi and See 1995). The dimensionless counterpart of c^*, the volume fraction (ϕ^*), i.e., the actual volume occupied by the polymer chains in the solution of volume V, can be estimated by the following expression:

$$\phi^* \sim \frac{L}{V} \times d^2 \tag{2.3}$$

For instance, a 16-mer MUC5AC chain will begin to form entanglements at ϕ^* values between ~0.0003 and 0.0005, meaning that oligomer overlapping starts when the mucin strands only fill ~0.03–0.05% of the available volume. Hence, at the concentration found in the mucus, mucin oligomer/multimers form intricate entangled networks due to their high flexibility and large sizes and, indeed, experimental studies have long supported this notion (e.g., Verdugo 1990; Raynal et al. 2002; Yakubov et al. 2007). Moreover, since mucins are more concentrated in the mucin granule lumen than in the mucus (see above), intragranular mucins must also be forming a highly entangled, condensed network (Perez-Vilar 2007, 2008; see Section 2.4.1).

2.5.2 Other Interchain Links

Although entanglements can govern and control the elasticity of polymer networks with long strands, the question arises whether in complex biological secretions such as intracellular compartments or the mucus gels other interchain links (e.g., mucin–mucin interactions, ions, link proteins) stabilize the entangled mucin network. In this respect, rheological studies suggest that mucus is formed by a mucin network mainly held together by weak non-covalent bonds with the likely intervention of certain ions (e.g., Madsen et al. 1998; Bansil et al. 1995; Raynal et al. 2003). Ca^{2+}-dependent, especially in the case of intracellular mucins (see Section 2.4.1), lectin-like, hydrophobic or protein-mediated interactions have been proposed to hold the mucin network. For instance, early reports documented the tendency of mucins in solution to form aggregates at low salt concentrations (Slayter et al. 1984; Rose et al. 1984), which could be mediated by interchain hydrophobic interactions (Bromberg and Barr 2000). It has been suggested that the CS-domains in pMuc5ac, the porcine counterpart of human MUC5AC, mediate pH-dependent, hydrophobic protein–protein interactions critical for gastric mucus gel formation at low pH (Cao et al. 1999; Hong et al. 2007). Interestingly, purified mucin CS-domains are able to interact non-covalently, but very weakly, at neutral or slightly acidic pH, although assays at lower pH have not yet been carried out (Perez-Vilar et al. 2004).

Members of the trefoil factor family of small proteins are solid candidates to be mucin interchain cross-linkers (Otto and Thim 2005). For example, TFF1 is mainly expressed in the goblet cells of gastric mucosa, where it co-localizes with and binds to MUC5AC (Ruchaud-Sparagano et al. 2004). Yeast two-hybrid screenings suggest that the COOH-terminal C-domain in MUC5AC is the domain interacting with MUC1 (Tomasetto et al. 2000). Consistent with these findings, TFF1 does not bind to MUC6 (Ruchaud-Sparagano et al. 2004), a mucin that lacks C-domains (see Section 2.2.1.4). Addition of TFF2, which is expressed in gastric mucus neck cells, to a mucin solution increases its viscosity and elasticity (Thim et al. 2002), which suggests an increase in the degree of cross-linking among the mucin chains.

2.5.3 Mucinases

An entangled network or gel is by definition a dynamic system, in which the polymeric chains continually diffuse. In fact, like other flexible polymers, the diffusion of mucin chains in the mucus can be explained by a reptation-like mechanism (De Gennes 1979), i.e., the chains move forward and backward until new entanglements are formed while others disappear. Because the reptative motion of a polymer is inversely proportional to N^2, the shorter mucin chains diffuse faster than their largest counterparts. Moreover, since the viscosity of such polymeric solutions is directly proportional to $\sim N^3$, mucus secretions with a higher proportion of shorter chains would be less viscous, more dispersible, and hence penetrable, than mucus containing larger mucin chains. Hence, it is not surprising that some microorganisms synthesize proteases that cleavage the mucin polypeptide and eventually disorganize the mucus barrier. Such is the case of *Entamoeba histolytica*, an intestinal protozoan parasite, which synthesizes cysteine proteases able to cleave the COOH-terminal region of MUC2 in a site within the D4-domain (Lidell et al. 2006). The resulting mucin chains are shorter as they lack continuity through their COOH-regions and accordingly form less intricate and penetrable mucus networks, allowing this parasite to traverse the intestinal mucus barrier.

The existence of bacterial proteases, glycosidases, and sulfatases acting on mucins, known collectively as mucinases, has been long recognized (e.g., Corfield et al. 1993;). Glycosidases and sulfatases are expected to alter the hydration shells embedding the mucin network by eliminating hydrophilic and negatively charged sugar residues. This would lead to increased diffusion of water molecules throughout the mucin chains, formation of less extended mucin chains, and perhaps the disruption of Ca^{2+}-mediated interchain interactions. Mucus in these conditions would be expected to be less viscous and more penetrable.

2.5.4 Emerging Notion

It is increasingly clear that mucins in mucus gels form a tri-dimensional network in which large macromolecules can diffuse through their pores. The mucin network likely originates in the goblet/mucous cell, i.e., inside the compartments of the secretory pathway, as mucin properties, large sizes, high intracellular concentration, and the sizes of the different Golgi complex subcompartments strongly argue in favor of the existence of intraluminal entangled networks (Perez-Vilar 2007, 2008). Simultaneously, the accumulation of mucin chains inside mucin granules cannot be understood without a role of the abundant intragranular Ca^{2+}, which presumably serves as a cross-linker of the polyanionic mucins (Perez-Vilar 2007). This is necessary not only to condense the mucin chains in a relatively small compartment, compared to mucin contour lengths, but also to minimize the osmotic pressure associated with mobile charged proteins. Hence, the existence of interchain interactions, in addition to topological entanglements, seems to depend on the environment in which the mucin is located and its function. For instance, gastric mucus would be expected to primarily protect the gastric mucosa from acidic pH and pepsin digestion. In this respect, the gastric mucus is able to cope with the gastric juice by (a) absorbing the bicarbonate secreted into the lumen, which forms a gradient ranging from near neutral at the epithelial surfaces to pH $\sim$ 2 at the boundaries with the stomach lumen (e.g., Allen and Flemstrom 2005); (b) forming pH-dependent mucin aggregates (i.e., increasing the cross-linking degree of the mucin network gel whenever acidic pH is encountered) (e.g., Hong et al. 2005), which permits the efflux of acid juices from the gastric mucosa (Bhaskar et al. 1992) while preventing these juices from penetrating the mucus from the lumen; (c) exploiting the hydrodynamic and pH-dependent properties of the gastric mucins (Bhaskar et al. 1992); (d) synthesizing MUC5AC and MUC6, a mucin lacking CS-domains and hence more resistance to proteases; and (e) maintaining a thick (>400 μm) mucus barrier likely by keeping a high rate of mucin synthesis and secretion.

On the other hand, it is important to consider that the mucus covers almost the entire gastrointestinal epithelia, which means mucin chains must be able to flow and anneal, especially considering the fact that mucins are secreted by a fraction of the gastrointestinal epithelial cells. Moreover, the permeability of the mucus must change in different areas of the gastrointestinal tract. These properties are better served by an entangled dynamic network, where non-covalent interactions, if any, are weak and reversible. The degree of cross-linking, whether entanglements or non-covalent interactions, can differ depending on the amount and types of gel-forming mucins present, the type of *O*-glycans, the ionic conditions, or the presence of cross-linking proteins. It can be anticipated that in each area of the gastrointestinal tract there are a specific combination of these factors and accordingly a mucus layer tailored for the specific needs of such area. Indeed, the gastrointestinal mucus layer varies in thickness from a very thin layer in the small intestine to more than 280 and 400 μm in the rectum and stomach, respectively (Deplancke and Gaskins 2001). Moreover, reports suggesting that the gastric mucus layer is not a homogeneous structure but comprises two different layers enriched in MUC5AC and MUC6, respectively (Hidaka et al. 2001), can be rationalized by considering the continuous diffusion of the chains inside the entangled gel and the fact that mixtures of polymers in solution will be fractionated over time into different phases, each enriched in one polymer.

2.6 Other Mucus Components

In the preceding sections we have summarized the most important properties of gel-forming mucins and how they are organized into a dynamic entangled network embedded in an aqueous medium. Mucus water content ranges from 70 to 95% of the total weight. Intermingled with the mucin network, there are many other mucus components, including the crucial microbiota and derived products (proteins, DNA, etc.), dislodged epithelial cells and side products (proteins, DNA, cell debris, etc.), proteoglycans, proteins (e.g., albumin), proteinases, antiproteinases, lipids (e.g., phospholipids, cholesterol, and free fatty acids), members of the innate and adaptive immunity systems (e.g., antimicrobial peptides, lysozyme, complement components, sIgA), and digestive enzymes. Quantitatively, non-mucin compounds can surpass mucins. For example, the dry weight of pig intestinal mucus contains, among other unidentified constituents, the following components: 5% (w/w) mucin, 39% (w/w) proteins, 27% (w/w) lipids, and 6% (w/w) DNA (Larhed et al. 1998).

Some mucus compounds such as TFFs (see Section 2.5.2) likely have a stable association with the mucus mucin framework. This could also be the case of protective compounds in general and defensins in particular (Otte et al. 2003), a family of antimicrobial peptides, which due to its strong cationic charge likely interact with the polyanionic mucin domains in the mucus mucin network. Other mucus (transient or permanent) compounds, including DNA, proteoglycans, and glycosylated subunits of membrane/tethered mucins likely contribute to the viscoelastic properties of the gastrointestinal mucus due to the fact that they are large polyanionic macromolecules and, in some cases (e.g., DNA), extraordinarily flexible. Moreover, non-mucin components likely have a significant role during the diffusion of macromolecules in mucus (see below).

2.7 Diffusion of Macromolecules in Mucus

2.7.1 Influence of Mesh Size

In an entangled polymer network the interchain length (i.e., mesh or pore size) can be assessed by determining the correlation length (ξ) (Doi and See 1995; Graessley 2004), which for linear polymers with cylindrical Kuhn segments such as mucins can be approximated by the following expression:

$$\xi \sim \phi^{-0.76} \times l \times \left(\frac{l}{d}\right)^{0.23} \tag{2.4}$$

where ϕ, l, and d are the volume fraction, *Kuhn* segment, and chain width, respectively (see Section 2.5.1). For instance, the mesh size for mucin gels with ϕ, l, and d values between 0.01–0.1, 30–60 nm, and 5–10 nm, respectively, would range from 220 to 2,800 nm, which is not far from the range of values obtained empirically (Table 2.1). Of course, differences in the concentration of mucins (e.g., mucus water content), O-glycosylation pattern (e.g., glycan chain length and charge), degree of cross-linking (e.g., Ca^{2+}- or protein–protein-dependent mucin–mucin interactions, see Section 2.5.2), and/or the specific mucin expressed (e.g., mucin chain length) will result in native mucin meshworks with characteristic average pore sizes, viscoelasticity, and eventually permeability. That the pore size physically constrains the diffusion of macromolecules is attested by the following observations: (a) in general, the larger the macromolecule, the lower its intramucus rate of diffusion (e.g., Desai et al. 1992); (b) mucolytic agents, including disulfide-reducing agents and proteases, augment protein diffusion in GI mucus (e.g., Bernkop-Schnürch et al. 1999); and (c) transfection agents, which compact the DNA conformation, increase supercoiled DNA diffusion in mucus by a factor of 2 (Shen et al. 2006).

Table 2.1 Average mesh size in different mucus

Mucus	Mesh size (nm)	References
Cervical (bovine)	1,000–12,000	Shen et al. (2006)
Cervical (human)	100–300	Saltzman et al. (1994)
Cervical (rat)	20–800	Olmsted et al. (2001)
Cystic fibrosis sputum (human)	>200	Dawson et al. (2003)
Gastric (porcine)	200–650	Celli et al. (2005)
Gastric (porcine)	>650	Celli et al. (2005)
Tracheobronchial – 2.5% (w/w) – (human)	200–1000	Matsui et al. (2006)
Tracheobronchial – 8.0% (w/w) – (human)	<200	Matsui et al. (2006)

When considering the influence of mesh size it is important to recognize the effect of other mucus components (see below) and also that the GI mucus appears to comprise two continuous but distinct layers: the outer, loosely adherent mucous layer and the inner, firmly adherent mucus layer (Atuma et al. 2001). These mucus layers differ in thickness (Table 2.2) and perhaps organization (pore mesh), although the underlying reason for this intramucus compartmentalization has not yet been determined. The fact that the inner mucus layer is most abundant and the ratio of inner layer/ outer layer thickness is larger in the stomach supports the view that the former has a protective function while the latter would mainly function as a lubricant (Atuma et al. 2001).

Table 2.2 Thickness of the two mucus layers along the rat GI system. Shown are the mean values (standard error) in micrometers. Adapted from Atuma et al. (2001)

	Corpus (*n* = 6)	Antrum (*n* = 6)	Duodenum (*n* = 7)	Jejunum (*n* = 6)	Ileum (*n* = 6)	Colon (*n* = 5)
Total	189 (11)	274 (41)	170 (38)	123 (4)	480 (47)	830 (110)
Inner	80 (5)	154 (16)	16 (3)	15 (2)	29 (8)	116 (51)
Outer	109 (12)	120 (38)	154 (39)	108 (5)	447 (47)	714 (109)
Inner/outer	0.73	1.28	0.10	0.13	0.06	0.16

2.7.2 Influence of Other Factors

Since most proteins and viruses have hydrodynamic sizes under 100 nm, and irrespective of differences due to mucus source, sampling, etc., other factors besides mucus mesh size must contribute to explain the restrictive nature of the GI mucus (e.g., Bernkop-Schnürch and Fragner 1996; Flemstrom et al. 1999). First, mucus layer thickness varies along the GI tract (Table 2.2) and accordingly, it is likely that mucus thickness influences the diffusion of luminal macromolecules. Second, mucus is an aqueous solution and mucins are highly hydrophilic proteins (see Section 2.2) and, accordingly, diffusion of hydrophilic compounds would be favored over diffusion of lipophilic ones. This has been corroborated by studying the diffusion rate of relatively small molecular weight model drugs in intestinal mucus (e.g., Larhed et al. 1997). Lipophilicity, and not charge, was the critical property determining the rate of drug diffusion within the GI mucus. However, as the drug molecular weight increases, size is the predominant parameter. Third, mucus components other than mucins (see Section 2.6) could contribute to mucus permeation. Indeed, the lipophilicity–-diffusion relationship mentioned above is no longer observed when the diffusion rate of the same drugs is measured in purified mucins (Larhed et al. 1997), which suggests that other mucus components also contribute to reduce the effective pore size of mucus. Such an effect could explain, at least in part, why large macromolecules or even nanoparticles are able to freely diffuse in mucin-enriched solutions (e.g., Matsui et al. 2006; Lai et al. 2007). Moreover, the chemical nature of non-mucin mucus components likely contributes as well to restrict the diffusion of certain compounds. For example, lipids appear to be the critical mucus components that reduce diffusion of lipophilic drugs in GI mucus (Larhed et al. 1996).

2.8 Conclusions

Mucus is basically a semi-dilute aqueous solution in which very long, highly hydrophilic and flexible mucin macromolecules form a tri-dimensional network or gel that keeps the water molecules and other components together. Mucins are among the largest and more complex proteins known, and even the most primitive metazoans have gel-forming mucins similar to those found in humans (Lang et al. 2007). This conservancy suggests that mucus organization and function do not permit much variability and likely depends on maintaining a thin balance between the mucin network and the aqueous medium where the network is embedded. The network is highly dynamic and responds not only to the water content (swelling vs. shrinking) but also to the particular environment (Ca^{2+}, pH, etc.), adapting its macromolecular structure accordingly (e.g., increasing interchain cross-linking in the stomach lumen). Other mucus components seem to be critical for certain mucus functional properties such as its role as a diffusion barrier against macromolecules.

Any experimental or pathological condition that directly or indirectly changes the amount of extracellular mucin (i.e., gene expression, biosynthesis, and/or secretion), mucin structure (post-translational modification, length, etc.) or mucus water, or ionic, content will result in an (structurally and functionally) altered mucus barrier. Gastric ulcer (Slomiany and Slomiany 2002), colitis (e.g., van der Sluis et al. 2006), or even colon cancer (Velcich et al. 2002) are likely examples of the consequences of a faulty mucin expression and/or mucus composition. Moreover, microorganisms have evolved mechanisms to disrupt the GI mucus barrier and it is tempting to envision us being able to do the same for pharmacological purposes as our knowledge on mucin biochemistry, biophysics, and cellular biology keeps growing.

Acknowledgments The author's studies mentioned in this review were supported by grants from the University of North Carolina Research Council, the North American Cystic Fibrosis Foundation, and the National Institute of Heath (NIDDK).

References

Allen A. and Flemstrom G. (2005) Gastroduodenal mucus bicarbonate barrier: protection against acid and pepsin. Am J Physiol Cell Physiol 288, C1–C19

Atuma C., Strugala V., Allen A. and Holm L. (2001) The adherent gastrointestinal mucus gel layer:thickness and physical state in vivo. Am J Physiol Gastrointest Liver Physiol 280,G922–G929

Bansil R., Stanley E. and LaMont J.T. (1995). Mucin biophysics. Annu Rev Physiol. 57, 635–657

Bansil R. and Turner B.S. (2006). Mucin structure, aggregation, physiological functions and biomedical applications. Curr Opin Colloid Interf Sci. 11, 164–170

Bell S.L., Xu G., Khatri I.A., Wang R., Rahman S. and Forstner J.F. (2003). N-linked oligosaccharides play a role in disulphide-dependent dimerization of intestinal mucin Muc2. Biochem J. 373, 893–900

Bernkop-Schnürch A. and Fragner R. (1996) Investigations into the diffusion behavior of polypeptides in native intestinal mucus with regard to their peroral administration. Pharmaceutical Sci 2, 361–363

Bernkop-Schnürch A., Valenta C. and Daee S.M. (1999) Peroral polypeptide delivery. A comparative in vitro study of mucolytic agents. Arzneimittelforschung 49, 799–803

Bhaskar K.R., Gank P., Turner B.S., Bradley, J.D., Bansil R., Stanley H.E. and LaMont J.T. (1992). Viscous fingering of HCl through gastric mucin. Nature 360, 458–461

Bi L.C. and Kaunitz J.D. (2003) Gastroduodenal mucosal defense: an integrated protective response. Curr Opin Gastroenterol. 19, 526–532

Bradbury N.A. (2000) Protein kinase A-mediated secretion of mucin from human colonic epithelial cells. J Cell Physiol 185, 408–415

Brockhausen I. (2003). Sulphotransferases acting on mucin-type oligosaccharides. Biochem Soc Trans. 31, 318–325

Bromberg L.E. and Barr D.P. (2000) Sel-association of Mucin. Biomacromolecules 1, 325–334

Burgoyne R.D. and Morgan A. (2003). Secretory granule exocytosis. Physiol Rev. 83, 581–632

Cao X., Bansil R., Bhaskar K.R., Turner B.S., LaMont J. T., Niu N. and Afdhal N.H. (1999). pH-dependent conformational change of gastric mucin leads to sol-gel transition. Biophys J. 76, 1250–1258

Celli J., Gregor B., Turner B., Afdhal N.H., Bansil R. and Erramilli S. (2005) Viscoelastic Properties and Dynamics of Porcine Gastric mucin. Biomacromolecules 6, 1329–1333

Chen Y., Zhao Y. H., Kalaslavadi T.B., Hamati E., Nehrke K., Le A.D., Ann D.K. and Wu R. (2004). Genome-wide search and identification of a novel gel-forming mucin MUC19/Muc19 in glandular tissues. A J Respir Cell Mol Biol. 30, 155–165

Chin W.C., Quesada I., Nguyen T. and Verdugo P. (2002). Oscillations of pH inside the secretory granule control the gain of Ca2+ release for signal transduction in goblet cell exocytosis. Novartis Found Symp. 248, 132–141

Corfield A.P., Wagner S.A., O'Donnell L.J., Durdey P., Mountford R.A. and Clamp J.R. (1993) The roles of enteric bacterial sialidase, sialate O-acetyl esterase and glycosulfatase in the degradation of human colonic mucin. Glycoconj J 10, 72–81

Dawson M., Wirtz D. and Hanes J. (2003) Enhanced viscoelasticity of human cystic fibrosic sputum correlates with increasing microheterogeneity in particle transport. J Biol Chem. 278, 50393–50401

De Gennes P.G. (1979) Scaling concepts in polymer physics. USA: Cornell University Press

Dekker J., Rossen J.W., Buller H.A. and Einerhand, A.W. (2002). The MUC family: an obituary. Trends Biochem Sci. 27, 126–131

Deplancke B. and Gaskins H.R. (2001) Microbial modulation of innate defense: goblet cells and the intestinal mucus layer. Am J Clin Nutr 73, 1131S–1141S

Desai M.A., Mutlu M. and Vadgama P. (1992) A Study of Macromolecular Diffusion through Native Porcine Mucus. Experientia 48, 22–26

Dessein J.L., Guyonnet-Duperat V., Porchet N., Aubert J.P. and Laine A. (1997). Human mucin gene MUC5B., the 10.7 kb large central exon encodes various alternate subdomains resulting in a super-repeat. Structural evidence for a 11p15.5 gene family. J Biol Chem. 272, 3168–3178

Doi M. and See H. Introduction to polymer physics. USA: Oxford University Press; 1995

Eckhardt A.E., Timpte C.S., Abernethy J.L., Toumadje A., Johnson W.C. and Hill R.L. (1987). Structural properties of porcine submaxillary mucin. J Biol Chem. 262, 11339–11344

Eckhardt A.E., Timpte C.S., Abernethy J.L., Zhao Y. and Hill R.L. (1991). Porcine submaxillary mucin contains a cystine-rich, carboxyl-terminal domain in addition to a highly repetitive, glycosylated domain. J Biol Chem. 266, 9678–9686

Ehre C., Zhu Y., Abdullah L.H., Olsen J., Nakayama K.I., Nakayama K., Messing R.O. and Daviss C.W. (2007) nPKCϵ, a P3Y2-R downstream effector in regulated mucin secretion from airway goblet cells. Am J Physiol Cell Physiol 293, C1445–C1454

El Homsi M., Ducroc R., Claustre J., Jourdan G., Gertler A., Estienne M., Bado A., Scoazec J.Y. and Plaisancie P. (2007) Leptin modulates the expression of secreted and membrane-associated mucins in colonic epithelial cells by targeting PKC, PI#K, and MAP.K. Am J Physiol Gastrointest Liver Physiol 293,G365–G373

Escande F., Aubert J.P., Porchet N. and Buisine M.P. (2001). Human mucin gene MUC5AC: organization of its 5'-region and central repetitive region. Biochem J. 358, 763–772

Flemstrom G., Hallgren A., Nylander O., Engstrans L., Wilander E. and Allen A. (1999) Adherent surface mucus gel restricts diffusion of macromolecules in rat duodenum in vivo. Am J Physiol Gastrointest Liver Physiol 277, 375–382

Forstner G. (1995) Signal transduction, packaging and secretion of mucins. Annu Rev Physiol 57, 585–605

Gerken T.A. (1993) Biophysical approaches to salivary mucin structure, conformation and dynamics. Crit Rev Oral Biol Med 4, 261–270

Gold K., Johansson M.E., Lidell M.E., Morgelin M., Karlsson H., Olson F. J., Gum J.R., Kim, Y.S. and Hansson G.C. (2002). The N terminus of the MUC2 mucin forms trimers that are held together within a trypsin-resistant core fragment. J Biol Chem. 277, 47248–47256

Graessley W.W. (2004) Polymeric liquids and networks: structure and properties. USA: Garland Sciences

Ham M. and Kaunitz J.D. (2007) Gastroduodenal defense. Curr Opin Gastroentrol 23, 607–616

Hidaka E., Ota H. and Hidaka H. (2001) Helicobacter pylori and two ultrastructurally distinct layers of gastric mucous cell mucins in the surface mucous gel layer. Gut 49, 474–480

Hong D.H., Forstner J. and Forstner G. (1997) Protein kinase C-ϵ is the likely mediator of mucin exocytosis in human colonic cell lines. Am J Physiol Gastrointest Liver Physiol 272,G31–G37

Hong D.H., Petrovics G., Anderson W.B., Forstner J. and Forstner G. (1999) Induction of mucin gene expression in human colonic cell lines by PMA is dependent on PKC-ϵ. Am J Physiol Gastrointest Liver Physiol 277,G1041–G1047

Hong Z., Chasan B., Bansil R., Turner B.S., Bhaskar K.R. and Afdhal N.H. (2005) Atomic force microscopy reveals aggregation of gastric mucin at low pH. Biomacromolecules 6, 3458–3466

Jentoft N. (1991) Why are proteins O-glycosylated?. Trends Biochem Sci. 15, 291–294

Kawakubo M., Ito Y., Okimura Y., Kobayashi M., Sakura K., Kasama S., Fukuda M.N., Fukuda M., Katsuyama T. and Nakayama J. (2004). Natural antibiotic function of a human gastric mucin against Helicobacter pylori infection. Science 305, 1003–1006

Klein A. and Roussel P. (1998) O-Acetylation of sialic acids. Biochimie 80, 49–57

Kuver R., Klinkspoor J.H., Osborne W.R.A. and Lee S.P. (2000) Mucous granule exocytosis and CFTR expression in gallbladder epithelium. Glycobiology 10, 149–157

Lai S.K., O'Hanlon E., Harrold S., Man S.T., Wang Y-Y., Cone R. and Hanes J. (2007) Rapid transport of large polymeric nanoparticles in fresh undiluted human mucus. Proc Natl Acad Sci USA 104, 1482–1487

Lang T., Alexandersson M., Hansson GC and Samuelsson T. (2004). Bioinformatic identification of polymerizing and transmembrane mucins in the puffer fish Fugu rubripes. Glycobiology 14, 521–527

Lang T., Hansson G.C. and Samuelsson T. (2007) Gel-forming mucins appeared early in metazoan evolution. Proc Natl Acad Sci USA 104, 16209–16214

Larhed A.W., Artursson P. and Bjork E. (1998) The influence of intestinal mucus components on the diffusion of drugs. Pharm Res 15, 66–71

Larhed A.W., Artursson P., Grasjo J. and Bjork E. (1997) Diffusion of drugs in native and purified gastrointestinal mucus. J Pharm Sci 86, 660–665

Li Y. and Tanaka T. (1992). Volume phase transitions of gels. Annu Rev Mater Sci. 22, 243–277

Li Y., Martin L.D., Spizz G. and Adler K.B. (2001). MARCKS protein is a key molecule regulating mucin secretion by human airway epithelial cells in vitro. J Biol Chem. 276, 40982–40990

Lidell M.E., Johansson M.E. and Hansson G.C. (2003). An autocatalytic cleavage in the C terminus of the human MUC2 mucin occurs at the low pH of the late secretory pathway. J Biol Chem. 278, 13944–13951

Lidell M.E. and Hansson G.C. (2006) Cleavage in the GDPH sequence of the C-terminal cysteine-rich part of the human MUC5AC mucin. Biochem J. 399, 121–129

Lidell M.E., Moncada D.M., Chadee K. and Hansson G.C. (2006). Entamoeba histolytica cysteine proteases cleave the MUC2 mucin in its C-terminal domain and dissolve the protective colonic mucus gel. Proc Natl Acad Sci U S A. 103, 9298–9303

Lievin-Le Moal V. and Servin A.L. (2006) The front line of enteric host defense against unwelcome intrusion of harmful microorganisms: mucins, antimicrobial peptides, and microbiota. Clin Microbiol Rev 19, 315–337

Lin J., Haruta A., Kawano H., Ho S.B., Adams G.L., Juhn S.K. and Kim Y. (2000) Induction of mucin gene expression in middle ear of rats by tumor necrosis factor-alpha: potential cause for mucoid otitis media. J Infect Dis 182, 882–887

Madsen F., Eberth K. and Smart J.D. (1998) A Rheological examination of the Mucoadhesive/Mucus Interaction; the Effect of Mucoadhesive Type and Concetration. J. Control Release 50, 167–178

Matsui H., Wagner V.E., Hill D.B., Schwab U.E., Rogers T.D., Buttom B., Taylor R.M., Superfine R., Rubinstein M., Iglewski R.H. and Boucher R.C. (2006) A physical linkage between cystic fibrosis airway surface dehydration and Pseudomonas aeruginosa biofilms. Proc Natl Acad Sci USA 103, 18131–18136

McCool D.J., Forstner J.F. and Forstner G.G. (1995) Regulated and unregulated pathways for MUC2 mucin secretion in human colonic LS180 adenocarcinoma cells are distinct. Biochem J 312, 125–133

McCool D.J., Okada Y., Forstner J.F. and Forstner G.G. (1999) Roles of calreticulin and calnexin during mucin synthesis in LS180 and HT29/A1 human colonic adenocarcinoma cells. Biochem J 341, 593–600

Meitinger T., Meindl A., Bork P., Rost B. and Hassemann M.M. (1993) Molecular modeling of the Norrie disease protein predicts a cystine knot growth factor tertiary structure. Nat Genet 5, 376–380

Miyake K., Tanaka T. and McNeil P.L. (2006) Disruption-induced mucus secretion: repair and protection. PLoS Biol 4, e276

Neutra M.R., Phillips T.L. and Phillips T.E. (1984). Regulation of intestinal goblet cells in situ, in mucosal explants and in the isolated epithelium. Ciba Found. Symp. 109, 20–39

Nguyen T., Chin W-C. and Verdugo P. (1998). Role of Ca^{2+}/K^{+} ion exchange in intracellular storage and release of Ca^{2+}. Nature 395, 908–912

Nochi T. and Kiyono H. (2006) Innate immunity in the mucosal immune system. Current Pharmaceutical Design 12,4203–4213

Noiva R. (1994) Enzymatic catalysis of disulfide formation. Protein Expr Purif. 5, 1–13

Olmsted S.S., Padgett J.L., Yudin A.I., Whaley K.J., Moench T.R. and Gone R.A. (2001) Diffusion of macromolecules and virus-like particles in human cervical mucus. Biophys J 81, 1930–1937

Otte J.M., Kiehne K. and Herzig K.H. (2003) Antimicrobial peptides in innate immunity of the human intestine. J Gastoenterol 38, 717–726

Otto W.R. and Thim L. (2005) Trefoil factor family-interacting proteins. Cell Mol Life Sci. 62, 2939–2946

Park J.A., Crews A.L., Lampe W.R., Fang S., Park J. and Adler K.B. (2007) Protein kinase Cδ regulates airway mucin secretion via phosphorylation of MARCKS protein. Am J Pathol. 171, 1822–1830

Perez-Vilar J., Eckhardt, A.E. and Hill R.L. (1996). Porcine submaxillary mucin forms disulfide-bonded dimers between its carboxyl-terminal domains. J Biol Chem. 271, 9845–9850

Perez-Vilar J. and Hill R.L. (1997). Norrie disease protein (norrin) forms disulfide-linked oligomers associated with the extracellular matrix. J Biol Chem 272, 33410–33415

Perez-Vilar J., Eckhardt, A.E., DeLuca A. and Hill R.L. (1998). Porcine submaxillary mucin forms disulfide-linked multimers through its amino-terminal D domains. J Biol Chem. 273, 14442–14449

Perez-Vilar J. and Hill R.L. (1998a). The carboxyl-terminal 90 residues of porcine submaxillary mucin are sufficient for forming disulfide-bonded dimers. J Biol Chem 273, 6982–6988
Perez-Vilar J. and Hill R.L. (1998b). Identification of the half-cystine residues in porcine submaxillary mucin critical for multimerization through the D-domains. Roles of the CGLCG motif in the D1 and D3-domains. J Biol Chem 273, 34527–34534
Perez-Vilar J. and Hill R.L. (1999). The structure and assembly of secreted mucins. J Biol Chem 274, 31751–31754
Perez-Vilar J. and Boucher R.C. (2004). Mucins form disulfide-linked multimers through a self-catalyzed pH-dependent mechanism. Pediatr Pulmonol. 38, 238(144)
Perez-Vilar J., Randell S.H. and Boucher R.C. (2004). C-Mannosylation of MUC5AC and MUC5B Cys subdomains. Glycobiology 4, 325–337
Perez-Vilar J., Olsen J.C., Chua M. and Boucher R.C. (2005a). pH-dependent intraluminal organization of mucin granules in live human mucous/goblet cells. J. Biol. Chem. 280, 16868–16881
Perez-Vilar J., Ribeiro C.M., Salmon W.C., Mabolo R. and Boucher R.C. (2005b). Mucin granules are in close contact with tubular elements of the endoplasmic reticulum. J Histochem Cytochem. 53, 1305–1309
Perez-Vilar J., Mabolo R. McVaugh C.T., Bertozzi C.R. and Boucher R.C. (2006). Mucin granule intraluminal organization in living mucous/goblet cells. Roles of protein post-translational modifications and secretion. J Biol Chem. 281, 4844–4855
Perez-Vilar J. (2007). Mucin granule intraluminal organization. Am J Respir Cell Mol Biol, 36, 183–190
Perez-Vilar J. and Mabolo R. (2007) Gel-forming Mucins. Notions from *in vitro* studies. Histol Histopathol. 22 455–464
Perez-Vilar J. (2008) Formation of Mucin Granules. In "The Golgi Apparatus. State of the Art 110 years After Camilo Golgi's Discovery" (Mironov A. & Pavelka M., eds.) Springer-Verlag, Viena/New York, pp. 535–562.
Plisancie P., Ducroc R., El Homsi M., Tsocas A., Guilmeau S., Zoghbi S., Thibaudeau O. and Bado A. (2006) Luminal: eptin Activates mucin-secreting Goblet Cells in the Large Bowel. Am. J. Physiol. Gastrointest. Liver Physiol. 290, G805–G812
Raynal B.D., Hardingham T.E., Thornton D.J. and Sheehan J.K. (2002) Concentrated solutions of salivary MUC5B mucin do not replicate the gel-forming properties of saliva. Biochem J. 362, 289–296
Raynal B.D., Hardingham T.E., Sheehan J.K. and Thornton D.J. (2003). Calcium-dependent protein interactions in MUC5B provide reversible cross-links in salivary mucus. J Biol Chem. 278, 28703–28710
Rose M.C., Voter W.A., Sage H., Brown C.F. and Kaufman, B. (1984). Effects of deglycosylation on the architecture of ovine submaxillary mucin glycoprotein. J Biol Chem. 259, 3167–3172
Roth J., Wang Y., Eckhardt A.E. and Hill R.L. (1994). Subcellular localization of the UDP-N-acetyl-D-galactosamine: polypeptide N-acetylgalactosaminyltransferase-mediated O-glycosylation reaction in the submaxillary gland. Proc Natl Acad Sci USA 91, 8935–8939
Rousseau K., Byrne C., Kim Y.S., Gum J.R., Swallow D.M. and Toribara N.W. (2004). The complete genomic organization of the human MUC6 and MUC2 mucin genes. 83 936–939
Ruchaud-Sparagano M.H., Westley B.R. and May F.E. (2004) The threfoil protein TFF1 is bound to MUC5AC in human gastric mucosa. Cell Mol Life Sci 61, 1946–1954
Sadler J.E. (1998). Biochemistry and genetics of von Willebrand factor. Annu Rev Biochem 67, 395–424
Saltzman W.M., Radomsky M.L., Whaley K.J. and Cone R.A. (1994) Antibody diffusion in human cervical mucus. Biophys J 66, 508–515
Shen H., Hu Y. and Saltzman W.M. (2006) DNA diffusion in mucus: effect of size, topology of DNAs, and transfection reagents. Biophys J 91, 639–644

Shogren R., Gerken T.A. and Jentoft N. (1989) Role of glycosylation on the conformation and chain dimensions of O-linked glycoproteins: light-scattering studies of ovine submaxillary mucin. Biochemistry 28, 5525–5536

Slayter H.S., Lamblin G., LeTreut A., Galabert C., Houdret N., Degand P. and Roussel P. (1984) Complex Structure of Human Bronchial Mucus Glycoprotein. Eur. J. Biochem. 142, 209–218

Slomiany B.L. and Slomiany A. (2002) Disruption in gastric mucin synthesis by Helicobacter pylori lipopolysaccharide involves ERK and P38 mitogen-activated protein kinase participation. Biochem Biophys Res Commun 294, 220–224

Snapp E., Altan N. and Lippincott-Schwartz J. (2003) Fluorescence Recovery After Photobleaching. In *Current Protocols in Cell Biology*. (Bonifacino, J., Dasso, M., Harford, J., Lippincott-Schwartz, J., Yamada K., Morgan K.S., eds.) Unit 21.1 John Wiley & Sons, Inc., New York

Spiro R.G. (2002). Protein glycosylation: nature, distribution, enzymatic formation, and disease implications of glycopeptide bonds. Glycobiology 12:43R–56R

Strous G.J. and Dekker J. (1992) Mucin-type glycoproteins. Crit Rev Biochem Mol Biol 27, 57–92

Sun P.D. and Davies D.R. (1995). The cystine-knot growth-factor superfamily. Annu Rev Biophys. Biomol. Struct. 24, 269–291

Thim L., Madsen F. and Poulsen S.S. (2002) Effect of trefoil factors on the viscoelastic properties of mucus gels. Eur J Clin Invest 32, 519–527

Tomasetto C., Masson R., Linares J.L., Wendling C., Levebvre O. and Chenard M.P. (2000) pS2/TFF1 interacts directly with the VWFC cysteine-rich domains of mucins. Gastroenterology 118, 70–80

Van der Sluis M., De Koning B.A., De Bruijin A.C., Velcich A., Meijerink J.P., Van Goudoever J.B., Buller H.A., Dekker J., Van Seuginen I., Renes I.B. and Einerhand A.W. (2006). Muc2-2-deficient mice spontaneously develop colitis, indicating that MUC2 is critical for colonic protection. Gastroenterology 131, 117–129

Velcich A., Yang W., Heyer J., Fragale A., Nicholas C., Viani S., Kucherlapati R., Lipkin M., Yang K. and Augenlicht L. (2002). Colorectal cancer in mice genetically deficient in the mucin Muc2. Science 295, 1726–1729

Veerman E.C., Van den Keijbus P.A., Nazmi K., Vos W., Van der Wal J.E., Bloemena E.J.G. and Amerongen A.V. (2003). Distinct localization of MUC5B glycoforms in the human salivary glands. Glycobiology 13, 363–366

Verdugo P. (1990). Goblet cells secretion and mucogenesis. Annu Rev Physiol 52, 157–176

Verdugo P. (1991). Mucin exocytosis. Am Rev Respir Dis 144, S33–S37

Vinall L.E., Pratt W.S. and Swallow D.M. (2000). Detection of mucin gene polymorphism. Methods Mol Biol 125, 337–350

Wickstrom C. and Carlstedt I. (2001). N-terminal cleavage of the salivary MUC5B mucin. Analogy with the Von Willebrand propolypeptide?. J. Biol. Chem. 276, 47116–47121

Yakubov G.E., Papagiannopoulos A., Rat E., Easton R.L. and Waigh T.A. (2007) Molecular structure and rheological properties of short-side-chain heavily glycosylated porcine stomach mucin. Biomacromolecules 8, 3467–3477

Zen Y., Harada K., Sasaki M., Tsuneyama K., Katayanagi K., Yamamoto Y. and Nakanuma Y. (2002) Lipopolysaccharide induces overexpression of MUC2 and MUC5AC in cultured biliary epithelial cells. Am J Pathol 161, 1475–1484

Zoghi S., Trompette K., Claustre J., El Homsi M., Garzon J., Jourdan G., Scoazee J.Y. and Plaisancie P. (2006) β–Casomorphin-7 regulates the secretion and expression of gastrointestinal mucins through a μ-opioid pathway. Am J Physiol Gastrointes Liver Physiol 290, G1105–G1113

Chapter 3
The Absorption Barrier

Gerrit Borchard

I have finally cum to the konklusion that a good reliable set ov bowels iz worth more to a man than enny quantity of brains.
Josh Billings *aka* Henry Wheeler Shaw (1818–1885)

Contents

Abstract This chapter attempts to give an overview on the properties of the intestinal epithelium with regard to both, barriers to transcellular (transporter and efflux systems) and paracellular (tight junctional complex) drug absorption and transport systems and tight junction modulation. A short introduction into the relation between the innate immune system and modulation of paracellular permeability is equally given.

3.1 Introduction

The intestinal mucosa serves two purposes: being constantly exposed to exogenous substances and pathogens, it represents a formidable barrier against the diffusion of toxins and against invasion and infection by pathogenic

G. Borchard (✉)
Laboratory of Pharmaceutics and Biopharmaceutics, School of Pharmaceutical Sciences, University of Geneva, Geneva, Switzerland
e-mail: gerrit.barchard@unige.ch; gerrit.borchard@pharm.unige.ch

A. Bernkop-Schnürch (ed.), *Oral Delivery of Macromolecular Drugs*,
DOI 10.1007/978-1-4419-0200-9_3, © Springer Science+Business Media, LLC 2009

organisms. It is equally responsible for the uptake and – to a certain extent – the metabolization of nutrients, cofactors, and vitamins. Only few active transport systems of interest for the delivery of pharmacologically active compounds are expressed in intestinal epithelial cells, such as the di/tripeptide transporter PEPT-1, and the neonatal Fc receptor (FcRn). Efflux systems expressed at the mucosal side of the epithelial cells contribute to the barrier function of the intestinal epithelium, as are the intercellular tight junctional complexes (TJ). These latter interconnect adjacent cells across the paracellular space, allowing only water and solutes of small molecular weight to permeate. Therefore, the vast majority of drugs applied by the peroral route today is of small molecular weight and pass the intestinal barrier by passive diffusion. While extrinsic absorption barriers of the gastrointestinal tract, such as mucus and secreted enzymes, are discussed in other chapters of this book, this chapter is merely focusing on the intrinsic barriers and absorptive systems of the digestive tract.

3.2 Anatomy of the Digestive Tract

The digestive tract from mouth to colon is often referred to as the digestive tube, whose biological and physical parameters are summarized in Table 3.1 (adapted from Daugherty and Mrsny 1999). In mammals, this tube has a four-layered

Table 3.1 Biological and physical parameters of the human gastrointestinal tract. Adapted from Daugherty and Mrsny (1999)

Segment	Surface area	Segment length	Residence time	pH of segment	Catabolic activities
Oral cavity	100 cm^2		Seconds to minutes	6.5	Polysaccharides
Esophagus	200 cm^2	23–25 cm	Seconds		
Stomach	3.5 m^2 (variable)	0.25 cm (variable)	1.5 h (variable)	1–2	Proteases, lipases
Duodenum	1.9 m^2	0.35 m	0.5–0.75 h	4–5.5	Polysaccharides, oligosaccharides, proteases, peptidases, lipases, nucleases
Jejunum	184 m^2	2.8 m	1.5–2.0 h	5.5–7.0	Oligosaccharides, peptidases, lipases
Ileum	276 m^2	4.2 m	5–7 h	7.0–7.5	Oligosaccharides, peptidases, lipases, nucleases, nucleotidases
Colon and rectum	1.3 m^2	1.5 m	1–60 h (35 h average)	7.0–7.5	Broad spectrum of bacterial enzymes

architecture. The outermost layer, called the *tunica serosa* in the abdominal cavity, and *tunica adventitia* in the esophagus and rectum areas, is represented by a loose connective tissue covered by mesothelium. Within the abdominal cavity the digestive tube is suspended from the mesentery, containing vasculature and nerves supplying the intestinal tract. Next, two layers of smooth muscles form the *tunica muscularis*, with the myenteric plexus, an important part of the digestive tract's nervous system, localized in between the two muscle layers. Connective tissue containing the vasculature and vessels of the lymphatic system forms the *tunica submucosa*. This layer also contains the submucous plexus, which provides nervous control to the mucosa. The innermost layer of the digestive tube is formed by the *tunica mucosa*, with its outermost layer of smooth muscle, called the *lamina muscularis mucosae*. The *lamina propria* is located directly beneath the epithelium, containing blood and lymphatic vessels and nodes responsible for the immune function of the intestinal mucosa. The composition of cell types is the most complex within the epithelial layer (*lamina epithelialis*), due to the various functions of the different parts of the digestive tract. In the stomach, a stratified squamous and a columnar epithelial type can be distinguished. The latter contains specialized cells for the secretion of mucus (mucous cells), hydrochloric acid (parietal cells), pepsin (chief cells), and gastrin (G-cells). Except for a limited number of lipid-soluble compounds, such as acetylic–salicylic acid and non-steroidal drugs, very few substances of therapeutic importance are absorbed through the gastric mucosa.

Internal folding and projections of the small intestine (circular folds, villi, and microvilli) increase the internal surface area by 3-fold, 30-fold, and 600-fold, respectively (Daugherty and Mrsny 1999). Epithelial cells of the small intestine are derived from stem cells located at the bottom of the villi, the crypts. From there, intestinal stem cells differentiate into enterocytes covered by a thick glycocalix (brush border), enteroendocrine cells, mucus-secreting goblet cells, and Paneth cells that are involved in the intestinal host defense (Bry et al. 1994). Enterocytes undergo several divisions as they move up the villi, differentiating further into mature absorptive cells expressing a broad range of absorptive and enzymatic systems. This migration takes about 3–6 days, and cells stay at the villi tip only for a few days. The intestinal epithelium is thus constantly replenished with new cells migrating up from the crypt area. Intestinal epithelial cells, through stimulation of submucosal lymphocytes, can differentiate into so-called M cells covering the Peyer's patches (Kraehenbuhl and Neutra 2000). M cells are able to sample antigens from the intestinal lumen, and, by presentation to lymphocytes, activate the innate immune system. M cells are not covered by glycocalix and can macroscopically be identified as interruptive structures of the intestinal brush border.

The gastrointestinal epithelium forms an extrinsic and an intrinsic barrier against diffusion of toxins and pathogens. The extrinsic barrier is characterized by secretion of mucus, which hinders colonization and accelerates clearance of pathogenic organisms. The importance of mucus as a barrier to drug absorption is discussed in Chapter 2 of this book. In addition, the intestinal immune system and antibiotic peptides, secreted by Paneth cells, assist to protect the body from infection with the gastrointestinal tract as port-of-entry.

The intrinsic barrier of the gastrointestinal epithelium is characterized by intercellular junctions at the apical (luminal) side of differentiated epithelial cells, the so-called tight junctions (TJ), and the maintenance of epithelial integrity based on the balance between cellular proliferation and cell death, as described above. Barriers against drug absorption by the intracellular and paracellular routes, their modulation, and maintenance will be discussed in the following, with the focus on the intestinal epithelium.

3.3 Barriers to Transcellular Absorption

Nutrients and drugs are absorbed at the intestinal epithelia via several pathways, as illustrated in Fig. 3.1. Depending on their physicochemical properties, including molecular weight, lipophilicity, and hydrogen-binding potential, molecules may pass the intestinal barrier by transcellular or paracellular passive diffusion (Fig. 3.1A and C).

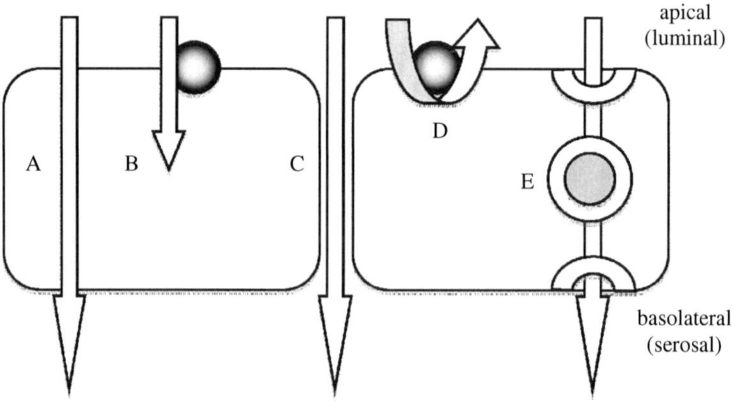

Fig. 3.1 Schematic presentation of absorption pathways through the intestinal epithelium. A: passive transcellular; B: active, carrier-mediated; C: passive, paracellular; D: efflux transporters; E: transcytosis

3.3.1 Passive Diffusion

As outlined by Lipinski's rule-of-5 (Lipinski et al. 1997), absorption of a drug can be estimated based on its physicochemical properties. For example, poor intestinal absorption is predicted for a compound of the following properties:

- Molecular weight > 500 Da
- Number of hydrogen bond donors (hydroxyl or amine groups) > 5

- Number of hydrogen bond acceptors (O or N) > 10
- log $P > 5.0$

In addition, the value of the polar surface areas (PSA) of a given molecule (Clark and Grootenhuis 2003), defined as the molecule's surface area able to undergo van-der-Waals interaction (hydrogen bond donors and acceptor groups), is used to estimate passive diffusion. A good correlation between PSA values and actual human intestinal absorption data was found for a number of compounds (Clark 1999), with good permeability found for compounds of PSA values of $\leq 61 A^2$ and poor absorption at values of $\geq 140\ A^2$.

3.3.2 Transporter Systems

Exceptions from Lipinski's rule, i.e., molecules of PSA values $> 140\ A^2$ are found to be actively absorbed by carrier-mediated transport systems (Wessel et al. 1998), as shown in Fig. 3.1B. As further detailed in Fig. 3.2, the intestinal epithelium expresses a number of such transport systems for amino acids, organic anions and cations, nucleosides, and hexoses. Among these systems are the apical sodium-dependent bile acid transporter (ASBT; Annaba et al. 2007), the monocarboxylate transporter (MCT; Halestrap and Price 1999), the sodium-D-glucose co-transporter (SLGT1; Kipp et al. 2003), and the nucleotide transporter SPNT1 (Balimane and Sinko 1999). In addition, the expression of a specialized transporter system for small peptides has been found in the intestinal epithelium with the di/tripeptide transporter, PepT1 (Tsuji 2002), after previous functional studies by Hu et al. (1989), and the cloning of PepT1

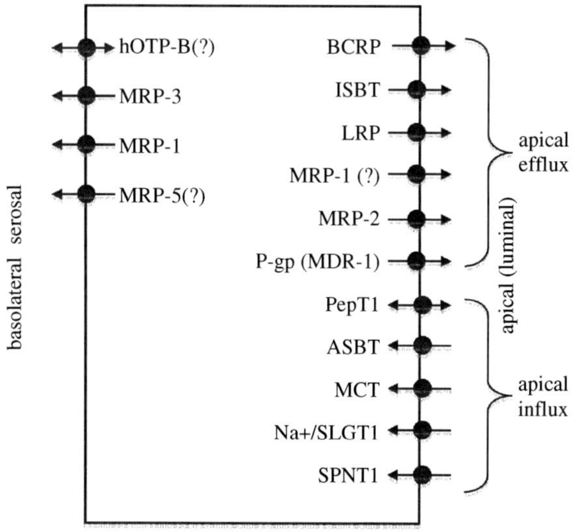

Fig. 3.2 Transport systems expressed at the intestinal epithelium

cDNA was accomplished (Boll et al. 1994; Fei et al. 1994). Apart from peptides derived from protein digestion in the intestinal lumen, several drugs (e.g., peptidomimetics, β-lactam antibiotics) have been identified as substrates for PepT1 (Han et al. 1999; Brodin et al. 2002), and methods based on cell-based assays to predict the extent of PepT1-mediated oral absorption were suggested (Shimizu et al. 2008).

3.3.3 Pinocytosis

Macromolecules, such as proteins and peptides, are taken up by polarized epithelial cells by random pinocytosis, i.e., sampling of luminal fluid containing such molecules into apical early endosomes (AEE) as shown for MDCK cells in vitro (Leung et al. 2000). Further intracellular sorting is depicted in general terms in Fig. 3.3. Recycling to the apical surface or transcytosis is then mediated

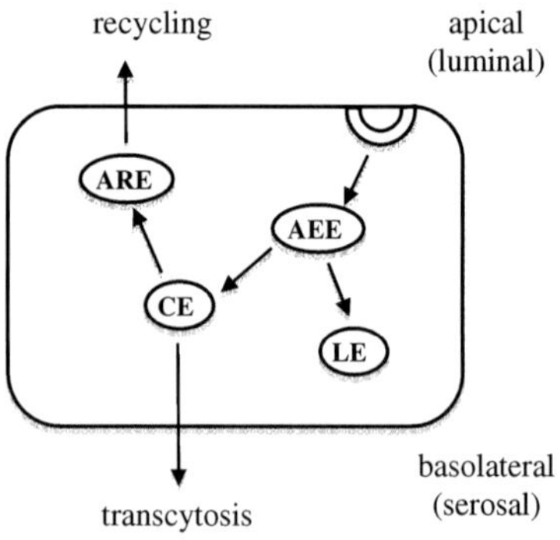

Fig. 3.3 Intracellular sorting in epithelial cells after clathrin-mediated endocytosis. AEE: apical early endosome; ARE: apical recycling endosome; CE: common endosome; LE: late endosome

through the common endosome (CE) and/or the apical recycling endosome (ARE). CEs are oriented in part along the apical–basolateral axis, and contain both transcytosing proteins, such as IgA and its receptor (pIgR), as well as recycling proteins (transferrin and its receptor). CEs do not contain fluid-phase markers and compounds to be degraded, such as epidermal growth factor (EGF) and low-density lipoprotein (LDL) (Mostov et al. 2000). AREs also contain molecules such as IgA, pIgR, transferrin and its receptor and are located directly at the apical membrane. While a low pH value of 5.8 was shown for CE, the pH of ARE is described at pH 6.5 (Wang et al. 2000), underlining the suggestion of a distinct nature of ARE. Degradation of pinocytosed material takes place in late endosomes (LEs), which may fuse with lysosomes, by intravesicular decrease in pH to about a value of 5.5 and release of oxidative species.

Pinocytosis at epithelial cells can occur by four distinct mechanisms, macropinocytosis, clathrin-mediated endocytosis, caveolae-mediated endocytosis, and clathrin/caveolin-independent endocytosis (Conner and Schmid 2003). Macropinocytosis is unlikely to occur in enterocytes, as it involves sampling of large volumes of luminal contents. Clathrin-mediated endocytosis represents the major route of uptake by enterocytes. Specialized receptors, localized in clathrin-coated pits, bind to their substrate (receptor-mediated pinocytosis), upon which coated vesicles are formed (Conner and Schmid 2003). Following uncoating, the vesicle fuses with an AEE. Examples for this process are the receptors of transferrin, epidermal growth factor (EGF), and immunoglobulins. Expression and distribution of the receptors of these compounds may vary depending on the region of the intestinal tract and the age of the patient (Sugiyama and Kato 1994). Molecules can also be taken up by adsorptive pinocytosis (non-specific binding) or dissolved in the solution sampled by the forming coated vesicle (Sanderson and Walker 1993).

3.3.4 Transport by Caveolae

Caveolae were first observed in heart endothelial cells (Palade 1953) and suggested to have the capability to transport molecules across cells (Simionescu 1983). Caveolae are dynamic structures that can take on different shapes (Thomas and Smart 2008) and are involved in endo/exocytosis, cholesterol homeostasis, and signal transduction (Hommelgaard et al. 2005). Structurally, caveolae are lipid rafts (Simons and Ikonen 1997), ordered microdomains of the plasma membrane containing phospholipids, sphingolipids, and cholesterol (Rietveld and Simons 1998). Other compounds, such as acylated proteins, are suggested to take part in lipid raft formation (Parton and Richards 2003). Caveolae can be distinguished from other lipid rafts by the presence of molecules of the caveolin protein family (caveolin-1, caveolin-2, and caveolin-3), which can directly interact with signaling molecules (Garcia-Gardena et al. 1996) and regulate the cholesterol content of caveolae (Uittenbogaard et al. 1998). Intracellular sorting appears to be distinct from clathrin-coated pits, as shown in Fig. 3.4. Previously described as a slow and

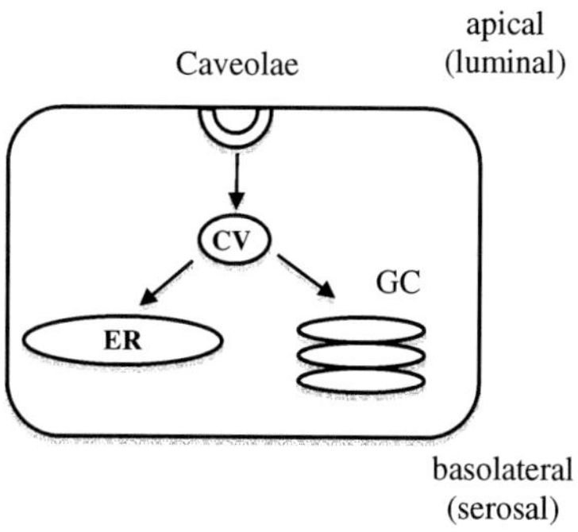

Fig. 3.4 Intracellular sorting by caveolae.CV: caveosome; ER: endoplasmatic reticulum; GC: Golgi complex

inefficient endocytosis and transcytosis process (Snoeck et al. 2005), recent in vivo kinetics data suggest rapid shuttling by caveolae, at least in endothelial cells of the rat lung (Oh et al. 2007).

3.3.5 Efflux Systems

A last type of transporter systems involved in the intrinsic barrier function of intestinal epithelium is the group of active efflux drug transporters (Fig. 3.1D) belonging to the family of ATP-binding cassette (ABC) transporters (Borst and Oude Elferink 2002). ABC transporters are large proteins showing several transmembrane domains (Table 3.2) and are able to transport a large range of substrates

Table 3.2 Intestinal epithelial efflux systems

Name	Mw (kDa)	Structure	Substrates	Human intestinal expression
P-glycoprotein, P-gp	170	1280 aa 2 NBD 12 TMD	Neutral and positively charged compounds	Superficial epithelial cells, increasing proximal to distal
Multidrug resistance-associated protein 2 (MRP2)	190	1545 aa 2 NBD 17 TMD	Hydrophilic compounds and conjugates	Villus cells, small intestine (BBM), decreasing jejunum to ileum
Breast cancer resistance protein (BCRP)	72 (homodimer)	655 aa 1 NBD 6 TMD	Relatively hydrophilic cytostatics	BBM of small intestine, colon muscle layers

aa: amino acids; BBM: brush border membrane; NBD: nucleotide-binding domains; TMD: transmembrane domains.

against a concentration gradient. While playing a large role in multidrug resistance (MDR) in cancer cells, the excretion of endogenous substrates from mammalian secretory epithelia and from the liver, and antigen presentation in immune cells, a variety of such transporters is also expressed both at the apical (luminal) and the basolateral (serosal) side of the epithelium (Fig. 3.2) (Takano et al. 2006).

Efflux systems of major importance in the intestinal epithelium are P-glycoprotein (P-gp) (Mizuno et al. 2003), multidrug resistance-associated protein 2 (MRP2) (Jansen et al. 1993), and breast cancer resistance protein (BCRP) (Doyle et al. 1998, Doyle and Ross 2003). The latter is described as a "half-transporter" and possibly functions as a homodimer (Schinkel and Jonker 2003). Details on the molecular weight, structure, substrates, and expression of P-gp, MRP2, and BCRP are listed in Table 3.2. It needs to be mentioned that expression of these intestinal efflux transporter systems shows high interindividual variation and is subject to drug interaction. Expression of, e.g., P-gp was found to be inducible by an extract of St. John's wort, and bioavailability of co-administered digoxin, a typical P-gp substrate, was reduced by 18% (Dürr et al. 2000). Therefore, pharmacokinetics of drugs applied by the oral route, which are substrates to these efflux systems, are equally fraught with high variation. Strategies to overcome this part of the intrinsic intestinal barrier function are further discussed in Chapter 7 of this book.

3.4 Barriers to Paracellular Absorption

3.4.1 The Tight Junctional Complex

Epithelial cells are interconnected at the apical (mucosal) side by a complex network of proteins, called the tight junctions (TJ). First thought to have merely a static role in restricting access of compounds present in the luminal fluid to the underlying subepithelial tissue and systemic circulation by the paracellular pathway, TJ are known today to be dynamic structures involved in cellular differentiation, cell signaling (Harder and Margolis 2008), polarized vesicle trafficking, and protein synthesis.

A growing number of proteins have been identified to be involved in the formation of TJ (Fig. 3.5). These proteins may be divided into three groups, namely the *integral TJ proteins* involved in the cell–cell contacts, the *TJ plaque proteins*, which connect integral TJ proteins to the actin network of the cytoskeleton, and various *cytosolic and nuclear proteins* interacting with TJ plaque

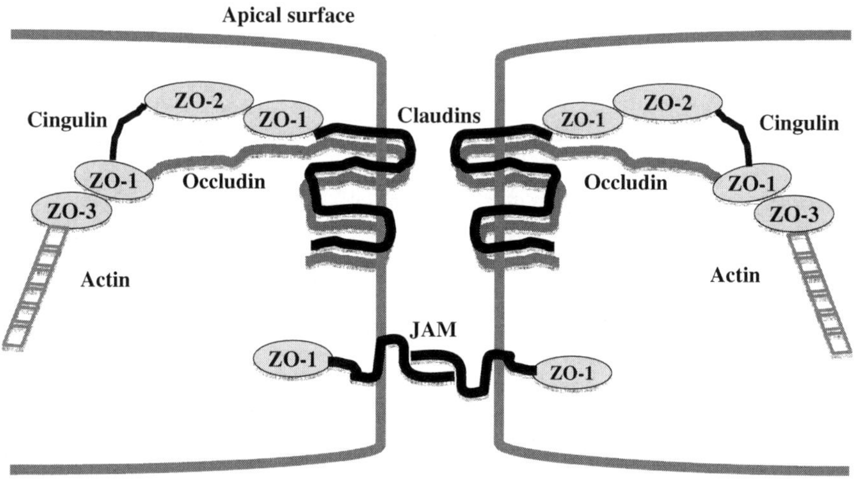

Fig. 3.5 Schematic representation of proteins involved in tight junction formation

proteins and involved in the regulation of TJ solute permeability, cell polarity and proliferation, and tumor suppression (Hartsock and Nelson 2008).

Major *integral TJ proteins* identified are occludin, the family of claudin proteins, and junction adhesion molecule 1 (JAM-1). Occludin is a 60-kDa membrane protein of four TMD and two extracellular loops, with both the amino and carboxy terminal groups residing in the cytosol (Furuse et al. 1993). The extracellular loops are rich in tyrosine and glycine and are suggested to be uncharged at physiological pH. In TJ, occludin is found in its phosphorylated state, whereas unphosphorylated protein is found along the basolateral membrane (REF). Phosphorylation of occludin is mediated by both classical (cPKC) and atypical (aPKC, phorbol ester insensitive) protein kinase C (PKC). Occludin is expressed in TJ of vertebrates, however, transfection of insect cells with occludin cDNA did not result in a typical TJ phenotype (Furuse et al. 1996). In addition, occludin null mice were found to express phenotypic alterations, however, in the absence of structural or functional TJ abnormalities (Saitou et al. 2000). Table 3.3 lists phenotypic aberrations of TJ protein-deficient mice (Schneeberger and Lynch 2004). As TJ function appears, at least to a certain extent, to be independent on the presence of occludin, its function needs to be further defined. Most likely, occludin plays a major role in the co-ordination of the actin cytoskeleton ("scaffolding") through different signaling pathways, by which it might be essential in the maintenance of the cell's phenotype.

As shown in Table 3.3, the absence of claudins results in disruption of TJ functioning and has severe changes in (mouse) phenotype as consequence. Members of the claudin family, of which 24 have been identified, are proteins of a molecular weight in the range of 20–27 kDa, showing 4 TMD, and cytosolic amino and carboxy terminal groups (Mitic and van Itallie 2001). Values for the isoelectric points (pK_i) of extracellular loops of claudins range, in contrast to occludin, from 4.17 to 10.5 for the different members of the claudin family (Mitic and van Itallie 2001). At the C terminus, claudins (with the exception of claudin-12) are connected to TJ plaque proteins bearing a PDZ domain (see below), such as ZO-1, ZO-2, and ZO-3 (Itoh et al. 1999).

Possibly due to the variety of their $pK_{i,}$values, claudins appear to confer selectivity in the paracellular transport of ions, with TJ in most epithelia resembling cation-selective ion channels (Powell 1981). As the extracellular loops of occludin remain uncharged at physiological pH, it is assumed that this selectivity is caused primarily by claudins (Yu et al. 2003). In recent in vitro studies, TJ were shown to have size and charge selectivity, dependence on ion concentration and competition, and pH (Tang and Goodenough 2003).

JAM-1 is a glycoprotein of the IgG superfamily of 43 kDa, one TMD, and two extracellular V-type Ig domains (Martin-Padura et al. 1998). The carboxy terminus is characterized by a PDZ-binding domain, which binds to the PDZ motif of ZO-1 (Hamazaki et al. 2002). Other members of the JAM family have been identified, such as JAM-4, which is expressed in renal glomerular and intestinal epithelial cells, where it was shown to induce Ca^{2+}-independent intercellular adhesion (Hirabayashi et al. 2003). JAM-1 is further connected to cytosolic

Table 3.3 Alterations of phenotype of occludin- or claudin-deficient mice (modified from Schneeberger and Lynch 2004)

Mouse	TJ protein	Phenotype
Null	Occludin	Growth retardation, male sterility, chronic inflammation, hyperplasia of gastric epithelium, brain calcification
Null	Claudin-1	Death at birth, failure to form barrier against a 600-kDa tracer molecule
Null	Claudin-5	Permeability increase of BBB (<800-kDa tracer)
Null	Claudin-11	Male sterility, slowed CNS conductance
Tg	Claudin-6	Death within 2 days of birth, aberrant expression of late differentiation markers, modified epidermal claudin profile

BBB: blood–brain barrier, CNS: central nervous system, Tg: transgenic.

proteins cingulin and calcium/calmodulin-dependent serine protein kinase (CASK) (Martinez-Estrada et al. 2001), which serves in cellular signaling.

Major *TJ plaque proteins* have in common the presence of PDZ motifs in their structure. PDZ domains are sequences of about 80–90 amino acids, forming hydrophobic pouches bound to the C terminus of claudins (Fanning and Anderson 1999). PDZ proteins expressed at TJ are *zonula occludens* (ZO) proteins, belonging to a family of so-called membrane-associated guanylyl kinase (MAGUK) homologues (Fanning and Anderson 1999). As an example, ZO-1 is suggested to play a role in cell differentiation and proliferation (Balda et al. 2003), whereas ZO-2 is involved in transcriptional regulation (Betanzos et al. 2004). Other examples of TJ plaque proteins are membrane-associated guanylyl kinase inverted proteins (MAGI) (Laura et al. 2002), and multi-PDZ domain protein 1 (MUPP-1) (Hamazaki et al. 2002), among others. For an in-depth review please refer to Citi (2001).

3.4.2 Modulation of Intestinal Tight Junctions

Permeation enhancers modulating the paracellular permeability of drugs across the intestinal epithelium will be discussed elsewhere in this book. Epithelial barrier function, and thus TJ integrity, is disturbed in inflammatory diseases of the intestine, e.g., inflammatory bowel disease (IBD) (Siccardi et al. 2005). Enhanced intestinal epithelial permeability in IBD may be caused by any, or any combination of, the following factors in conjunction with lesions of the intestinal mucosal tissue. In IBD, increased levels of, e.g., thrombin and trypsin have been observed and correlated to the disease, which may also act through apoptose-inducing pathways by the activation of proteinase-activated receptors (PAR) (Chin et al. 2003). Other factors modulating TJ function are pro-inflammatory cytokines, including tumor-necrosis factor alpha (TNF-α) and interferon-gamma (IFN-γ), which act synergistically to down-regulate expression of occludin (Mankertz et al. 2000) and upregulate myosin II light chain phosphorylation (Zolotarevsky et al. 2002). In addition, commensal bacteria or pathogens present in the gut lumen may interact with TJ components, resulting in enhanced permeability (Johanson et al. 1989). As an example, *Vibrio cholerae* has been described to produce a protein, zonula occludens toxin (Zot), that reversibly perturbs TJ integrity, allowing the pathogen to penetrate into the submucosa (Fasano et al. 1991). The mechanism of action of Zot on epithelial tight junctions, not only in the intestine, is thought to be connected to the protein kinase C alpha-dependent F-actin polymerization, resulting in a rearrangement of the cytoskeleton (Fasano et al. 1995). Zot is therefore being considered as a mucosal adjuvant, able to increase access of vaccines to sub-epithelial tissue (Marinaro et al. 2003). Recently (Wang et al. 2000), a mammalian analog to Zot, zonulin, has been described as an intestinal TJ regulator protein. Tissue lesions or bacterial load triggers the luminal release of zonulin, which regulates TJ by binding to the same receptor as Zot (Di Pierro et al. 2001). Zonulin has been shown to be involved in the innate immunity system in the gut (El Asmar et al. 2002). Another study in diabetic-prone rats (Watts et al. 2005) also revealed that zonulin might be responsible for the increase in small intestinal permeability at the onset of type I diabetes.

Another indication that the mucosal innate immune system is connected to the modulation of TJ was shown in reports examining the role of Toll-like receptor 2 (TLR2) on the regulation of intestinal epithelial barrier function in mucosal inflammation (Cario et al. 2007). Recent studies have revealed the importance of this family of pattern recognition receptors (PRR) for the bridging of the innate and acquired immune systems. Several agonists and antagonists of this receptor family have been found and partially tested in pre-clinical and clinical settings (Vasselon and Detmers 2002). TLR signaling at the intestinal epithelium was found to be altered in inflammatory disease (such as inflammatory bowel disease, IBD) or aged patients (Abreu et al. 2005). Apparently, the absence of TLR2 in mice alone did not cause IBD; however, the expression of interferon-gamma (IFN-γ) and other cytokines was suppressed, resulting in severe inflammation upon insult, especially in aged animals (Albert and Marshall 2008). In addition, the secretion of TFF3-peptide in the cecum and mid-colon of TLR2$^{-/-}$ mice was significantly reduced. Trefoil family factor (TFF)3-peptide, expressed in the gastrointestinal mucosa (Hoffmann 2005), is involved in the protection and repair of the mucosal tissue. A recent study in HT29/B6 cells in vitro showed that TFF3-peptide increased the expression of TJ protein claudin-1, and decreased expression of claudin-2, which is involved in the formation of cation-selective TJ ion channels (Meyer zum Büschenfelde et al. 2006). Overall, there appears to exist an intimate interplay between the innate immune system of the gastrointestinal tract and the modulation of paracellular permeability and barrier function in the intestine. Studies examining these interactions are posed to yield therapeutic strategies to restore or maintain intestinal epithelial barrier function in inflammatory or infectious diseases.

References

Abreu MT, Fukata M, Arditi M (2005) TLR signaling in the gut in health and disease. J Immunol 174:4453–4460

Albert EJ, Marshall JS (2008) Aging in the absence of TLR2 is associated with reduced IFN-g responses in the large intestine and increased severity of induced colitis. J Leukoc Biol 3:833–842

Annaba F, Sarwar Z, Kumar P, Saksena S, Turner JR, Dudeja PK, Gill RK, Alrefai WA (2007) Modulation of ileal bile acid transporter (ASBT) activity by depletion of plasma membrane cholesterol: association with lipid rafts. Am J Physiol Gastrointest Liver Physiol 294:489–497

Balda MS, Garrett MD, Matter K (2003) The ZO-1-associated Y-box factor ZONAB regulates epithelial cell proliferation and cell density. J Cell Biol 160:423–432

Balimane PV, Sinko PJ (1999) Involvement of multiple transporters in the oral absorption of nucleoside analogues. Adv Drug Deliv Rev 39:183–209

Betanzos A, Huerta M, Lopez-Bayghen E, Azuara E, Amerena J, Gonzalez-Mariscal L (2004) The tight junction protein ZO-2 associates with Jun, Fos and C/EBP transcription factors in epithelial cells. Exp Cell Res 292:51–66

Boll M, Markovich D, Weber WM, Korte H, Daniel H, Murer H (1994) Expression cloning of a cDNA from rabbit small intestine related to proton-coupled transport of peptides, beta-lactam antibiotics and ACE-inhibitors. Plugers Arch 429:146–149

Borst P, Oude Elferink R (2002) Mammalian ABC transporters in health and disease. Annu Rev Biochem 71:537–592

Brodin B, Nielsen CU, Steffansen B, Frokjaer S (2002) Transport of peptidomimetic drugs by the intestinal di/tripeptide transporter, PepT1. Pharmacol Toxicol 90:285–296

Bry L, Falk P, Huttner K, Ouellette A, Midtvedt T, Gordon JI (1994) Paneth cell differentiation in the developing intestine of normal and transgenic mice. Proc Natl Acad Sci 91:10335–10339

Cario E, Gerken G, Podolsky DK (2007) Toll-like receptor 2 controls mucosal inflammation by regulating epithelial barrier function. Gastroenterol 132:1359–1374

Chin AC, Vergnolle N, MacNaughton WK, Wallace JL, Hollenberg MD, Buret AG (2003) Proteinase-activated receptor 1 activation induces epithelial apoptosis and increases intestinal permeability. Proc Natl Acad Sci USA 100:11104–11109

Citi S (2001) The cytoplasmic plaque proteins of the tight junction. In: Anderson JM, Cereijido M (eds) Tight Junctions (2nd edn.): CRC, Boca Raton 2001, pp 231–264

Clark DE (1999) Rapid calculation of polar molecular surface area and its application to the prediction of transport phenomena: 1. Prediction of intestinal absorption. J Pharm Sci 88:807–814

Clark DE, Grootenhuis PD (2003) Predicting passive transport in silico – history, hype, hope. Curr Top Med Chem 3:1193–1203

Conner SD, Schmid SL (2003) Regulated portals of entry into the cell. Nature 422:37–44

Daugherty AL, Mrsny RJ (1999) Transcellular uptake mechanisms of the intestinal epithelial barrier. Part one. PSTT 2:144–151

Di Pierro M, Lu R, Uzzau S, Wang W, Margaretten K, Pazzani C, Maimone F, Fasano A (2001) Zonula occludens toxin structure-function analysis. J Biol Vhem 276:19160–19165

Doyle LA, Yang W, Abruzzo LV, Krogmann T, Gao Y, Rishi AK, Ross DD (1998) A multidrug resistance transporter from human MCF-7 breast cancer cells. Proc Natl Acad Sci USA 95:15665–15670

Doyle LA, Ross DD (2003) Multidrug resistance mediated by the breast cancer resistance protein BCRP (ABCG2). Oncogene 22:7340–7358

Dürr D, Stieger B, Kullak-Ublick GA, Rentsch KM, Steinert HC, Meier PJ, Fattinger K (2000) St. John's Wort induces intestinal Pglycoprotein/MDR1 and intestinal and hepatic CYP3A4. Clin Pharmacol Ther 68:598–604.

El Asmar R, Panigrahi P, Bamford P, Berti I, Not T, Coppa GV, Catassi C, Fasano A (2002) Host-dependent zonulin secretion causes the impairment of the small intestin barrier function after bacterial exposure. Gastroenterol 123:1607–1615

Fanning AS, Anderson JM (1999) PDZ domains: fundamental building blocks in the organization of protein complexes at the plasma membrane. J Clin Invest 103:767–772

Fasano A, Baudry B, Pumplin DW, Wassermann SS, Tall BD, Ketley JM, Kaper JB (1991) *Vibrio cholerae* produces a second enterotoxin, which affects intestinal tight junctions. Proc Natl Acad Sci USA 88:5242–5246

Fasano A, Fiorentini C, Donelli G, Uzzau S, Kaper JB, Margaretten K, Ding X, Guandalini S, Comstock L, Goldblum SE (1995) Zonula occludens toxin modulates tight junctions through protein kinase C-dependent actin reorganization, in vitro. J Clin Investig 96:710–720

Fei YJ, Kanai Y, Nussberger S, Ganapathy V, Leibach FH, Romero MF, Singh SK, Boron WF, Hediger MA (1994) Expression cloning of a mammalian proton-coupled oligopeptide transporter. Nature 368:563–566

Furuse M, Hirase T, Itoh M, Nagafuchi A, Yonemura S, Tsukita S, Tsukita S (1993) Occludin: a novel integral membrane protein localizing at tight junctions. J Cell Biol 123:1777–1788

Furuse M, Fujimoto K, Sato N, Hirase T, Tsukita S, Tsukita S (1996) Overexpression of occludin, a tight junction-associated integral membrane protein, induces the formation of intracellular multilamellar bodies bearing tight junction-like structures. J Cell Sci 109:429–435

Garcia-Gardena G, Oh P, Liu J, Schnitzer JE, Sessa WC (1996) Targeting of nitric oxide synthase to endothelial cell caveolae via palmitoylation: implications for nitric oxide signaling. Proc Natl Acad Sci USA 93:6448–6453

Halestrap AP, Price NT (1999) The proton-linked monocarboxylate transporter (MCT) family: structure, function and regulation. Biochem J 343:281–299

Hamazaki Y, Itoh M, Sasaki H, Furuse M, Tsukita S (2002) Multi-PDZ domain protein 1 (MUPP1) is concentrated at tight junctions through its possible interaction with claudin-1 and junctional adhesion molecule. J Biol Chem 277:455–461

Han HK, Rie JK, Oh DM, Saito G, Hsu CP, Stewart BH, Amidon GL (1999) CHO/hPEPT1 cells overexpressing the human peptide transporter (hPEPT1) as an alternative in vitro model for peptidomimetic drugs. J Pharm Sci 88:347–350

Harder JL, Margolis B (2008) SnapShot: Tight and adherens junction signaling. Cell 133:1118

Hartsock A, Nelson WJ (2008) Adherens and tight junctions: Structure, function and connections to the actin cytoskeleton. Biochim Biophys Acta 1778:660–669

Hirabayashi S, Tajima M, Yao I, Nishimura W, Mori H, Hata Y (2003) JAM-4, a junctional cell adhesion molecule interacting with a tight junction protein, MAGI-1. Mol Cell Biol 23:4267–4282,

Hoffmann W (2005) Trefoil factors TFF (trefoil factor family) peptide-triggered signals promoting mucosal restitution. Cell Mol Life Sci 62:2932–2938

Hommelgaard AM, Roepstorff K, Vilhardt F, Torgersen ML, Sandvig K, van Deurs B (2005) Caveolae: stable membrane domains with a potential for internalization. Traffic 6:720–724

Hu M, Subramanian P, Mosberg HI, Amidon GL (1989) Use of the peptide carrier system to improve the intestinal absorption of L-alpha-methydopa: carrier kinetics, intestinal permeabilities, and in vitro hydrolysis of dipeptidyl derivatives of L-alpha-methyldopa. Pharm Res 6:66–70

Itoh M, Furuse M, Morita K, Kubota K, Saitou M, Tsukita S (1999) Direct binding of three tight junction-associated MAGUKs, ZO-1, ZO-2 and ZO-3, with the COOH termini of claudins. J Cell Biol 147:1351–1363

Jansen PL, van Klinken JW, van Gelder M, Ottenhoff R, Elferink RP (1993) Preserved organic anion transport in mutant TR-rats with a hepatobiliary secretion defect. Am J Physiol 265:G445–G452.

Johanson K, Stintzing G, Magnusson KE, Sundqvist T, Jalil F, Murtaza A, Khan SR, Lindblad BS, Mollby R, Orusil E, Svensson L (1989) Intestinal permeability assessed with poly-ethylene glycols in children with diarrhea due to rotavirus and common bacterial pathogens in a developing community. J Pediatr Gastroenterol Nutr 9:307–313

Kipp H, Khoursandi S, Scharlau D, Kinne RKH (2003) More than apical: distribution of SLGT1 in Caco-2 cells. Am J Physiol Cell Physiol 54:C737–C749

Kraehenbuhl J-P, Neutra MR (2000) Epithelial M cells: differentiation and function. Annu Rev Cell Dev Biol 16:301–332

Laura RP, Ross S, Koeppen H, Lasky LA (2002) MAGI-1: A widely expressed, alternatively spliced tight junction protein. Exp Cell Res 275:155–170

Leung S-M, Ruiz WG, Apodaca G (2000) Sorting of membrane and fluid at the apical pole of polarized MDCK cells. Mol Biol Cell 11:2131–2150

Lipinski CA, Lombardo F, Dominy BW, Feeney PJ (1997) Experimental and computational approaches to estimate solubility and permeability in drug discovery and development settings. Adv Drug Del Rev 23:3–25

Mankertz J, Tavalali S, Schmitz H, Mankertz A, Riecken EO, Fromm M, Schulzke JD (2000) Expression from the human occludin promoter is affected by tumor necrosis factor alpha and interferon gamma. J Cell Sci 113:2085–2090

Marinaro M, Fasano A, De Magistris MT (2003) Zonula occludens toxin acts as an adjuvant through different mucosal routes and induces protective immune responses. Infect Immun 71:1897–1902

Martin-Padura I, Lostaglio S, Schneemann M, LW, Romano M, Fruscella P, Panzeri C, Stoppacciaro A, Ruco L, Villa A, Simmons D, Dejana E (1998) Junctional adhesion molecule, a novel member of the immunoglobulin superfamily that distributes at intercellular junctions and modulates monocyte transmigration. J Cell Biol 142:117–127

Martinez-Estrada OM, Villa A, Breviario F, Orsenigro F, Dejana E, Bazzoni G (2001) Association of junctional adhesion molecule with calcium/calmodulin-dependent serine protein kinase (CASK/LIN-2) in human epithelial Caco-2 cells. J Biol Chem 276:9291–9296

Meyer zum Büschenfelde D, Tauber R, Huber O (2006) TFF3-peptide increases transepithelial resistance in epithelial cells by modulating claudin-1 and -2 expression. Peptides 27:3383–3390

Mitic LL, van Itallie CM (2001) Occludin and claudins: transmembrane proteins of the tight junction. In: Cereijido M, Anderson JM (eds) Tight Junctions (2nd edn.). CRC, Boca Raton, pp 213–230

Mizuno N, Niwa T, Yotsumoto Y, Sugiyama Y (2003) Impact of drug transporter studies on drug discovery and development. Pharmacol Rev 55:425–461

Mostov KE, Verges M, Altschuler Y (2000) Membrane traffic in polarized epithelial cells. Curr Opin Cell Biol 12:483–490

Oh P, Borgström P, Witkiewicz H, Li Y, Borgström BJ, Chrastina A, Iwata K, Zinn KR, BAldwin R, Testa JE, Schnitzer JE (2007) Live dynamic imaging of caveolae pumping targeted antibody rapidly and specifically across endothelium in the lung. Nature Biotechnol 25:327–337

Palade GE (1953) Fine structure of blood capillaries. J Appl Physiol 24:1424

Parton RG, Richards AA (2003) Lipid rafts and caveolae as portals for endocytosis: new insights and common mechanisms. Traffic 4:724–738

Powell DW (1981) Barrier function of epithelia. Am J Physiol Gastrointest Liver Physiol 241:G275–G288

Rietveld A, Simons K (1998) The differential miscibility of lipids as basis for the formation of functional membrane rafts. Biochim Biophys Acta 1376:467–479

Saitou M, Furuse M, Saski H, Schulzke JK, Fromm M, Takano H, Noda T, Tsukita S (2000) Complex phenotype of mice lacking occludin, a component of tight junction strands. Mol Biol Cell 11:4131–4142

Sanderson IR, Walker WA (1993) Uptake and transport of macromolecules by the intestine: possible role in clinical disorder (An update). Gastroenterol 104:622–629

Schneeberger EE, Lynch RD (2004) The tight junction: a multifunctional complex. Am J Physiol Cell Physiol 286:1213–1228

Schinkel AH, Jonker JW (2003) Mammalian drug efflux transporters of ATP binding cassette (ABC) family: an overview. Adv Drug Del Rev 55:3–29

Shimizu R, Sukegawa T, Tsuda Y, Itoh T (2008) Quantitative prediction of oral absorption of PEPT1 substrates based on in vitro uptake into Caco-2 cells. Int J Pharm 354:104–110

Siccardi D, Turnerb JR, Mrsny R (2005) Regulation of intestinal epithelial function: a link between opportunities for macromolecular drug delivery and inflammatory bowel disease. Adv Drug Deliv Rev 57:219–235

Simionescu N (1983) Cellular aspects of transcapillary exchange. Physiol Rev 63:1536–1579

Simons K, Ikonen E (1997) Functional rafts in cell membranes. Nature 387:569–572

Snoeck V, Goddeeris B, Cox E (2005) The role of enterocytes in the intestinal barrier function and antigen uptake. Microbes Infect 7:997–1004

Sugiyama Y, Kato Y (1994) Pharmacokinetic aspects of peptide delivery and targeting: importance of receptor-mediated endocytosis. Drug Dev Ind Pharm 20:591–614

Takano M, Yumoto R, Murakami T (2006) Expression and function of efflux transporters in the intestine. Pharmacol Therap 109:137–161

Tang VW, Goodenough DA (2003) Paracellular ion channel at the tight junction. Biophys J 84:1660–1673

Thomas CM, Smart EJ (2008) Caveolae structure and function. J Cell Mol Med 12:796–809

Tsuji A (2002) Transporter-mediated drug interactions. Drug Metab Pharmacokinet 17:253–274

Uittenbogaard A, Ying Y, Smart EJ (1998) Characterization of a cytosolic heat-shock protein-caveolin chaperone complex. Involvement in cholesterol trafficking. J Biol Chem 273:6525–6532.

Vasselon T, Detmers PA (2002) Toll receptors: a central element in innate immune responses. Infect Immun 70:1033–1041

Wang E, Brown PS, Aroeti B, Chapin SJ, Mostov KE, Dunn KD (2000) Apical and basolateral endocytic pathways of MDCK cells meet in acidic common endosomes distinct from a nearly-neutral apical recycling endosome. Traffic 1:480–493

Wang W, Uzzau S, Goldblum SE, Fasano A (2000) Human zonulin, a potential modulator of intestinal tight junctions. J Cell Sci 113:4435–4440

Watts T, Berti I, Sapone A, Gerarduzzi T, Not T, Zielke R, Fasano A (2005) Role of the intestinal tight junction modulator zonulin in the pathogenesis of type I diabetes in BB diabetic-prone rats. Proc Natl Acad Sci USA 102:2916–2921

Wessel MD, Jurs PC, Tolan JW, Muskal SM (1998) Prediction of human intestinal absorption of drug compounds from molecular structure. J Chem Inf Comput Sci 38:726–735

Yu AS, Enck AH, Lencer WI, Schneeberger EE (2003) Claudin-8 expression in Madin-Darby canine kidney cells augments the paracellular barrier to cation permeation. J Biol Chem 278:17350–17359

Zolotarevsky Y, Hecht G, Koutsouris A, Gonzalez DE, Quan C, Tom J, Mrsny RJ, Turner JR (2002) A membrane-permeant peptide that inhibits MLC kinase restores barrier function in in vitro models of intestinal disease. Gastroenterol 123:163–172

Chapter 4
Strategies to Overcome the Enzymatic Barrier

Martin Werle and Hirofumi Takeuchi

Contents

Abstract Enzymatic degradation of various hydrophilic macromolecules including peptide- or protein drugs by enzymes present in the gastrointestinal tract can be regarded as one main reason for their poor bioavailability after peroral administration. Within the current chapter, strategies to overcome the so-called enzymatic barrier are described. Besides formulations that can protect the drug from enzymatic digestion via, e.g., drug encapsulation and chemical modifications of the drug itself, an emphasis has been put forward on enzyme inhibitors.

M. Werle (✉)
Laboratory of Pharmaceutical Engineering, Department of Drug Delivery Technology and Science, Gifu Pharmaceutical University, 5-6-1 Mitahora-Higashi, Gifu, Japan
e-mail: martin.werle@uibk.ac.at

A. Bernkop-Schnürch (ed.), *Oral Delivery of Macromolecular Drugs*,
DOI 10.1007/978-1-4419-0200-9_4, © Springer Science+Business Media, LLC 2009

4.1 Introduction

The enzymatic barrier, which has been comprehensively described in Chapter 1, can be regarded as one of the main barriers for orally administered macromolecular drugs. The impact of this barrier on the delivery of therapeutic peptides and proteins has been thoroughly investigated; however, due to GI secretion of various other enzymes apart from peptidases and proteases, the enzymatic barrier also constitutes a hurdle for oligosaccharides, genes, and other drugs. The main focus of this chapter will be on the oral delivery of peptides and proteins, though the discussed strategies might be applied in general to any drug susceptible to enzymatic degradation.

Basically, there are two main approaches to overcome the enzymatic barrier. The first approach is based on a chemical modification of the drug itself, whereas the second one is based on the delivery of the unmodified drug in combination with auxiliary agents. These auxiliary agents can be either a proper formulation such as nanoparticles, liposomes, and patches or compounds capable of inhibiting enzymes. Nanoparticles and liposomes are discussed in detail in Chapters 9 and 10. An emphasis has been placed on enzyme inhibitors that can be co-administered with the drug and which are capable of effectively protecting it from enzymatic digestion.

4.2 Formulations That Can Protect Drugs from Enzymatic Degradation

Formulations that can protect drugs from GI enzymes include nanoparticles, liposomes, multifunctional carrier matrices, patch systems as well as formulations targeting GI segments with low enzyme concentrations. Their advantage is that they can improve the uptake of already approved drugs without the co-administration of any additional pharmacologically active compounds. Disadvantages might include problems in the scale-up process, limited stability, and high production costs.

Micro- and nanoparticulate drug delivery systems as well as delivery systems based on liposomes are described in detail in Chapters 9 and 10. Besides other advantages which are discussed in the mentioned chapters, these formulations are capable of protecting incorporated drugs from enzymatic degradation.

Multifunctional carrier matrices offer various advantages for oral drug delivery. Their application for oral drug delivery is described in Chapter 8. Generally, such carriers can be used either to protect the drug from enzymatic degradation via steric hindrance or to directly inhibit enzymes. Multifunctional polymers that can inhibit GI proteolytic enzymes per se are polyacrylates and modifications thereof as well as polymer–inhibitor conjugates. The involved mechanisms are described in subdivision Section 4.4.6.3 of the current chapter.

Patch systems have gained lots of attention for transdermal drug delivery. In addition to transdermal delivery, this approach might also be applied to oral drug delivery. Especially patches which display a mucoadhesive matrix allowing controlled drug release on one side and a closed layer that can protect the drug from luminally secreted enzymes on the backside are believed to improve oral drug absorption by a combination of different mechanisms.

It has been demonstrated that delivery systems targeting the stomach as potential absorption site can be beneficial for drugs which are highly susceptible to degradation mediated by intestinal occurring enzymes. Besides the low pH of the gastric juice, the only occurring proteolytic enzyme is pepsin. Another absorption site with very low enzymatic activity is the colon. One more advantage of colonic delivery systems is that via this absorption route, the first pass effect can be minimized, because only two of the four veins located in the stomach transport the blood via the liver to the blood circulation.

With the exception of multifunctional carrier matrices, the above-discussed formulations can protect the drug to a certain extent from enzymatic degradation in the GI tract, but they are not capable of directly inhibiting the according enzymes. Although promising data from animal studies are available, it must always be considered if the evaluated dosage form and administration route can be used in humans too. This refers to dosage forms, such as solutions, which have in some cases be evaluated using high volumes, or to intraduodenal administered delivery systems as well as to possible stability problems of particulate systems.

4.3 Chemical Modification

Chemical modification of a drug is one of the most important approaches in order to circumvent GI digestion. A targeted modification should be based on an exact knowledge of the substructures susceptible to enzymatic degradation. To investigate this, simple enzyme assays can be performed. Various GI enzymes are commercially available. Incubating, e.g., a peptide or protein in solutions containing different isolated proteases such as trypsin, chymotrypsin, or elastase at physiological pH allows an identification of the enzymes responsible for enzymatic degradation. Due to a narrow cleavage specificity of those proteases, the susceptible amino acids can be identified and modified. Another option is to characterize the cleavage products via, e.g., MS. Regarding peptides and proteins, some common modifications to improve the enzymatic stability are discussed below. Although most of them have been used to prolong peptide/protein plasma half-life times, they can in general also be applied to oral drug delivery. However, it always has to be taken into consideration that such modifications will most likely cause alterations in the pharmacological profile of the drug. Such modifications generally lead to the generation of new chemical entities, for which extensive toxicological data are necessary if drug approval is intended. Therefore, such alterations – with the exception of PEGylation – are not of great interest for pharmaceutical technologists working in the field of oral drug delivery and whose aim is to improve or to make feasible the oral uptake of an already approved drug.

4.3.1 Modification of N and C Terminus

Various exopeptidases including aminopeptidase N or carboxypeptidase B occur in the intestine. Therefore, a modification of the terminal amino acids of a therapeutic peptide or protein can protect the drug from exopeptidase degradation. One common way of terminal modification is N-acetylation and C-amidation. Moreover, the attachment of various compounds to the terminal amino acids can improve stability. Various fatty acids of chain lengths ranging from 4 to 18 were conjugated to RC-160, a somatostatin analog with antiproliferative activity. The novel compounds exhibited greater resistance toward trypsin and serum degradation in comparison to unmodified RC-160 (Dasgupta et al. 2002).

4.3.2 Replacement of Labile Amino Acids

As described in Chapter 1, GI endopeptidases such as trypsin and chymotrypsin have narrow substrate specificity. The preferred theoretical cleavage sites of most proteolytic enzymes are known, so if the amino acid sequence of a drug is known as well, the amino acid sequence of the drug can be screened for these potential cleavage sites. If the positions within a therapeutic peptide/protein that are susceptible to enzymatic degradation can be identified, a substitution of the labile amino acids can improve peptide/protein stability. It has, for example, been demonstrated that by substituting the single chymotrypsin cleavage site of a 29-amino acid cystine-knot microprotein, stabilization toward this protease could be achieved. The time-dependent degradation in the presence of chymotrypsin of the cystine-knot microprotein before and after substituting the theoretical cleavage sites is shown in Fig. 4.1 (Werle et al. 2006).

Fig. 4.1 Time-dependent degradation profile of a cystine-knot microprotein in the presence of chymotrypsin before (-○-) and after (-●-) substitution of the theoretical chymotrypsin cleavage site; each point represents the mean ± SD of at least three experiments. Figure adapted from Werle et al. (2006)

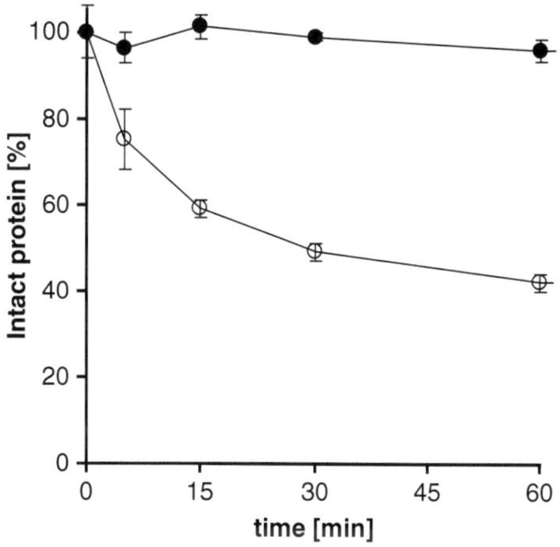

4.3.3 PEGylation

Another approach is the specific covalent attachment of PEG to either or both termini of a peptide or a protein drug. The N-terminal modification of glucose-dependent insulinotropic polypeptide (GIP1-30) with 40-kDa PEG abrogated functional activity, whereas C-terminal PEGylation of GIP1-30 maintained full agonism at the GIP receptor and conferred a high level of dipeptidyl peptidase IV (DP IV) resistance. Moreover, the dual modification of N-terminal palmitoyl and C-terminal PEGylation resulted in a full agonist of comparable potency to native GIP that was stable to DP IV cleavage (Salhanick et al. 2005). The stability of INF-α2b toward trypsin-caused degradation was strongly improved by the conjugation of $PEG_{2,40\ K}$ to the native protein (Ramon et al. 2005). The results of this study are provided in Fig. 4.2. Poly(ethylene glycol) (PEG) exhibits

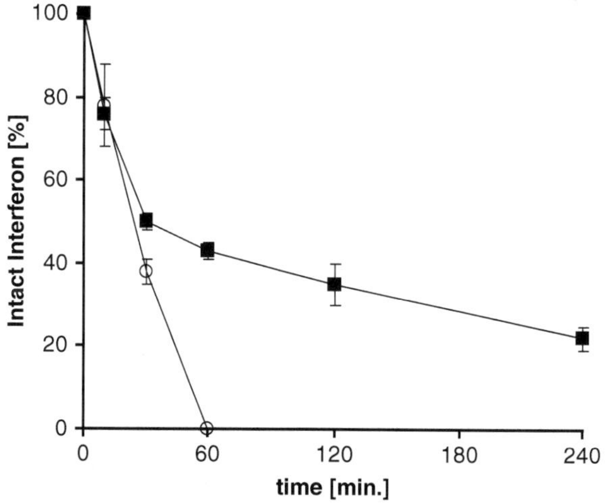

Fig. 4.2 Degradation of (-●-) PEGylated interferon α2b ($PEG_{2,40\ K}$-IFN-α2b) and (-○-) non-PEGylated interferon α2b (IFN-α2b) in the presence of trypsin at 37°C. Figure adapted from Ramon et al. (2005)

several properties that are of relevance for pharmaceutical applications: high water solubility, high mobility in solution, lack of toxicity and immunogenicity, and ready clearance from the body (Delgado et al. 1992). Interestingly, many of these properties are transferred to PEG–protein or PEG–peptide conjugates. The extent of these features is dependent on the molecular weight of the attached PEG. As demonstrated, for example, by He et al., only minor changes in immunogenicity of trichosanthin after modification with $PEG_{5\ k}$ was observed, whereas modification with $PEG_{20\ k}$ led to significantly reduced immunogenicity (He et al. 1999).

4.4 Enzyme Inhibitors

The efficacy of enzyme inhibitors to improve the oral absorption of therapeutic macromolecular drugs has been studied extensively. The use of enzyme inhibitors should be based on an exact knowledge regarding the enzymes responsible for drug degradation. In vitro methods to identify these enzymes have been summarized previously and include drug stability studies in presence of collected gastric and/or intestinal juices, isolated proteases, or intestinal mucosa (Werle 2007). Based on the information gained in such studies, an appropriate enzyme inhibitor can be chosen. Although the potential of enzyme inhibitors for oral drug delivery has been clearly demonstrated in numerous studies, their use remains questionable; in particular because of concerns regarding a potential toxicity. Therefore, great effort has been put in the development of enzyme inhibitors such as polymer–inhibitor conjugates, which are believed to be a safer alternative.

4.4.1 Protease Inhibitors Which Are Not Based on Amino Acids

The drawback of this highly potent class of inhibitors is their toxicity (Stryer 1988). Only few representatives are known to be non-toxic. Therefore, enzyme inhibitors that are not based on amino acids are only of theoretical interest for oral drug delivery and other applications in human. Nevertheless, they might be interesting as lead compounds to develop novel potent and non-toxic inhibitors.

Representatives of this class of inhibitors are DFP (diisopropylfluorophosphate), PMSF (phenylmethylsulfonyl fluoride), APMSF (4-aminophenyl)-methanesulfonyl, AEBSF (4-(2-aminoethyl)-benzenesulfonyl fluoride), FK-448 (4-(4-isopropylpiperadinocarbonyl) phenyl 1,2,3,4,-tetrahydro-1-naphthoate methanesulfonate), camostat mesilate (*N*,*N*′-dimethyl carbamoylmethyl-*p*-(*p*′-guanidino-benzoyloxy)phenylacetate methanesulfonate), and Na-glycocholate. The organophosphorus inhibitors DFP and PMSF are potent irreversible inhibitors of serine proteases. Due to their additional inhibition of acetylcholine-esterase these compounds are highly toxic (Stryer 1988). APMSF is a toxic trypsin inhibitor. In contrast, AEBSF is markedly less toxic than DFP and PMSF but displays comparable enzyme inhibitor activity. Also FK-448 is a low toxic inhibitor. It is capable of inhibiting chymotrypsin. Moreover, FK-448 has already been investigated in rats and dogs regarding its potential for oral drug delivery. Co-administration of FK-448 led to an enhanced absorption of insulin. It was found that the inhibition of chymotrypsin was mainly responsible for the observed decrease in blood glucose (Fujii et al. 1985). Other representatives exhibiting low toxicity are camostat mesilate (Yamamoto et al. 1994) and Na-glycocholate (Okagawa et al. 1994, Yamamoto et al. 1994).

4.4.2 Protease Inhibitors Which Are Based on Amino Acids and Modified Amino Acids

Amino acids and modified amino acids offer some advantages to protease inhibitors that are not based on amino acids. In general they are non-toxic and can be easily produced at low costs. Disadvantage is their small size, which leads to a rapid dilution and absorption in the GI tract. Therefore, high concentrations of this class of inhibitors would be necessary to achieve a sufficient improvement of oral drug uptake in vivo. However, they gained some interest as protease inhibitors in other non-invasive routes, especially in nasal delivery. Although it has been demonstrated that unmodified amino acids are reversible and are competitive inhibitors of exopeptidases (McClellan and Garner 1980), their low inhibitory activity renders them unsuitable as protease inhibitors for oral drug delivery. Contrary to unmodified amino acids, modified amino acids exhibit a much stronger inhibitory activity. This can be explained, because modified amino acids belong to the group of "transition-state inhibitors." These transition state inhibitors are generally reversible and competitive inhibitors with high inhibitory activity. The underlying mechanism of transition state inhibitors is based on the theory that inhibitors which resemble the geometry of a substrate in its transition state has a much higher affinity to the active site of the enzyme than the substrate itself. The aminopeptidase transition state inhibitor boroleucine for example exhibits a 100-fold higher enzyme inhibitory activity than bestatin. In comparison to puromycin, the inhibitory activity of boroleucine is even 1000-fold higher (Hussain et al. 1989). Boro-valine and boroalanine are two more representatives of α-aminoboronic acid derivatives acting as aminopeptidase transition state inhibitors. All these α-aminoboronic acid derivatives have been so far investigated only for nasal delivery. Another modified amino acid which is capable of inhibiting proteolytic enzymes, in particular aminopeptidase N, is *N*-acetylcysteine (Bernkop-Schnürch and Marschütz 1997). *N*-Acetylcysteine is a mucolytic agent which can reduce the diffusion barrier and which displays low toxicity (Bernkop-Schnürch and Fragner 1996).

4.4.3 Protease Inhibitors Which Are Based on Peptides and Modified Peptides

A representative of the group of peptide-based protease inhibitors is bacitracin. This docapeptide, which is almost exclusively used in veterinary medicine as well as for the topical antibiotic treatment in human, is known to interfere with a variety of biological processes, including inhibition of bacterial peptidoglycan synthesis, mammalian transglutaminase activity, and proteolytic enzymes such as aminopeptidase N. Due to these properties, it has therefore been used to inhibit the degradation of various therapeutic (poly)peptides, such as insulin, metkephamid, LH-RH, and buserelin (Langguth et al. 1994, Raehs et al. 1988, Yamamoto et al. 1994). It has also been demonstrated that bacitracin exhibits absorption-enhancing effects without leading to serious intestinal mucosal damage (Gotoh et al. 1995).

Nevertheless, the use of bacitracin as an enzyme inhibitor in order to improve drug uptake in vivo remains questionable, because bacitracin-mediated nephrotoxicity has been reported previously (Drapeau et al. 1992). However, it was demonstrated recently that bacitracin can be covalently bond to a mucoadhesive, non-absorbable polymer and that this polymer–bacitracin conjugate still retains its inhibitory activity (Bernkop-Schnürch and Marschütz 1997). The reason for using such a polymer–inhibitor conjugate instead of unmodified bacitracin is that the polymer–bacitracin conjugate is believed to remain in the GI tract which would exclude systemic side effects. More information about polymer–inhibitor conjugates are provided in Section 4.4.6.3 of this article.

According to amino acids, also di- and tripeptides display weak and unspecific activity toward some exopeptidases (Langguth et al. 1994). However, chemical modification of di- and tripeptides can lead to analogs with an enhanced inhibitory activity. Phosphinic acid dipeptides, which also belong to the group of "transition state inhibitors," are potent inhibitors of aminopeptidase. The phosphinate inhibitor VI, for instance, is 10 times more potent than bestatin and even 100-fold more potent than (Hussain et al. 1992). The use of a phosphinic acid dipeptide analog for the nasal delivery of the model drug leu-enkephalin was investigated. It could be demonstrated in rats that even very low concentration of this inhibitor can reduce the enzymatic degradation of nasally administered leu-enkephalin and that the effects were reversible (Hussain et al. 1992). Therefore, inhibitors based on dipeptides might be interesting for oral drug delivery too.

Pepstatin, another representative of the group of "transition state inhibitors" and a very potent inhibitor of pepsin, is a modified pentapeptide (McConnell et al. 1991). The inhibition of the gastric-secreted pepsin is of practical relevance for two reasons: the first reason is that there are several drugs that must be liberated in the stomach. An example is the epidermal growth factor in the treatment of gastric ulcer (Itoh and Matsuo 1994). The second reason is that drugs which are extremely susceptible to intestinally secreted proteases might be liberated and absorbed in the stomach. An enteric coating of the dosage form with a gastric fluid resistant layer would disable the liberation of the drug in the stomach. Therefore, pepsin inhibitors can in certain cases be useful tools for oral drug delivery. Unfortunately, it was reported that co-administration of pepstatin led to several side effects mediated by an inhibition of physiologically essential enzymes (Carmel 1994, McCaffrey and Jamieson 1993, Plumpton et al. 1994).

Peptides that display a terminally located aldehyde function in their structure constitute another group of modified peptide enzyme inhibitors. The sequence benzyloxycarbonyl-Pro-Phe-CHO fulfils the known primary and secondary specificity requirements of chymotrypsin and has been found to be a potent reversible inhibitor of this proteolytic enzyme (Walker et al. 1993). Further, protease inhibitors comprising terminally located aldehyde function are antipain, leupeptin, chymostatin, and elastatinal. In addition, also phosphoramidon, bestatin, puromycin, and amastatin represent modified peptides which can reversibly inhibit enzymes.

4.4.4 Protease Inhibitors Which Are Based on Polypeptides

This class of protease inhibitors constitutes representatives such as aprotinin or Bowman-Birk inhibitor (BBI), which have been extensively investigated regarding their potential for oral drug delivery. The 58 amino acid polypeptide which is derived from bovine tissues is known to inhibit various enzymes including chymotrypsin, kallikrein, plasmin, and trypsin. Moreover, it was one of the first protease inhibitors used as an auxiliary agent for oral (poly)peptide delivery. Co-administration of aprotinin led to an increased bioavailability of peptide and protein drugs. The underlying mechanism of this increase has been attributed to the chymotrypsin and trypsin inhibitory properties of aprotinin (Kimura et al. 1996, Saffran et al. 1988, Yamamoto et al. 1994). The soybean trypsin inhibitors can be divided into two main families: the Bowman-Birk inhibitor (71 amino acids, 8 kDa) and the Kunitz trypsin inhibitor (184 amino acids, 21 kDa). Both are known to inhibit trypsin, chymotrypsin, and elastase, whereas carboxypeptidases A and B cannot be inhibited (Reseland et al. 1996, Ushirogawa 1992). Other representatives of this class of polypeptide inhibitors are the chicken egg white trypsin inhibitor (chicken ovomucoid; 186 amino acids) (Ushirogawa 1992), the chicken ovoinhibitor (449 amino acids) (Scott et al. 1987), and the human pancreatic trypsin inhibitor (56 amino acids).

Due to the comparably high mass of polypeptide-based protease inhibitors, a controlled release of these inhibitors out of a drug carrier matrix can be achieved. The efficacy of a drug delivery systems for the slow release of the protease inhibitors aprotinin and bacitracin to improve the oral bioavailability of insulin has been demonstrated previously (Kimura et al. 1996). In this study, poly(vinyl alcohol)-gel spheres were used as a mucoadhesive drug delivery matrix. The release rate of the aprotinin was only around 10% per hour. This release rate was almost synchronous with the release rate of the polypeptide insulin. In vivo studies carried out with this delivery system showed an improved oral drug bioavailability (Kimura et al. 1996). Polypeptide inhibitors often exhibit low toxicity as well as strong inhibitory activity and have therefore been used to the greatest extent as auxiliary agents to overcome the enzymatic barrier of perorally administered therapeutic peptides and proteins.

4.4.5 Protease Inhibitors That Can Complex Ions

Divalent cations, in particular Zn^{2+} and Ca^{2+}, are essential co-factors for most proteolytic enzymes. Therefore, depletion of divalent cations consequently leads to reduced protease activity. This can be achieved by using complexing agents, which are capable of removing ions from the enzymatic structure. Representatives of complexing agents are for instance ethylenediamine tetraacetic acid (EDTA), ethylene glycol tetraacetic acid (EGTA), 1,10-phenanthroline, and hydroxychinolin (Garner and Behal 1974, Ikesue et al. 1993, Sangadala et al. 1994). The concept of inhibiting enzymes via ion-complexation generally works in vitro, however, some important points must be considered regarding the in vivo situation: in order to inhibit enzymes by using complexing agents, high concentrations thereof are necessary. It has been demonstrated in vitro that a concentration of 7.5% (w/v) of the chelating agent EDTA was not sufficient to inhibit the calcium-dependent endopeptidase trypsin (Luessen et al. 1996). Taking into account that the calcium ion concentration in gastric and intestinal fluids has been determined to be in the range of 0.4–0.7 mM (Lindahl et al. 1997), it can be anticipated that such high calcium concentrations will negatively influence the efficacy of orally administered complexing agents in terms of protease inhibition. Hence, complexing agents do not appear to be the most suitable auxiliary agents to successfully inhibit calcium-dependent endoproteases such as trypsin, chymotrypsin, and elastase. However, the potential of complexing agents to inhibit various zinc-dependent exopeptidases including carboxypeptidases A and B as well as aminopeptidase N has already been demonstrated (Bernkop-Schnürch et al. 1997, Ikesue et al. 1993, Sanderink et al. 1988). Maybe the most important point in order to achieve sufficient protease inhibition and minimizing toxic side effects in vivo by using complexing agents is to avoid the extensive dilution during the GI passage. Regarding oral drug delivery it must be considered that co-administration of certain complexing agents can improve drug permeation (Lee 1990).

4.4.6 Protease Inhibitors Based on Multifunctional Mucoadhesive Polymers

The main drawback of low molecular mass enzyme inhibitors for oral drug delivery, namely the necessity of using high inhibitor concentrations due to extensive dilution effects during GI passage, can be excluded by using mucoadhesive polymers that display enzyme inhibitory activity. Multifunctional polymers for oral drug delivery exhibiting mucoadhesive properties as well as enzyme inhibitory activity are poly(acrylates), thiolated polymers, and polymer–enzyme inhibitor conjugates. Whereas the inhibitory mechanism of polymer–enzyme inhibitor conjugates depends on the immobilized inhibitor, the inhibitory mechanism of poly(acrylates) and thiolated polymers is based on the complexing properties of these polymers. Besides a direct complexation of divalent cations such as Zn^{2+} or Ca^{2+} membrane-bound enzymes can also be inhibited via a "far distance inhibitory effect" (Luessen et al. 1996). Results of this study are shown in Fig. 4.3. The GI mucosa is covered by a mucus layer, which separates the mucoadhesive polymer from the brush border membrane (Bernkop-Schnürch and Marschütz 1997). Although there is no direct contact between polymer and membrane-bound enzymes, it was demonstrated that membrane-bound proteolytic enzymes such as aminopeptidase N can be inhibited by mucoadhesive polymers. However, as mentioned above, a general problem of complexing agents in oral drug delivery is the high ion concentrations in the GI tract.

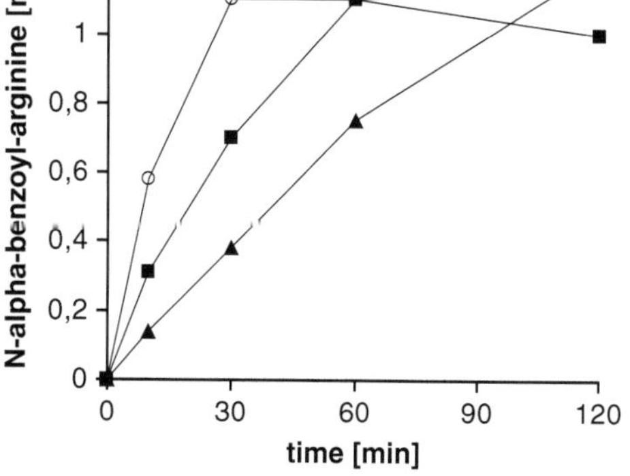

Fig. 4.3 Formation of BA during incubation with BAEE and trypsin in the presence of -○- Ca^{2+}, -■- 0.35% PCP and Ca^{2+} added directly before trypsin incubation, and -▲- 0.35% PCP and Ca^{2+} added after 10 min of incubation. Figure adapted from Luessen et al. (1995)

Still, mucoadhesive polymers offer various advantages in comparison to low molecular weight complexing agents. The first advantage is that the polymeric matrix itself can to a certain amount protect the embedded drug from degradation. Second, the drug can diffuse directly from the matrix to the mucus layer when using mucoadhesive polymers. This limits the contact time of the drug with luminally secreted proteases. Another advantage is that the enzyme inhibition takes place at the intestinal segment where the drug is released. Therefore, lower concentrations of inhibitors can be used and the likeliness of an unintended absorption of harmful compounds is limited.

4.4.6.1 Poly(acrylates)

Regarding poly(acrylates) and derivatives, detailed studies focusing on the inhibitory effect toward luminally secreted and brush border membrane enzymes are available (Luessen et al. 1996). It was demonstrated that poly(acrylate) derivatives are capable of inhibiting trypsin, chymotrypsin, and carboxypeptidases A and B (Fig. 4.3). However, whether the protective effect of these polymers is sufficient to prevent luminal enzymatic drug degradation will mainly depend on the dosage form used. Simple formulations of poly(acrylate) derivatives are not believed to be sufficient for enzyme inhibition in vivo (Bernkop-Schnürch and Göckel 1997). However, poly(acrylates) are of great importance for oral drug delivery. Besides their enzyme inhibitory activity and their mucoadhesiveness (Junginger 1990, Lehr 1994), they can generally be regarded as safe and furthermore, they exhibit additional permeation-enhancing properties (Brochard et al. 1996, Luessen et al. 1996). In summary, the combination of all these properties can lead to improved drug absorption from the GI tract.

4.4.6.2 Thiolated Polymers

The use of thiolated polymers or designated thiomers is investigated for various routes of drug delivery, including the oral, nasal, buccal, vaginal, and ocular routes. A detailed description of this technology is provided in Chapter 8. In brief, a thiol group bearing small molecule is covalently bound to a polymeric backbone such as chitosan or poly(acrylates). This modification leads to significant improvements of mucoadhesive- as well as permeation-enhancing properties toward the unmodified polymer (Bernkop-Schnürch et al. 2003, Bernkop-Schnürch and Krajicek 1998). Moreover, certain thiomers are capable of inhibiting proteins including proteolytic enzymes and transmembrane-located glycoproteins such as P-glycoprotein (see Chapter 7). Thiomers are useful excipients for the delivery of peptides/proteins, genes, and efflux pump substrates. In comparison to unmodified polymer polycarbophil, the thiomer polycarbophil-cysteine significantly improves the inhibition of carboxypeptidases A and B as well as chymotrypsin

(Bernkop-Schnürch and Thaler 2000). Moreover, the model drug leu-enkephalin was protected from degradation caused by enzymes of excised native porcine intestinal mucosa and aminopeptidase N (Bernkop-Schnürch et al. 2001). Another poly(acrylate) derivative which was found to exhibit significantly higher enzyme inhibitory properties against aminopeptidase N as well as vaginal mucosa compared to the unmodified polymer is the thiomer carbopol(974 P)-cysteine. Valenta et al. demonstrated that the inhibitory activity was strongly dependent on the amount of immobilized thiol groups. Increasing amounts of free thiol group on the polymeric backbone led to increased inhibitory activity (Valenta et al. 2002).

4.4.6.3 Polymer–Enzyme Inhibitor Conjugates

In order to avoid the absorption of enzyme inhibitors, polymer–inhibitor conjugates have been developed. Well-known enzyme inhibitors including EDTA, Bowman-Birk inhibitor (BBI), or aprotinin are covalently bound to mucoadhesive polymers such as poly(acrylic acid) or chitosan. Besides the advantage that these conjugates are not absorbed, additional features are believed to contribute to a safer toxicological profile in comparison to the administration of the corresponding unbound inhibitor. Due to the mucoadhesive properties of the polymeric backbone, enzyme inhibition takes place at the same intestinal segments as where the drug is released. In comparison to polymer–inhibitor conjugates, enzyme inhibitors which are not connected to mucoadhesive polymers inhibit enzymes in the entire intestine, facilitating the uptake of harmful compounds. Moreover, polymer–inhibitor conjugates are much less affected by dilution effects in comparison to unbound enzyme inhibitors. Therefore, the targeted inhibition of polymer–inhibitor conjugates allows lower concentrations of enzyme inhibitors, leading to an overall reduced risk regarding toxicity mediated by feedback mechanisms. Additional properties of the polymer itself including mucoadhesive properties, enzyme inhibitory properties per se, or permeation-enhancing properties can further improve the oral uptake of hydrophilic macromolecular drugs.

Maybe the first and still the most important representative of polymer–inhibitor conjugates is chitosan–EDTA. This conjugate has been synthesized around 10 years ago and has been evaluated for the oral delivery of peptides (Bernkop-Schnürch et al. 1997, Wu et al. 2004) and pDNA (Loretz et al. 2006) as well as for transdermal drug delivery (Biruss and Valenta 2006, Valenta et al. 1998). The underlying idea of developing chitosan–EDTA was to combine the mucoadhesive properties of chitosan and the well-known ion-complexing properties of EDTA. It has been demonstrated that chitosan–EDTA is capable of inhibiting zinc-dependent proteases including aminopeptidases A, N, P, and W as well as carboxypeptidases A, B, M, and P. Although the ability of chitosan–EDTA to bind calcium ions is higher than that of various polyacrylates (Bernkop-Schnürch and Krajicek 1998), chitosan–EDTA conjugates do not display any inhibitory effects toward calcium-dependent proteases such as elastase, chymotrypsin, or trypsin. However, chitosan–EDTA which can additionally be used as a matrix for controlled drug release is an important tool for the delivery of enzymatic susceptible drugs.

Other so far developed polymer–inhibitor conjugates are sodium carboxymethylcellulose–BBI and sodium carboxymethylcellulose–elastatinal (Marschütz and Bernkop-Schnürch 2000), various polymer–inhibitor conjugates based on polycarbophil (Marschütz et al. 2001), chitosan–BBI and chitosan–elastatinal (Guggi and Bernkop-Schnürch 2003), chitosan–pepstatin (Guggi et al. 2003), and chitosan–aprotinin (Werle et al. 2007). A comparison of the in vitro efficacy of chitosan–aprotinin and chitosan to protect the trypsin and chymotrypsin substrates benzoyl-arginine-*p*-nitroanilide (trypsin) and benzoyl-tyrosine-*p*-nitroanilide (chymotrypsin) from degradation in the presence of the corresponding protease is shown in Fig. 4.4 (Werle et al. 2007) (Table 4.1).

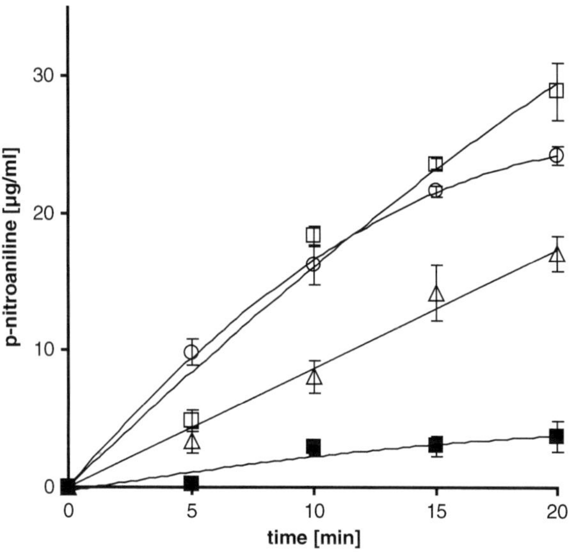

Fig. 4.4 Increase of *p*-nitroaniline in a solution containing the trypsin substrate benzoyl-arginine-*p*-nitroanilide and trypsin at 37°C and phosphate buffer pH 7.6; -○- buffer only, -□- 0.1% chitosan, -△- 0.025% chitosan–aprotinin, -■- 0.1% chitosan–aprotinin; each point represents the mean ± SD of at least three experiments. Figure adapted from Werle et al. (2007)

For an intended drug uptake in the stomach, the only proteolytic enzyme which has to be inhibited is pepsin. It has been demonstrated that the polymeric pepsin–inhibitor conjugate chitosan–pepstatin was capable to efficiently protect the peptide drug calcitonin in the presence of pepsin. These results have been additionally confirmed in vivo. The administration of minitablets comprising of a mixture of calcitonin, chitosan, and chitosan–pepstatin to rats led to a reduction of the plasma calcium level, whereas no effect was observed when

Table 4.1 Summary of various protease inhibitors

Inhibitor group	Inhibitor name and reference
Not based on amino acids	DFP, PMSF, APMSF, AEBSF, phenyl 1,2,3,4,-tetrahydro-1-naphthoate methanesulfonate), (Stryer 1988), FK-448 (4-(4-isopropylpiperadinocarbonyl) (Fujii et al. 1985), camostat mesilate (Yamamoto et al. 1994), Na-glycocholate (Okagawa et al. 1994, Yamamoto et al. 1994)
Amino acids and modified amino acids	Amino acids having hydrophobic side chains such as L-phenylalanine (McClellan and Garner 1980), boro-leucine, boro-valine, and boro-alanine (Hussain et al. 1989), *N*-acetylcysteine (Bernkop-Schnürch and Marschütz 1997)
Peptides and modified peptides	Bacitracin (Langguth et al. 1994, Raehs et al. 1988, Yamamoto et al. 1994), di- and tripeptides (Langguth et al. 1994), phosphinate inhibitor VI (Hussain et al. 1992). Pepstatin, (McConnell et al. 1991), benzyloxycarbonyl-Pro-Phe-CHO (Walker et al. 1993), antipain, leupeptin, chymostatin, and elastatinal, phosphoramidon, bestatin, puromycin, and amastatin (Bernkop-Schnürch 1998)
Polypeptides	Aprotinin (Kimura et al. 1996, Saffran et al. 1988, Yamamoto et al. 1994), Bowman-Birk inhibitor, Kunitz trypsin inhibitor, chicken egg white trypsin inhibitor (Reseland et al. 1996, Ushirogawa 1992), chicken ovoinhibitor (Scott et al. 1987), human pancreatic trypsin
Complexing agents	EDTA, EGTA, 1,10-phenanthroline, hydroxychinolin (Garner and Behal 1974, Ikesue et al. 1993, Sangadala et al. 1994)
Multifunctional polymers	Poly(acrylates) (Luessen et al. 1996), polycarbophil-cysteine (Bernkop-Schnürch and Thaler 2000, Bernkop-Schnürch et al. 2001), carbopol(974 P)-cysteine (Valenta et al. 2002), carboxymethylcellulose–BBI, and carboxymethylcellulose–elastatinal (Marschütz and Bernkop-Schnürch 2000), PCP–tetramethylenediamine–chymostatin, PCP–poly(ethylene glycol))–chymostatin conjugate, PCP–tetramethylenediamine–antipain, and PCP–tetramethylenediamine–elastatinal (Marschütz et al. 2001), chitosan–BBI and chitosan–elastatinal (Guggi and Bernkop-Schnürch 2003), chitosan–pepstatin (Guggi et al. 2003), chitosan–aprotinin (Werle et al. 2007)

administering tablets omitting chitosan–pepstatin (Guggi et al. 2003). Even more pronounced effects were achieved with tablets additionally containing the thiomer chitosan–TBA. The results of this study are provided in Fig. 4.5. More recently, the potential of a chitosan–aprotinin conjugate was evaluated in vitro as well as in vivo. Even concentration of 0.025% of the polymer–inhibitor conjugates displayed inhibitory properties toward trypsin and chymotrypsin. Moreover, the blood glucose level after oral administration of a chitosan–aprotinin-based delivery system containing insulin could be decreased for several hours (Werle et al. 2007).

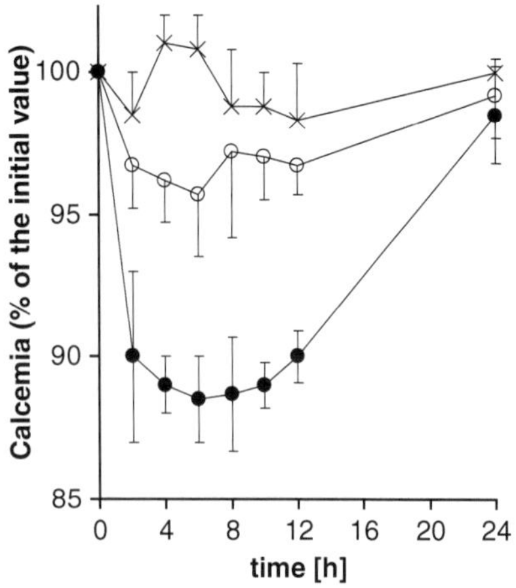

Fig. 4.5 Decrease in plasma calcium level after oral administration of calcitonin in tablets based on -x- chitosan, -○- chitosan–pepstatin, and -●- chitosan–pepstatin/chitosan–TBA; each point represents the mean ± SD of at least three experiments. Figure adapted from Guggi et al. (2003)

4.5 Conclusion and Future Trends

The enzymatic barrier can be regarded as a main obstacle for orally administered hydrophilic macromolecular drugs. Therefore, strategies to overcome this barrier are highly on demand. A modification of the drug itself in order to avoid enzymatic degradation is always connected with possible changes in the pharmacokinetic profile and challenges regarding regulatory issues. The efficacy of formulations that can protect the drug from enzymatic degradation as well as of enzymes inhibitors has been demonstrated in various studies and a list of so far used enzyme-inhibitors is provided in Table 4.1. Especially for enzyme inhibitors, possible toxic side effects must be taken into consideration. Although systemic side effects can be minimized by using polymer–inhibitor conjugates, the risk of side effects mediated by feedback mechanisms remains. Therefore, an effective protection of the drug by appropriate formulations might be a safer alternative. For future applications, especially the emerging field of nanotechnology will play a major role.

References

Bernkop-Schnürch A, Fragner R (1996) Investigations into the diffusion behaviour of polypeptides in native intestinal mucus with regard to their peroral administration. Pharm Sci 2:361–363

Bernkop-Schnürch A, Göckel NC (1997) Development and analysis of a polymer protecting from luminal enzymatic degradation caused by a-chymotrypsin. Drug Dev Ind Pharm 23:733–740

Bernkop-Schnürch A, Kast CE, Guggi D (2003) Permeation enhancing polymers in oral delivery of hydrophilic macromolecules: thiomer / GSH systems. J Control Release. 2003 Dec 5; 93(2):95–103.

Bernkop-Schnürch A (1998) The use of inhibitory agents to overcome the enzymatic barrier to perorally administered therapeutic peptides and proteins. J Control Release 52(1-2):1–16.
Bernkop-Schnürch A, Krajicek ME (1998) Mucoadhesive polymers as platforms for peroral peptide delivery and absorption: synthesis and evaluation of different chitosan-EDTA conjugates. J Control Rel 50:215–223
Bernkop-Schnürch A, Marschütz MK (1997) Development and in vivo evaluation of systems to protect peptide drugs from aminopeptidase N. Pharm Res 14:181–185
Bernkop-Schnürch A, Paikl C, Valenta C (1997) Novel bioadhesive chitosan-EDTA conjugate protects leucine enkephalin from degradation by aminopeptidase N. Pharm Res 14:917–922
Bernkop-Schnürch A, Thaler S (2000) Polycarbophil-cysteine conjugates as platforms for oral (poly)peptide delivery systems. J Pharm Sci 89:901–909
Bernkop-Schnürch A, Walker G, Zarti H (2001) Thiolation of polycarbophil enhances its inhibition of intestinal brush border membrane bound aminopeptidase N. J Pharm Sci 90:1907–1914
Biruss B, Valenta C (2006) Skin permeation of different steroid hormones from polymeric coated liposomal formulations. Eur J Pharm Biopharm 62:210–219
Brochard G, Luessen HL, Verhoef JC, Lehr CM, de Boer AG, Junginger HE (1996) The potential of mucoadhesive polymers in enhancing intestinal peptide drug absorption III: Effects of chitosan-glutamate and carbomer on epithelial tight junctions in vitro. J Control Rel 39:131–138
Carmel R (1994) In vitro studies of gastric juice in patients with food-cobalamin malabsorption. Dig Dis Sci 39:2516–2522
Dasgupta P, Singh A, Mukherjee R (2002) N-terminal acylation of somatostatin analog with long chain fatty acids enhances its stability and anti-proliferative activity in human breast adenocarcinoma cells. Biol Pharm Bull 25:29–36
Delgado C, Francis GE, D. F (1992) The uses and properties of PEG-linked proteins. Crit Rev Ther Drug Carrier Syst 9:249–304
Drapeau G, Petitclerc E, Toulouse A, Marceau F (1992) Dissociation of the antimicrobial activity of bacitracin USP from its renovascular effects. Antimicrob Agents Chemother 36:955–961
Fujii S, Yokohama T, Ikegaya K, Salo F, Yakoo N (1985) Promoting effect of the new chymotrypsin inhibitor FK-448 on the intestinal absorption of insulin in rats and dogs. J Pharm Pharmacol 37:545–549
Garner CWJ, Behal FJ (1974) Human liver aminopeptidase. Role of metal ions in mechanism of action. Biochemistry 13:3227–3233
Gotoh S, Nakamura R, Nishiyama M, Fujita T, Yamamoto S, Muranishi S (1995) Does bacitracin have an absorption-enhancing effect in the intestine? Biol Pharm Bull 18:794–796
Guggi D, Bernkop-Schnürch A (2003) In vitro evaluation of polymeric excipients protecting calcitonin against degradation by intestinal serine proteases. Int J Pharm 252:187–196
Guggi D, Krauland AH, Bernkop-Schnürch A (2003) Systemic peptide delivery via the stomach: in vivo evaluation of an oral dosage form for salmon calcitonin. J Control Release 92:125–135
He XH, Shaw PC, Tam SC (1999) Reducing the immunogenicity and improving the in vivo activity of trichosanthin by site-directed pegylation. Life Sci 65:355–368
Hussain MA, Lim MS, Raghavan KS, Rogers NJ, Hidalgo R, Kettner CA (1992) A phosphinic acid dipeptide analogue to stabilize peptide drugs during their intranasal absorption. Pharm Res 9:626–628
Hussain MA, Shenvi AB, Rowe SM, Shefter E (1989) The use of alpha-aminoboronic acid derivatives to stabilize peptide drugs during their intranasal absorption. Pharm Res 6:186–189
Ikesue K, Kopecková P, Kopecek J (1993) Degradation of proteins by guinea pig intestinal enzymes. Int J Pharm 95:171–179

Itoh M, Matsuo Y (1994) Gastric ulcer treatment with intravenous human epidermal growth factor: a double-blind controlled clinical study. J Gastroenterol Hepatol 9:78–83
Junginger HE (1990) Bioadhesive polymer systems for peptide delivery. Acta Pharm Technol 36:110–126
Kimura T, Sato K, Sugimoto K, Tao R, Murakami T, Kurosaki Y, Nakayama T (1996) Oral administration of insulin as poly(vinyl alcohol)-gel spheres in diabetic rats. Biol Pharm Bull 19:897–900
Langguth P, Bohner V, Biber J, Merkle HP (1994) Metabolism and transport of the pentapeptide metkephamid by brush-border membrane vesicles of rat intestine. J Pharm Pharmacol 46:34–40
Lee VHL (1990) Protease Inhibitors and Penetration Enhancers as Approaches to Modify Peptide Absorption. J Control Rel 213–223
Lehr C-M (1994) Bioadhesion technologies for the delivery of peptide and protein drugs to the gastrointestinal tract. Crit Rev Ther Drug 11:119–160
Lindahl A, Ungell A-L, Knutson L, Lennerna:s H (1997) Characterization of fluids from the stomach and proximal jejunum in men and women. Pharm Res 14:497–502
Loretz B, Föger FA, Werle M, Bernkop-Schnürch A (2006) Oral gene delivery: Strategies to improve stability of pDNA towards intestinal digestion. J Drug Targ 14:311–319
Luessen HL, Bohner V, Perard D, Langguth P, Verhoef JC, de Boer AG, Merkle HP, Junginger HE (1996) Mucoadhesive polymers in peroral peptide drug delivery.V. Effect of poly(acrylates) on the enzymatic degradation of peptide drugs by intestinal brush border membrane vesicles. Int J Pharm 141:39–52
Luessen HL, de Leeuw BJ, Langemeyer MW, de Boer AG, Verhoef JC, Junginger HE (1996) Mucoadhesive polymers in peroral peptide drug delivery. VI. Carbomer and chitosan improve the intestinal absorption of the peptide drug buserelin in vivo. Pharm Res 13:1668–1672
Luessen HL, de Leeuw BJ, Perard D, Lehr C-M, de Boer AG, Verhoef JC, Junginger HE (1996) Mucoadhesive polymers in peroral peptide drug delivery. I. Influence of mucoadhesive excipients on the proteolytic activity of intestinal enzymes. Eur J Pharm Sci 4:117–128
Luessen HL, Verhoef JC, Borchard G, Lehr CM, de Boer AG, Junginger HE (1995) Mucoadhesive polymers in peroral peptide drug delivery. II. Carbomer and polycarbophil are potent inhibitors of the intestinal proteolytic enzyme trypsin. Pharm Res 12:1293–1298
Marschütz MK, Bernkop-Schnürch A (2000) Oral peptide drug delivery: polymer-inhibitor conjugates protecting insulin from enzymatic degradation in vitro. Biomaterials 21:1499–1507
Marschütz MK, Veronese FM, Bernkop-Schnürch A (2001) Influence of the spacer on the inhibitory effect of different polycarbophil-protease inhibitor conjugates. Eur J Pharm Biopharm 52:137–144
McCaffrey G, Jamieson JC (1993) Evidence for the role of a cathepsin D-like activity in the release of gal beta 1-4 glnac alpha 2-6 sialyltransferase from rat and mouse liver in whole-cell systems. Comp Biochem Phys C 104:91–94
McClellan JBJ, Garner CW (1980) Purification and properties of human intestine alanine aminopeptidase. Biochim Biophys Acta 613:160–167
McConnell RM, Frizell D, Evans ACA, Jones W, Cagle C (1991) New pepstatin analogues: synthesis and pepsin inhibition. J Med Chem 34:2298–2300
Okagawa T, Fujita T, Murakami M, Yamamoto A, Shimura T, Tabata S, Kondo S, Muranishi S (1994) Susceptibility of ebiratide to proteolysis in rat intestinal fluid and homogenates and its protection by various protease inhibitors. Life Sciences 55:677–683
Plumpton C, Kalinka S, Martin RC, Horton JK, Davenport AP (1994) Effects of phosphoramidon and pepstatin A on the secretion of endothelin-1 and big endothelin-1 by human umbilical vein endothelial cells – measurement by two-site enzyme-linked immunosorbent assay. Clin Sci 87:245–251

Raehs SC, Sandow J, Wirth K, Merkle HP (1988) The adjuvant effect of bacitracin on nasal absorption of gonadorelin and buserelin in rats. Pharm Res 5:689–693

Ramon J, Saez V, Baez R, Aldana R, Hardy E (2005) PEGylated Interferon-alpha2b: A Branched 40K Polyethylene Glycol Derivative. Pharm Res 22:1375–1387

Reseland JE, Holm H, Jacobsen MB, Jenssen TG, Hanssen LE (1996) Proteinase inhibitors induce selective stimulation of human trypsin and chymotrypsin secretion. J Nutr 126:634–642

Saffran M, Bedra C, Kumar GS, Neckers DC (1988) Vasopressin: a model for the study of effects of additives on the oral and rectal administration of peptide drugs. J Pharm Sci 77:33–38

Salhanick AI, Clairmont KB, Buckholz TM, Pellegrino CM, Ha S, Lumb KJ (2005) Contribution of site-specific PEGylation to the dipeptidyl peptidase IV stability of glucose-dependent insulinotropic polypeptide. Bioorg Med Chem Lett 15:4114–4117

Sanderink G-J, Artur Y, Siest G (1988) Human aminopeptidase: a review of the literature. J Clin Chem Clin Biochem 26:795–807

Sangadala S, Walters FS, English LH, Adang MJ (1994) A mixture of Manduca sexta aminopeptidase and phosphatase enhances Bacillus thuringiensis insecticidal CryIA(c) toxin binding and 86Rb(+)-K+ efflux in vitro. J Biol Chem 269:10088–10092

Scott MJ, Huckaby CS, Kato I, Kohr WJ, Laskowski MJ, Tsai MJ, O'Malley BW (1987) Ovoinhibitor introns specify functional domains as in the related and linked ovomucoid gene. J Biol Chem 262:5899–5907

Stryer L (1988) Biochemistry. W.H. Freeman and Company, New York

Ushirogawa Y (1992) Effect of organic acids, trypsin inhibitors and dietary protein on the pharmacological activity of recombinant human granulocyte colony-stimulating factor (rhG-CSF) in rats. Int J Pharm 81:133

Valenta C, Christen B, Bernkop-Schnürch A (1998) Chitosan-EDTA conjugate: a novel polymer for topical gels. J Pharm Pharmacol 50:445–452

Valenta C, Marschutz M, Egyed C, Bernkop-Schnürch A (2002) Evaluation of the inhibition effect of thiolated poly(acrylates) on vaginal membrane bound aminopeptidase N and release of the model drug LH-RH. J Pharm Pharmacol 54:603–610

Walker B, McCarthy N, Healy A, Ye T, McKervey MA (1993) Peptide gyoxals-a novel class of inhibitor for serine and cysteine proteinases. Biochem J 193:321–323

Werle M (2007) Analytical methods for the characterisation of multifunctional polymers for oral drug delivery. Current pharmaceutical analysis 3:111–116

Werle M, Loretz B, Entstrasser D, Föger F (2007) Design and evaluation of a Chitosan-Aprotinin conjugate for the peroral delivery of therapeutic peptides and proteins susceptible to enzymatic degradation. J Drug Targ 15:327–333

Werle M, Schmitz T, Huang H, Wentzel A, Kolmar H, Bernkop-Schnürch A (2006) The potential of cystine-knot microproteins as novel pharmacophoric scaffolds in oral peptide drug delivery. J Drug Targ 14:137–146

Wu ZH, Ping QN, Song YM, Lei XM, Li JY, Cai P (2004) Studies on the insulin-liposomes double-coated by chitosan and chitosan-EDTA conjugates. Yao Xue Xue Bao 39:933–938

Yamamoto A, Taniguchi T, Rikyun K, Tsuji T, Fujita T, Murakami M, Muranishi S (1994) Effects of Various Protease Inhibitors on the Intestinal Absorption and Degradation of Insulin in Rats. Pharm Res 11:1496–1500

Chapter 5
Low Molecular Mass Permeation Enhancers in Oral Delivery of Macromolecular Drugs

Andreas Bernkop-Schnürch

Contents

Abstract Within the last decades strong evidence has been provided on the potential of various types of low molecular mass permeation enhancers to improve the oral uptake of macromolecular drugs. Although a considerable number of low molecular mass permeation enhancers have been developed, medium-chain fatty acids can still be regarded as gold standard. The development of more potent and less toxic low molecular mass permeation enhancers is therefore highly on demand. Moreover, drug delivery systems providing synchronized release properties of both drug and permeation enhancer and additional favourable features such as protective and mucoadhesive properties contribute to an improved oral drug uptake. Having the great potential of low molecular mass permeation enhancers in mind and taking all the

A. Bernkop-Schnürch (✉)
Institute of Pharmacy, University of Innsbruck, Innrain 52, 6020, Innsbruck, Austria
e-mail: andreas.bernkop@uibk.ac.at

A. Bernkop-Schnürch (ed.), *Oral Delivery of Macromolecular Drugs*,
DOI 10.1007/978-1-4419-0200-9_5,

opportunities ahead into consideration, this class of permeation enhancers will certainly further alter the landscape of drug delivery towards more efficient therapeutic systems.

5.1 Introduction

One of the most likely promising strategies in order to improve the bioavailability of orally administered macromolecular drugs is the co-administration of permeation enhancers. Generally, they can be divided into low molecular mass permeation enhancers, polymeric permeation enhancers (Chapter 6) and efflux pump inhibitors (Chapter 7). This chapter will focus on low molecular mass permeation enhancers. Although this group of permeation enhancers has by far the longest history going back in the 1970s medium-chain fatty acids can still be regarded as gold standard. The main reason for this situation is their insufficient permeation-enhancing properties when being applied in non-toxic concentrations. In contrast to polymeric permeation enhancers being un-absorbable because of their much higher molecular mass, most low molecular mass permeation enhancers are rapidly diluted in the GI tract and absorbed from the mucosa. Consequently, when considerably high amounts of low molecular mass permeation enhancers have to be administered local and systemic toxic side effects cannot be excluded. They seem therefore to be more appropriate for short-term treatments such as for treatment of acute diseases than for long-term treatments such as of chronic diseases.

Generally, low molecular mass permeation enhancers can be divided into transcellular and paracellular permeation enhancers. On the one hand the potential of permeation enhancers to open the paracellular route of uptake can be determined by the reduction in the transepithelial electrical resistance (TEER) (enhancement potential = EP). On the other hand the potential of permeation enhancers to open the transcellular route of uptake can be determined by the lactate dehydrogenase (LDH) assay (LDH potential = LP). The parameter $K = (\mathrm{EP}-\mathrm{LP})/\mathrm{EP}$ represents the relative contribution of the paracellular pathway. Consequently, a K value of 0 means predominantly transcellular and a K value of 1 means predominantly paracellular. Based on this classification system Whitehead and Mitragotri classified over 50 low molecular mass permeation enhancers showing that most of them are paracellular and only a few of them are transcellular permeation enhancers (2008).

Low molecular mass permeation enhancers include numerous classes of compounds with diverse chemical properties including detergents, surfactants, N-acetylated α-amino acids, N-acetylated non-α-amino acids, fatty acids, medium-chain mono- and diglycerides, acyl carnitine, alkanoyl cholines, Ca^{2+} chelating agents, zonula occludens toxins and NO donors.

Some of these low molecular mass permeation enhancers act as surfactants/detergents to increase the transcelluar transport of more lipophilic drugs by disrupting the structure of the lipid bilayer and rendering the cell membrane more permeable (Hellriegel et al. 1996). As most macromolecular drugs, however, are highly hydrophilic, this type of permeation enhancers is apart from a few exceptions of minor practical relevance. Furthermore, the potential lytic nature of surfactants may cause exfoliation of the intestinal epithelium, irreversibly compromising its barrier functions (Aungst et al. 1996).

As macromolecular drugs are primarily uptaken via the paracellular route, the use of low molecular mass permeation enhancers opening tight junctions is the likely more straightforward approach. Furthermore, evidence is provided that opening tight junctions – in particular in a reversible manner – is less damaging than a disruption of cell membrane structure. Low molecular mass permeation enhancers opening tight junctions include fatty acids, medium-chain mono- and diglycerides, acylcarnitines, alkanoylcholines, zonula occludens toxins and NO donors.

5.2 Transcellular Permeation Enhancers

5.2.1 Non-ionic Surfactants

Numerous non-ionic surfactants have been investigated as intestinal permeation enhancers. The size and structure of both the lipophilic and the hydrophilic part of the surfactant influence its permeation-enhancing properties. Their permeation-enhancing effect is likely based on interactions with the membrane by solubilizing membrane components. Studies with polyoxyethylene-ethers, -esters and -sorbitan esters revealed, for instance, a good correlation between the enhancement of colonic absorption of *p*-aminobenzoic acid in rats and lactic dehydrogenase (LDH) release from the intestine (Sakai et al. 1986). These results were confirmed by Swenson et al. (1994) showing a good correlation between enhancement of intestinal absorption of phenol red in rats and LDH release in the presence of nonylphenoxypolyoxyethylene surfactants.

Dodecylmaltoside increased the small intestinal and colonic absorption of phenol red in rats without causing membrane protein and phospholipid release (Yamamoto et al. 1996; Uchiyama et al. 1996). These results, however, are not in agreement with an in vitro study showing a significant protein and phospholipid release from rat jejunum and colon when dodecylmaltoside is applied in the same concentration (Yamamoto et al. 1997).

5.2.2 Steroidal Detergents

Steroidal detergents showing permeation-enhancing properties are bile salts, glycosylated bile acid analogues and saponins.

Bile salts are produced in the liver and secreted into the intestinal lumen forming mixed micelles with lecithin, monoglycerides, fatty acids and cholesterol. Because of the formation of mixed micelles the in vitro effect of bile salts on the intestinal mucosa can be quite different in comparison to in vivo. The permeation-enhancing effect of taurocholate on Caco-2 monolayers, for instance, was greatly reduced when incorporated with phospholipids or cholesterol in mixed micelles. In contrast, a mixed micelle composed of taurocholate and oleic acid showed a much greater permeation-enhancing effect than taurocholate alone (Werner et al. 1996). The permeation-enhancing effect of bile salts is likely based on their high capacity for phospholipid solubilization. This theory is supported by numerous studies showing a correlation between membrane permeation-enhancing effects and the release of membrane phospholipids and proteins. This effect of bile salts on the intestinal mucosa seems to be reversible, which was shown in various studies. Nevertheless, the safety of bile salts as intestinal permeation enhancers has not yet been resolved. 1% Chenodeoxycholate increased the uptake of octreotide on Caco-2 monolayer approximately threefold and in rats from 0.26 to 20%. Further studies in humans showed that the oral bioavailability of octreotide is 1.26% with chenodeoxycholate and 0.13% with urodeoxycholate. According to studies performed in rats with insulin used as model drug, the colon seems to be more sensitive than the small intestine to the permeation-enhancing effect of bile salts (Yamamoto et al. 1994).

Glycosylated bile acid analogues were shown to be more effective than taurocholate in increasing the intestinal uptake of calcitonin in rats (Bowe et al. 1997).

Saponins are found in plants. Their chemical structure is based on glycosylated steroid or triterpenoid. Glycyrrhizinate, for instance, contains two glucuronosyl moieties linked to a steroid. Glycyrrhizinate itself shows almost no permeation-enhancing effect on Caco-2 monolayer (Sakai et al. 1997), whereas the aglycone glycyrrhetinic acid being formed by glucuronidase in colonic flora increased Caco-2 permeability and enhanced colonic uptake of calcitonin in rats (Imai et al. 1999). Another saponin, DS-1, was also shown to increase the uptake of mannitol and a peptide drug from Caco-2 monolayer without influencing cell viability (Chao et al. 1998a).

5.2.3 N-Acetylated α-Amino Acids and N-Acetylated Non-α-Amino Acids

Initiated by work on proteinoid microspheres, which are formed from thermally condensed α-amino acid mixtures, various N-acetylated α-amino acids and N-acetylated non-α-amino acids were synthesized and evaluated regarding their permeation-enhancing properties. Among these compounds sodium *N*-[8-(2-hydroxybenzoyl)amino]caprylate (SNAC) turned out as the most likely promising permeation enhancer. SNAC was identified from a series of structurally related compounds on the basis of its permeation-enhancing properties on heparin in rats (Leon-Bay et al. 1998a). At doses of 300 mg/kg

SNAC increased the oral bioavailability of heparin (Rivera et al. 1997). In another study SNAC increased the ileal permeation of heparin in situ in rats (Brayden et al. 1997). The postulated mechanism of action is based on an association of SNAC to the therapeutic macromolecule rendering the drug more lipophilic. Consequently large or highly charged molecules can cross cell membranes via the passive transcellular route of uptake. Once the drug molecule crosses the membrane, SNAC dissociates from the therapeutic agent, which then re-establishes its natural conformation (Wu and Robinson 1999; Leon-Bay et al. 1998b; Milstein et al. 1998). How SNAC can be associated with a variety of different macromolecular drugs in the harsh environment of the GI tract being subsequently dissociated once having reached the blood stream, however, remains unclear.

The SNAC/heparin combination has been evaluated in phase I and phase II clinical trials. Furthermore, in phase I clinical trials dosing insulin in combination with SNAC a rapid elevation of plasma insulin and a subsequent decrease in plasma glucose levels were observed. In a phase II clinical trial in patients with type 2 diabetes, insulin was orally administered in combination with SNAC and metformin failing to achieve significant superior glycemic control over treatment with metformin alone (Hoffman and Qadri 2008).

5.3 Paracellular Permeation Enhancers

5.3.1 Fatty Acids

Numerous fatty acids were shown to have membrane permeation-enhancing properties. Among them salts of medium-chain fatty acids including caproates (C6), caprylates (C8), caprates (C10) and laureates (C12) were shown to boost the flux of hydrophilic agents at millimolar concentrations (Lindmark et al. 1995). Longer chain fatty acids are also effective as intestinal permeation enhancers, but because of their lower water solubility they have to be combined with emulsifying agents. Shorter chain fatty acids are much less effective and in many cases of unpleasant odour. Within medium-chain fatty acids in particular sodium caprate is the most thoroughly characterized for use as oral permeation enhancer. Sodium caprate was shown to significantly increase the permeability of PEGs and FITC-dextrans in vitro as well as in vivo (Lindmark et al. 1997). The enhancement by sodium caprate of in vitro permeation of a peptide was greater for rat colon than rat jejunum (Yamamoto et al. 1997).

In the presence of 10–24 mM sodium caprate on Caco-2 monolayers a dilatation of the tight junctions was observed (Anderberg et al. 1993). Based on Caco-2 studies Tomita et al. (1995) and Lindmark et al. (1998) proposed the activation of phospholipase C, followed by an increase in inositol triphosphate and intracellular calcium concentration leading to a contraction of calmodulin-dependent actin–myosin filaments and the opening of tight junctions. This likely mechanism was confirmed by further studies on isolated rat and human colon specimens (Shimazaki et al. 1998). In addition, sodium caprate seems to affect also transcellular permeation as it caused a release of phospholipids into rat colonic lumen in situ (Tomita et al. 1988). Furthermore, it was demonstrated that sodium caprate affects the permeability of brush border membrane vesicles (Tomita et al. 1992). The permeation-enhancing effect of sodium caprate on Caco-2 TEER and mannitol permeability strongly depends on the applied concentration (Anderberg et al. 1993). In concentrations > 24 mM sodium caprate was shown to be cytotoxic (Anderberg et al. 1993; Sakai et al. 1998). Chao et al. (1999) also demonstrated that sodium caprate enhances the oral bioavailability in a dose-dependent fashion utilizing a D-decapeptide as model drug in rats. Results of this study are shown in Fig. 5.1.

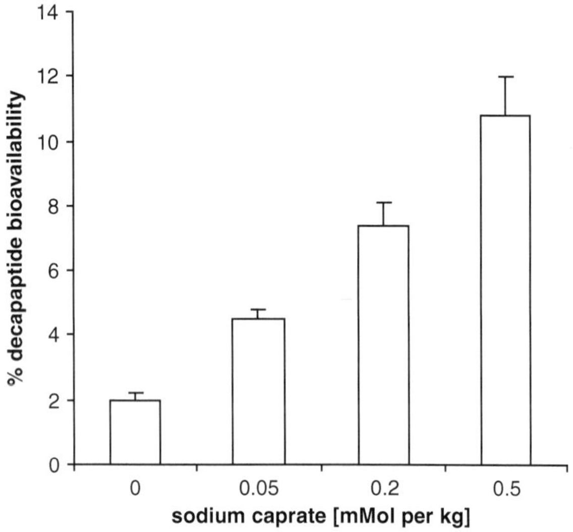

Fig. 5.1 Correlation between the amount of sodium caprate and the bioavailability of a D-decapeptide. Adapted from Chao et al. (1999)

The in vivo performance of sodium caprate depends on the delivery system and how the drug and sodium caprate are released. The oral bioavailability of a peptide, for instance, was enhanced in rats and dogs using capsules containing a semisolid matrix of sodium caprate, polyethylene glycol and water. In contrast, other formulations were less effective. The extent of absorption enhancement reported ranged from almost no improvement up to fivefold (Aungst et al. 1996; Burcham et al. 1995).

Longer chain fatty acids are also effective as intestinal permeation enhancers. Morishita et al. (1998), for instance, could significantly increase the in situ colonic absorption of insulin in rats utilizing emulsions (w/o/w) containing oleic, linoleic or linolenic acid.

5.3.2 Medium-Chain Mono- and Diglycerides

Medium-chain mono- and diglycerides exhibiting permeation-enhancing properties are mainly based on caprylic and capric acid. They are supplied as mixtures containing traces of triglycerides as well as monoglycerides and diglycerides of shorter and longer chain fatty acids. In the intestinal lumen medium-chain mono- and diglycerides are subjected to presystemic metabolism to form free fatty acids. Metabolism to free fatty acids, however, is not required for their intestinal permeation-enhancing properties.

Lohikangas et al. (1994) showed that a medium-chain glyceride (MCG)/phosphatidylcholine (3:1) mixture applied in a concentration of $\geq$4 mM reduces Caco-2 TEER and increases P_{app} of mannitol and low molecular weight heparin approximately tenfold. Irreversible increase in permeability and altered cell morphology were observed at concentrations $\geq$ 8 mM. The in vivo safety of MCG has been assessed in rats, rabbits and dogs showing that rectally administered MCGs lead to no remarkable morphologic changes to the rectal mucosa (Sekine et al. 1985).

The oral bioavailability of a peptide was improved from 0.5 to 27% due to the co-administration of MCG (Constantinides et al. 1994). Lundin et al. (1997) demonstrated that monohexanoin is even more effective than MCG in increasing oral absorption of the therapeutic peptide dDAVP in rats. Furthermore, due to the co-administration of a monoolein/sodium taurocholate combination the colonic uptake of polyethylene glycol 4000, calcitonin and horseradish peroxidase in rats was significantly increased without causing morphologic damage to the mucosa. The appearance of horseradish peroxidase in the cytoplasm suggested enhancement via the transcellular route of uptake (Hastewell et al. 1994).

5.3.3 Acylcarnitines and Alkanoylcholines

Medium- and long-chain fatty acid esters of carnitine and choline are another group of intestinal permeation enhancers. Palmitoyl-DL-carnitine chloride was shown to dilate paracellular spaces on Caco-2 monolayers leading to an improved uptake of fluorescent compounds via the paracellular route of uptake (Hochman et al. 1994; Chao et al. 1998b). In this connection it was also shown that the effective pore radius of Caco-2 monolayers increases with the concentration of applied palmitoyl-DL-carnitine chloride in the 0.05–0.35 mM range (Knipp et al. 1997). Furthermore, it was shown that this effect of palmitoyl-DL-carnitine chloride on tight junctions is calcium independent (Hochman et al. 1994). So far, to our knowledge the influence of fatty acids formed by the presystemic metabolism of palmitoyl-DL-carnitine chloride on its permeation-enhancing effect has not been addressed. Medium-chain alkanoylcholines, however, were shown to be rapidly hydrolyzed in rat intestine (Chelminska-Bertilsson et al. 1995).

In orientating toxicity studies palmitoyl-DL-carnitine chloride and lauroyl choline were shown to cause only slight alterations of the mucosal cell structure of jejunum and colon under prolonged exposure when being tested in permeation-enhancing concentrations (LeCluyse et al. 1993; Fix et al. 1996). In another study on Caco-2 monolayers, however, palmitoyl-DL-carnitine chloride caused LDH release, increased uptake of propidium iodide and reduced neutral red retention, in concentrations needed to provide an improved uptake of PEG 4000 (Duizer et al. 1998).

5.3.4 Chelating Agents

In comparison to other low molecular mass permeation enhancers chelating agents seem to be of minor practical relevance for oral macromolecular delivery. The mechanism of action is likely based on the depletion of Ca^{2+} ions leading to an increase in the paracellular permeability of epithelial cells (e.g. Noach et al. 1993). Chitosan–EDTA and chitosan–DTPA conjugates exhibiting a comparatively high Ca^{2+} ions binding affinity and binding capacity (Bernkop-Schnürch and Krajicek 1998), however, did not show any permeation-enhancing properties. Sodium salicylate and 5-methoxysalicylate, for instance, were shown to remarkably enhance the rectal absorption of insulin in rats (Nishihata et al. 1983; Aungst and Rogers 1988). Most of the chelating agents, however, cause functional damage to the intestinal mucosa (e.g. Yamashita et al. 1987).

5.3.5 Zonula Occludens Toxin

Although zonula occludens toxin (Zot) is a polypeptide of 44.8 kDa, it is described within this chapter of low molecular mass permeation enhancers, as also smaller fragments of this protein show permeation-enhancing properties. Zot is present in toxigenic strains of *Vibrio cholerae* (Baudry et al. 1992) with the ability to reversibly alter intestinal epithelial tight junctions allowing the passage of macromolecular drugs through the intestinal mucosal barrier. It was first identified by Uzzau and Fasano (2000) in the outer membrane of the bacterium. The toxin possesses multiple domains that allow a dual function as an enterotoxin and as a morphogenetic phage protein (Uzzau et al. 1999). Permeation studies on freshly excised mucosa have shown that the activity of Zot is reversible and sensitive to protease digestion (Fasano et al. 1991). To identify Zot domains being involved in the permeation-enhancing effect several Zot gene deletion mutants were constructed and tested (Di Pierro et al. 2001). The study revealed the carboxyl terminus of the protein as functionally and immunologically related endogenous modulator of epithelial tight junctions. Extensive studies on Zot were triggered by its ability to allow for the oral administration of insulin (Fasano and Uzzau 1997). Rats treated with Zot experienced neither diarrhoea, fever, nor other systemic symptoms, and no structural changes were demonstrated in the small intestine on histological examination (Fasano and Uzzau 1997). In rabbits it was shown that Zot induces a reversible increase of small intestinal permeability without any cytotoxic side effects (Fasano et al. 1991). In vitro experiments in the rabbit ileum demonstrated that Zot reversibly increased intestinal absorption of insulin and immunoglobulin G (Fasano and Uzzau 1997). In vivo (Fasano and Uzzau 1997; Uzzau et al. 1996) and in vitro (Fasano et al. 1991; Uzzau et al. 1996) studies demonstrated that the effect of Zot on membrane permeability occurs within 20 min and reaches its maximum within 80 min. In a primate model of diabetes mellitus it was shown that the oral bioavailability of insulin increased from 5.4% in controls to 10.7 and 18% when Zot 2 and 4 μg/kg were co-administered, respectively (Watts et al. 2000).

The transport-enhancing effect of Zot was shown to be reversible and non-toxic (Fasano et al. 1991; Cox et al. 2002). More recently a smaller fragment of Zot in the size of 12 kDa referred to as *ΔG* was identified (Di Pierro et al. 2001). *ΔG* displayed significant potential as permeation enhancer. In vitro studies showed that it is capable of significantly increasing the apparent permeability coefficients for a wide variety of drugs across Caco-2 monolayer (Salama et al. 2003, 2004). In the presence of peptidase inhibitors *ΔG* improved the bioavailability of mannitol, inulin and PEG 4000 after intraduodenal administration to rats (Salama et al. 2003, 2004). In another in vivo study the oral bioavailability of cyclosporin A was increased up to 50-fold due to the co-administration of *ΔG* when metabolic protection was provided (Salama et al. 2005). Results of this study are illustrated in Fig. 5.2.

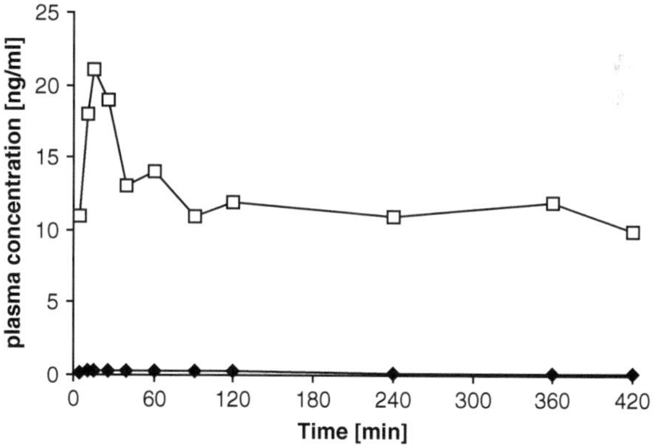

Fig. 5.2 Average plasma concentration versus time profile for cyclosporin A with (□) and without (◆) *ΔG* (720 μg/kg) intraduodenally administered to rats. Data are means of 3–6/group. Adopted from Salama et al. (2005)

Zonulin, an intestinal Zot analogue sharing an 8-amino acid motif with Zot (Di Pierro et al. 2001), was shown in a primate model to reversibly open intestinal tight junctions after engagement to the same receptor activated by Zot (Wang et al. 2000) The physiological role of Zonulin remains unclear but it is likely that it is involved in several functions, including tight junction regulation during developmental, physiological and pathological processes, including tissue morphogenesis, protection against microorganisms' colonization and the movement of fluid, macromolecules and leucocytes between the bloodstream and the intestinal lumen and vice versa (Fasano 2001). After a Zonulin induced opening of tight junctions, water is secreted into the intestinal lumen followed by a hydrostatic pressure gradient (Fasano et al. 1997). Consequently bacteria are flushed out from the small intestine (Fasano 2001).

5.3.6 NO Donors

Nitric oxide (NO) donors include a variety of compounds such as 3-(2-hydroxy-1-(methylethyl)-2-nitrosohydrazino)-1-propanamine (NOC5), *N*-ethyl-2-(1-ethyl-hydroxy-2-nitrosohydrazino)-ethanamine (NOC12), *S*-nitroso-*N*-acetyl-penicillamine and sodium nitroprusside. Their mode of action seems to be based on an opening of tight junctions. NOC12, for instance, was shown to significantly decrease the transepithelial electrical resistance value of the colonic membrane, suggesting that the absorption-enhancing mechanism is related to the dilation of the tight junctions (Yamamoto et al. 2001).

Furthermore, the permeation-enhancing effects of NO donors seem to be strongly dose-dependent. Utoguchi et al. (1998) reported that the absorption-enhancing effect of *S*-nitroso-*N*-acetyl-penicillamine for rectal insulin absorption was dose-dependent over the range of 0.25–4.0 mg in rats. Similarly, Salzman et al. (1995) demonstrated that incubation with sodium nitroprusside resulted in a concentration-dependent increase in the transepithelial transport of fluorescein sulphonic acid in Caco-2 cells. To our knowledge the potential of NO donors for oral macromolecular delivery, however, has so far not been tested in a valid animal model.

5.4 Oral Macromolecular Drug Delivery Systems Containing Low Molecular Mass Permeation Enhancers

Although a low molecular mass permeation enhancer may greatly increase intestinal permeability in vitro or in situ, this does not guarantee that the enhancer will significantly improve oral bioavailability in vivo. The in vivo performance of a permeation enhancer is mainly related to its concentration at the site of drug absorption. The design of appropriate oral delivery systems comprising a combination of a macromolecular drug and a low molecular mass permeation enhancer is therefore likely the key for sufficient high oral bioavailability. Apart from general points to be considered for a potent oral delivery system which are listed in Table 5.1, in case of delivery systems comprising low molecular mass permeation enhancers a further point needs to be addressed.

Table 5.1 Advantageous features of oral macromolecular drug delivery systems

Demanded features	Benefit	Additional information
Intimate contact of the delivery system with the intestinal mucosa	Exclusion of a presystemic metabolism	This intimate contact can be achieved by utilizing mucoadhesive polymers and/or nanoparticulate delivery systems
Prolonged intestinal residence time	Prolonged period of time being available for drug uptake	
Protective effect towards an enzymatic attack	Exclusion of a presystemic metabolism	
Controlled drug release	Best compromise between protective effect/intimate contact and concentration gradient on the absorption membrane	On the one hand, if the drug release is too rapid the beneficial properties of the delivery system such as protective effect or intimate contact with the absorption membrane are lost. On the other hand, if the drug release is too slow the concentration gradient on the absorption membrane representing the driving force for passive drug uptake and the absorption area being available for drug uptake is strongly reduced
Targeted drug release	Exclusion of drug inactivation in the acidic environment of the stomach	A targeted release of the macromolecular drug in the small intestine can be achieved by an enteric coating

The drug and the permeation enhancer must be delivered to the absorption site simultaneously and a sufficient concentration of the permeation enhancer must be achieved and maintained there. As low molecular mass permeation enhancers are rapidly diluted by the fluids of the gastrointestinal tract, spread over a large surface and are consequently rapidly absorbed leaving the macromolecular drug alone behind in the intestine, the release of both the drug and the permeation enhancer needs to be synchronized. Lin et al. (2006), for instance, investigated the influence of various release kinetics of a polysaccharide drug and sodium caprate on the oral bioavailability in rats. They compared three formulations which eroded within 1, 2 and 4 h providing a synchronous release of both the drug and the permeation enhancer (S-1, S-2 and S-4) with corresponding non-synchronous formulations (NS-1, NS-2 and NS-4). Generally formulations providing a controlled release over 2 h (S-2 and NS-2) showed the greatest potential supporting the advantage of a well-adjusted drug release rate as listed in Table 5.1. Furthermore, as listed in Table 5.2, all synchronous formulations led to comparatively higher bioavailability than the corresponding non-synchronous formulations.

Table 5.2 Pharmacokinetics parameters after intra-intestinal administration of synchronous (S) and non-synchronous (NS) formulations to rats (means ± SD, $n = 4$); adapted from Lin et al. (2006)

Formulation	T_{max} (h)	C_{max} (μg/mL)	AUC $_{0–4}$ (h·μg/mL)	F (%)
S-1	0.50	29.1 ± 5.9	46.4 ± 11.2	27.8 ± 6.7
S-2	1.00	21.3 ± 4.6	52.0 ± 7.3	31.2 ± 4.4
S-4	0.50	11.2 ± 2.4	20.5 ± 3.7	12.3 ± 2.2
NS-1	0.25	12.8 ± 2.4	19.8 ± 2.5	11.8 ± 1.5
NS-2	0.50	23.4 ± 4.8	42.0 ± 6.6	25.2 ± 3.9
NS-4	0.50	9.6 ± 2.5	14.2 ± 5.6	8.5 ± 3.4

Furthermore, permeation enhancers may differ in their delivery requirements in order to guarantee best performance. Sodium decanoate was more effective in permeation enhancement from the rat jejunum when administered at a concentration of 50 mM over 30 min than when administered at 100 mM over 15 min (Baluom et al. 1998).

Another aspect from the drug delivery point of view is the combined use of different types of permeation enhancers in the same formulation. Whitehead et al. (2008), for instance, could show a synergistic effect of binary combinations of hexylamine and chembetaine and ternary combinations of sodium laureth sulphate, decyltrimethyl ammonium bromide and chembetaine on the uptake of 70 kDa dextran from Caco-2 monolayer while inducing very little toxicity in Caco-2 cells.

5.5 Conclusion

Within the last decades strong evidence has been provided on the potential of various types of low molecular mass permeation enhancers to improve the oral uptake of macromolecular drugs. Nevertheless, so far no oral macromolecular drug delivery system comprising a low molecular mass permeation enhancer has entered the market indicating that the permeation-enhancing properties of these enhancers are still associated with too high toxic risks.

The development of more potent and less toxic low molecular mass permeation enhancers is therefore highly on demand. The design of shorter and more stable Zots, for instance, might be a promising approach in this direction. The identification of further synergistic acting combinations of permeation enhancers might also be a fruitful strategy. Moreover, improvements in drug delivery systems providing better synchronized release properties and additional favourable features such as protective and mucoadhesive properties will contribute to more potent formulations for oral macromolecular delivery. Last but not least, improved production techniques for most macromolecular drugs have significantly lowered the costs of these therapeutic agents within the last years. Consequently, comparatively lower oral up-take improvements are meanwhile sufficient justifying the development of oral delivery systems showing lower bioavailability of the macromolecular drug.

Having the great potential of low molecular mass permeation enhancers in mind and taking all the opportunities ahead into consideration, this class of permeation enhancers will certainly further alter the landscape of drug delivery towards more efficient therapeutic systems. This chapter should encourage and motivate scientists in academia and industry to move on or intensify their activities in this challenging research field.

References

Anderberg EK, Lindmark T, Artursson P (1993) Sodium caprate elicits dilatations in human intestinal tight junctions and enhances drug absorption by the paracellular route. Pharm Res 10:857–864

Aungst BJ, Rogers NJ (1988) Site dependence of absorption-promoting actions of Laureth-9, Na salicylate, Na_2EDTA, and aprotinin on rectal, nasal, and buccal insulin delivery. Pharm Res 5:305–308

Aungst BJ, Saitoh H, Burcham DL, Huang S-M, Mousa SA, Hussain MA (1996) Enhancement of the intestinal absorption of peptides and non-peptides. J Contr Rel 41:19–31

Baluom M, Friedman M, Rubinstein A (1998) The importance of intestinal residence time of absorption enhancer on drug absorption and implication on formulative considerations. Int J Pharm 176:21–30

Baudry B, Fasano A, Ketley J, Kaper JB (1992) Cloning of a gene (zot) encoding a new toxin produced by Vibrio cholerae. Infect Immun 60:428–434

Bernkop-Schnürch A, Krajicek ME (1998) Mucoadhesive polymers as platforms for peroral peptide delivery and absorption: synthesis and evaluation of different chitosan-EDTA conjugates J Control Rel 50:215–23

Bowe CL, Mokhtarzadeh L, Venkatesan P, Babu S, Axelrod HR, Sofia MJ, Kakarla R, Chan TY, Kim JS, Lee HJ, et al. (1997) Design of compounds that increase the absorption of polar molecules. Proc Natl Acad Sci USA 94:12218–12223

Brayden D, Creed E, O'Connell A, Leipold H, Argawal R, Leone-Bay A (1997) Heparin absorption across the intestine: effects of sodium N-[8-(2-hydroxybenzoyl)amino]caprylate in rat in situ intestinal instillations and in Caco-2 monolayers. Pharm Res 14:1772–1779

Burcham DL, Aungst BJ, Hussain M, Gorko MA, Quon CY, Huang S-M (1995) The effect of absorption enhancers on the oral absorption of the GPIIb/IIIa receptor antagonist, DMP 728, in rats and dogs. Pharm Res 12:2065–2070

Chao AC, Nguyen JV, Broughall M, Recchia J, Kensil CR, Daddona PE, Fix JA (1998a) Enhancement of intestinal model compound transport by DS-1, a modified Quillaia saponin. J Pharm Sci 87:1395–1399

Chao AC, Taylor MT, Daddona PE, Broughall M, Fix JA (1998b) Molecular weight-dependent paracellular transport of fluorescent model compounds induced by palmitoylcarnitine chloride across the human intestinal epithelial cell line Caco-2. J Drug Targeting 6:37–43

Chao AC, Nguyen JV, Broughall M, Griffin A, Fix JA, Daddona PE (1999) In vitro and in vivo evaluation of effects of sodium caprate on enteral peptide absorption and on mucosal morphology. Int J Pharm 191:15–24
Chelminska-Bertilsson M, Edebo A, Thompson RA, Edebo L (1995) Enzymatic hydrolysis of long-chain alkanoylcholines in rat intestinal loops. Scand J Gastroenterol 30:670–674
Constantinides PP, Scalart J-P, Lancaster C, Marcello J, Marks G, Ellens H, Smith PL (1994) Formulation and intestinal absorption enhancement evaluation of water-in-oil microemulsions incorporating medium-chain glycerides. Pharm Res 11:1385–1390
Cox D, Raje S, Gao H, et al. (2002) Enhanced permeability of molecular weight markers and poorly bioavailable compounds across Caco-2 cell monolayers using the absorption enhancer, Zonula Occludens toxin. Pharm Res 19:1680–1688
Di Pierro M, Lu R, Uzzau S, et al. (2001) Zonula Occludens toxin structure-function analysis: identification of the fragment biologically active on tight junctions and of the Zonulin receptor binding domain J Biol Chem 276:19160–19165
Duizer E, Van der Wulp C, Versantvoort CHM, Groten JP (1998) Absorption enhancement, structural changes in tight junctions and cytotoxicity caused by palmitoyl carnitine in Caco-2 and IEC-18 cells. J Pharmacol Exp Ther 287:395–402
Fasano A (2001) Intestinal Zonulin: open sesame! Gut 49:159–162
Fasano A, Uzzau S (1997) Modulation of intestinal tight junctions by Zonula Occludens toxin permits enteral administration of insulin and other macromolecules in an animal model. J Clin Invest 99:1158–1164
Fasano A, Baudry B, Pumplin D, et al. (1991) Vibrio cholerae produces a second enterotoxin, which affects intestinal tight junctions. Proc Natl Acad Sci USA 88:5242–5246
Fasano A, Uzzau S, Fiore C, Mararetten K (1997) The enterotoxic effect of Zonula Occludens toxin on rabbit small intestine involves the paracellular pathway. Gastroentrology 112:839–846
Fix JA (1996) Strategies for delivery of peptides utilizing absorption-enhancing agents. J Pharm Sci 85:1282–1285
Fix JA, Engle K, Porter PA, Leppert PS, Selk SJ, Gardner CR, Alexander J (1986) Acylcarnitines: drug absorption-enhancing agents in the gastrointestinal tract. Am J Physiol 251:G332–G340
Hastewell J, Lynch S, Fox R, Williamson I, Skelton-Stroud P, Mackay M (1994) Enhancement of human calcitonin absorption across the rat colon in vivo. Int J Pharm 101:115–120
Hellriegel ET, Bjornsson TD, Hauck WW (1996) Interpatient variability in bioavailability is related to the extent of absorption: implications for bioavailability and bioequivalence studies. Clin Pharmacol Ther 60:601–607
Hochman JH, Fix JA, LeCluyse EL (1994) In vitro and in vivo analysis of the mechanism of absorption enhancement by palmitoylcarnitine. J Pharmacol Exp Ther 269:813–822
Hoffman A, Qadri B (2008) Eligen insulin–a system for the oral delivery of insulin for diabetes. IDrugs 11:433–41
Imai T, Sakai M, Ohtake H, Azuma H, Otagiri M (1999) In vitro and in vivo evaluation of the enhancing activity of glycyrrhizin on the intestinal absorption of drugs. Pharm Res 16:80–86
Knipp GT, Ho NFH, Barsuhn CL, Borchardt RT (1997) Paracellular diffusion in Caco-2 cell monolayers: effect of perturbation on the transport of hydrophilic compounds that vary in charge and size. J Pharm Sci 86:1105–1110
LeCluyse EL, Sutton SC, Fix JA (1993) In vitro effects of long-chain acylcarnitines on the permeability, transepithelial electrical resistance and morphology of rat colonic mucosa. J Pharmacol Exp Ther 265:955–962
Leone-Bay A, Paton DR, Freeman J, Lercara C, O'Toole D, Gschneidner D, Wang E, Harris E, Rosado C, Rivera T, et al. (1998a) Synthesis and evaluation of compounds that facilitate the gastrointestinal absorption of heparin. J Med Chem 41:1163–1171

Leone-Bay A, Paton D, Variano B, Leipold H, Rivera T, Miura-Fraboni J, Baughman RA, Santiago N (1998b) Acylated non-α-amino acids as novel agents for the oral delivery of heparin sodium, USP. J Contr Rel 50:41–49

Lin X, Xu DS, Feng Y, Li SM, Lu ZL, Shen L (2006) Release-controlling absorption enhancement of enterally administered Ophiopogon japonicus polysaccharide by sodium caprate in rats. J Pharm Sci 95:2534–42

Lindmark T, Nikkilä T, Artursson P (1995) Mechanisms of absorption enhancement by medium chain fatty acids in intestinal epithelial Caco-2 cell monolayers. J Pharmacol Exp Ther 275:958–964

Lindmark T, Kimura Y, Artursson P (1998) Absorption enhancement through intracellular regulation of tight junction permeability by medium chain fatty acids in Caco-2 cells. J Pharmacol Exp Ther 284:362–369

Lindmark T, Schipper N, Lazorová L, de Boer AG, Artursson P (1997) Absorption enhancement in intestinal epithelial Caco-2 monolayers by sodium caprate: assessment of molecular weight dependence and demonstration of transport routes. J Drug Targeting 5:215–223

Lohikangas L, Wilen M, Einarsson M, Artursson P (1994) Effects of a new lipid-based drug delivery system on the absorption of low molecular weight heparin (Fragmin) through monolayers of human intestinal epithelial Caco-2 cells and after rectal administration to rabbits. Eur J Pharm Sci 1:297–305

Lundin PDP, Bojrup M, Ljusberg-Wahren H, Weström BR, Lundin S (1997) Enhancing effects of monohexanoin and two other medium-chain glyceride vehicles on intestinal absorption of desmopressin (dDAVP). J Pharmacol Exp Ther 282:585–590

Milstein SJ, Leipold H, Sarubbi D, Leone-Bay A, Mlynek GM, Robinson JR, Kasimova M, Freire E (1998) Partially unfolded proteins efficiently penetrate cell membranes – implications for oral drug delivery. J Contr Rel 53:259–267

Morishita M, Matsuzawa A, Takayama K, Isowa K, Nagai T (1998) Improving insulin enteral absorption using water-in-oil-in-water emulsion. Int J Pharm 172:189–198

Nishihata T, Rytting JH, Kamada A, Higuchi T, Routh M, Caldwell L (1983) Enhancement of rectal absorption of insulin using salicylates in dogs. J Pharm Pharmacol 35:148–151

Noach ABJ, Kurosaki Y, Blom-Roosemalen MCM, de Boer AG, Breimer DD (1993) Cell-polarity dependent effect of chelation on the paracellular permeability of confluent Caco-2 cell monolayers. Int J Pharm 90:229–237

Rivera TM, Leone-Bay A, Paton DR, Leipold H, Baughman RA (1997) Oral delivery of heparin in combination with sodium N-[8-(2-hydroxybenzoyl) amino]caprylate: pharmacological considerations. Pharm Res 14:1830–1834

Sakai K, Kutsuna TM, Nishino T, Fujihara Y, Yata N (1986) Contribution of calcium ion sequestration by polyoxyethylated nonionic surfactants to the enhanced colonic absorption of p-aminobenzoic acid. J Pharm Sci 75:387–390

Sakai M, Imai T, Ohtake H, Azuma H, Otagiri M (1997) Effects of absorption enhancers on the transport of model compounds in Caco-2 cell monolayers: assessment by confocal laser scanning microscopy. J Pharm Sci 86:779–785

Sakai M, Imai T, Ohtake H, Otagiri M (1998) Cytotoxicity of absorption enhancers in Caco-2 cell monolayers. J Pharm Pharmacol 50:1101–1108

Salama NN, Fasano A, Lu R, et al. (2003) Effect of the biologically active fragment of Zonula Occludens toxin, ΔG, on the intestinal paracellular transport and oral absorption of mannitol. Int J. Pharm 251:113–121

Salama NN, Fasano A, Thakar M, et al. (2004) The effect of ΔG on the transport and oral absorption of macromolecules. J Pharm Sci 93:1310–1319

Salama NN, Fasano A, Thakar M, et al. (2005) The effect of ΔG on the oral bioavailability of low bioavailable therapeutic agents. J Pharmacol Exp Ther 312:199–205

Salzman AL, Menconi MJ, Unno N, Ezzell RM, Casey DM, Gonzalez PK, Fink MP (1995) Nitric oxide dilates tight junctions and depletes ATP in cultured Caco-2BBe intestinal epithelial monolayers. Am J Physiol 268:G361–G373

Sekine M, Maeda E, Sasahara K, Okada R, Kimura K, Fukami M, Awazu S (1985) Improvement of bioavailability of poorly absorbed drugs. III. Oral acute toxicity and local irritation of medium chain glyceride. J Pharmacobio-Dyn 8:633–644

Shimazaki T, Tomita M, Sadahiro S, Hayashi M, Awazu S (1998) Absorption-enhancing effects of sodium caprate and palmitoyl carnitine in rat and human colons. Dig Dis Sci 43:641–645

Swenson ES, Milisen WB, Curatolo W (1994) Intestinal permeability enhancement: structure-activity and structure-toxicity relationships for nonylphenoxypolyoxyethylene surfactant permeability enhancers. Pharm Res 11:1501–1504

Tomita M, Hayashi M, Horie T, Ishizawa T, Awazu S (1988) Enhancement of colonic drug absorption by the transcellular permeation route. Pharm Res 5:786–789

Tomita M, Sawada T, Ogawa T, Ouchi H, Hayashi M, Awazu S (1992) Differences in the enhancing effects of sodium caprate on colonic and jejunal drug absorption. Pharm Res 9:648–653

Tomita M, Hayashi M, Awazu S (1995) Absorption-enhancing mechanism of sodium caprate and decanoylcarnitine in Caco-2 cells. J Pharmacol Exp Ther 272:739–743

Uchiyama T, Yamamoto A, Hatano H, Fujita T, Muranishi S (1996) Effectiveness toxicity screening of various absorption enhancers in the large intestine: intestinal absorption of phenol red and protein and phospholipid release from the intestinal membrane. Biol Pharm Bull 19:1618–1621

Utoguchi N, Watanabe Y, Shida T, Matsumoto M (1998) Nitric oxide donors enhance rectal absorption of macromolecules in rabbits. Pharm Res 15:870–876

Uzzau S, Fasano A (2000) Cross-talk between enteric pathogens and the intestine. Cell Microbiol 2:83–89

Uzzau S, Fiore C, Margaretten K, et al. (1996) Modulation of intestinal tight junctions: a novel mechanism of intestinal secretion. Gastroentrology 110:370

Uzzau S, Cappuccinelli P, Fasano A (1999) Expression of Vibrio cholerae Zonula Occludens toxin and analysis of its subcellular localization. Microb Pathog 27:377–385

Wang W, Uzzau S, Goldblum S, Fasano A (2000) Human Zonulin, a potential modulator of intestinal tight junctions. J Cell Sci 113:4435–4440

Watts TL, Alexander T, Hansen B, et al. (2000) Utilizing the paracellular pathway: a novel approach for the delivery of oral insulin in diabetic rhesus macaques. J Invest Med 48:185

Werner U, Kissel T, Reers M (1996) Effects of permeation enhancers on the transport of a peptidomimetic thrombin inhibitor (CRC 220) in a human intestinal cell line (Caco-2). Pharm Res 13:1219–1227

Whitehead K, Mitragotri S (2008) Mechanistic analysis of chemical permeation enhancers for oral drug delivery. Pharm Res 25:1412–1419

Whitehead K, Karr N, Mitragotri S (2008) Discovery of synergistic permeation enhancers for oral drug delivery. J Control Rel 128:128–33

Wu S-Y, Robinson JR (1999) Transcellular lipophilic complex-enhanced intestinal absorption of human growth hormone. Pharm Res 16:1266–1272

Yamashita S, Saitoh H, Nakanishi K, Masada M, Nadai T, Kimura T (1987) Effects of diclofenac sodium and disodium ethylenediaminetetraacetate on electrical parameters of the mucosal membrane and their relation to the permeability enhancing effects in the rat jejunum. J Pharm Pharmacol 39:621–626

Yamamoto A, Taniguchi T, Rikyuu K, Tsuji T, Fujita T, Murakami M, Muranishi S (1994) Effects of various protease inhibitors on the intestinal absorption and degradation of insulin in rats. Pharm Res 11:1496–1500

Yamamoto A, Uchiyama T, Nishikawa R, Fujita T, Muranishi S (1996) Effectiveness and toxicity screening of various absorption enhancers in the rat small intestine: effects of absorption enhancers on the intestinal absorption of phenol red and the release of protein and phospholipids from the intestinal membrane. J Pharm Pharmacol 48:1285–1289

Yamamoto A, Okagawa T, Kotani A, Uchiyama T, Shimura T, Tabata S, Kondo S, Muranishi S (1997) Effects of different absorption enhancers on the permeation of ebiratide, an ACTH analogue, across intestinal membranes. J Pharm Pharmacol 49:1057–1061

Yamamoto A, Tatsumi H, Maruyama M, Uchiyama T, Okada N, Fujita T (2001) Modulation of intestinal permeability by nitric oxide donors: implications in intestinal delivery of poorly absorbable drugs. J Pharmacol Exp Ther 296:84–90

Chapter 6
Polymeric Permeation Enhancers

Hans E. Junginger

Contents

Abstract There are enough low molecular permeation enhancers available to increase the absorption of hydrophilic drugs, but all of them have the disadvantage of strongly interfering with the phospholipid membranes and of damaging them. Polymeric permeation enhancers are a class of substances which are able to selectively trigger mechanisms to selectively open the watery channels of the tight junctions which allow the passage of hydrophilic drugs alongside the enterocytes. As these polymeric permeation enhancers are hydrophilic polymeric substances they are generally not absorbed and hence show basically no toxicity. They specifically act, e.g. by binding calcium ions (polyacrylates) and the reversible opening of the tight junctions is triggered. Others – like the chitosans and their quaternary analogues like trimethyl chitosan (TMC) – specifically interact with their positive charges of sialic acid or sulphuric acid of the mucosal linings to

H.E. Junginger (✉)
Department of Pharmaceutics, Faculty of Pharmaceutical Sciences, Naresuan University, Phitsanulok, 65000, Thailand
e-mail: hejunginger@yahoo.com

A. Bernkop-Schnürch (ed.), *Oral Delivery of Macromolecular Drugs*,
DOI 10.1007/978-1-4419-0200-9_6,

induce the same effect. As all the polymeric permeation enhancers have to show mucoadhesivity in order to be effective, their residence time on the mucosal surface is increased. This effect is even strongly increased when thiolated polymeric permeation enhancers are used, which are able to even further increase the mucus residence time of the permeation enhancer. Whereas delivery systems containing polymeric permeation enhancers for application at directly accessible mucous membranes are feasible the peroral application is still a challenge.

6.1 Introduction

In the early 1990s more or less by serendipity two new classes of polymeric permeation enhancers have been identified being polymers which have been used since long time and in huge amounts as 'neutral' excipients in formulation of drug medicines. These new classes are polyacrylates and chitosan salts which showed the ability to reversibly open the so-called tight junctions and allow the paracellular permeation of hydrophilic substances across mucosal tissues. Because of their different structure and ionic charge (polyacrylates being negatively charged and chitosans being positively charged), the mechanisms underlying this alteration of permeability have to be of different characters. A direct interaction of the cationic polymer molecule with the negatively charged cell membrane sugar residues is now established for chitosan and its derivatives, whereas depletion of extracellular Ca^{2+} probably plays the major role in triggering the opening of the tight junctions (Noach et al. 1993, Lueßen et al. 1996a, 1997, Borchard et al. 1996). Both classes of polymers are mucoadhesive. Additionally polyacrylates can also block intestinal enzymes to a certain extent (Lueßen et al. 1996b). Most importantly, however, is the fact that these classes of polymeric permeation enhancers only act on the tight junctions, thus allowing paracellular transport of hydrophilic molecules only, and not additionally interact, as low molecular penetration enhancer do, with the cell membrane components by inducing transcellular drug transport by membrane damage. As these polymeric permeation enhancers are hydrophilic macromolecular substances their absorption by enterocytes is negligible and their safety profile is very good (Junginger and Verhoef 1998).

6.2 Polyacrylates

The term polyacrylates includes synthetic, high molecular weight polymers of acrylic acid (polyacrylic acid or PAA) that are also known as carbomers. They are either linear or (weakly) cross-linked (either by allyl sucrose (carbomers) or by divinyl glycol (polycarbophils)) polymers that are broadly applied in pharmaceutical and cosmetic industry (mostly as excipient for controlled drug release from oral dosage forms and as stabilizers for gels). Cross-linked carbomers, distributed now by Lubrizol (www.lubrizol.com), formerly by B.F. Goodrich under the commercial name carbopols and polycarbophils (PCPs), are also used as mucoadhesive platforms for drug delivery. Carbopols and PCPs have received extensive review and toxicological evaluation. PCPs and calcium PCPs are classified as category 1 GRAS (generally recognized as safe) materials. Polyacrylates interact with mucus by hydrogen and van der Waals bonds, created between the carboxylic groups of the polyacrylates and the sulphate and sialic acid residues of mucin glycoproteins (Dodou et al. 2005). However, polyacrylates do possess properties of absorption enhancers for paracellular absorption of hydrophilic compounds as peptides and are additionally able to inhibit the activities of enzymes present in the intestinal fluid.

6.2.1 Polyacrylates as Absorption Enhancers

The absorption across rat intestinal tissue of the model peptide-drug 9-desglycinamide, 8-L-arginine vasopressin (DGAVP) from mucoadhesive formulations was studied by Lehr et al. (1992a) in vitro, in a chronically isolated internal loop in situ and after intraduodenal administration in vivo. Only the polycarbophil suspensions of the drug could show significant increases of bioavailabilities in all three models, whereas a controlled-release bioadhesive drug delivery system consisting of microspheres of poly(2-hydroxyethyl methacrylate) with a mucoadhesive polycarbophil coating was practically ineffective, because its mucoadhesive coating was de-activated by soluble mucins before reaching the intestinal mucosa. A prolongation of theabsorption phase in vitro and in the chronically isolated loop in situ suggested that the polymer was able to protect the peptide from proteolytic degradation.

In another study Lueßen and coworkers (1997) compared the absorption-enhancing effects of polycarbophil, chitosan and chitosan glutamate and found that all three mucoadhesive polymers were potent enhancers of the model peptide-drug DGAVP using Caco-2 cell layers and the vertically perfused intestinal loop model of the rat. The observed comparable transport effect of polycarbophil on the intestinal loop model was mainly ascribed to the protection of DGAVP against proteolytic degradation in the intestinal lumen, which allows for sufficient concentration and thus transport of the peptide drug when the polycarbophil-induced paracellular transport is less pronounced. In an in vivo study the same researchers also studied the intestinal absorption of buserelin using Carbomer 934P, the sodium salt of Carbomer (made in order to increase the solubility of Carbomer and chitosan hydrochloride as possible polymeric penetration enhancers (Fig. 6.1a)). Further results, however, presented in Fig. 6.1b showed a strong superiority of chitosan as absorption-enhancing polymer in comparison with polyacrylates (Lueßen et al. 1997).

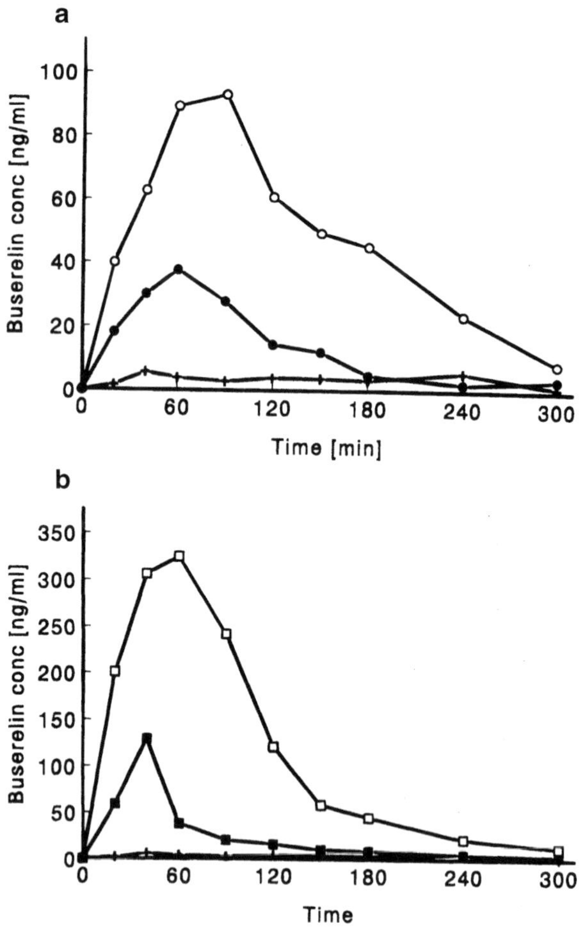

Fig. 6.1 **(a)** Serum concentrations after intraduodenal application of buserelin (500 μg/rat). +, control (MES/KOH buffer pH 6.7); •, 0.5% (w/v) FNaC934P; ○, 0.5% (w/v) C934P. Data are presented as the mean of 5 or 6 rats; **(b)** Serum concentrations after intraduodenal application of buserelin (500 μg/rat). +, control (MES/KOH buffer pH 6.7); ■, 0.5% (w/v) FNaC934P/1.5% (w/v) chitosan-HCl mixture (1:1); □, 1.5% (w/v) chitosan-HCl. Data are presented as the mean of 5 or 6 rats.

Carbopol 934P was also used by Thanou and collaborators (2001b) for the enhancement of the intestinal absorption of low molecular weight heparin (LMWH) in rats and pigs (Figs. 6.2 and 6.3, respectively). LMWH is a polyanion and does not interact with polycarbophil 934P. To both animal species LMWH was administered intraduodenally and the AntiXa levels were measured. Both studies showed a remarkably enhanced LMWH uptake after about 1 h and the effect for providing sufficient antithrombotic effect lasted for both animal species for about 7 h showing that this polyacrylate may be a good absorption enhancer for LMWH, provided a good delivery system can be developed based on polycarbophil 934P. The in vivo results of both the study in rats and in pigs are shown in Figs. 6.2 and 6.3, respectively.

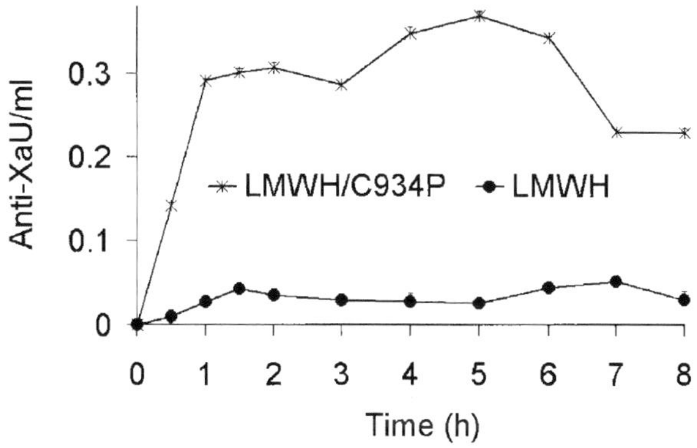

Fig. 6.2 Serum anti-Xa levels after intraduodenal administration of low molecular weight heparin (10,800 anti-Xa U/kg) with or without 1% (w/v) C934P in rats (mean ± standard error of the mean; $n = 6$) (Thanou et al. 2001b)

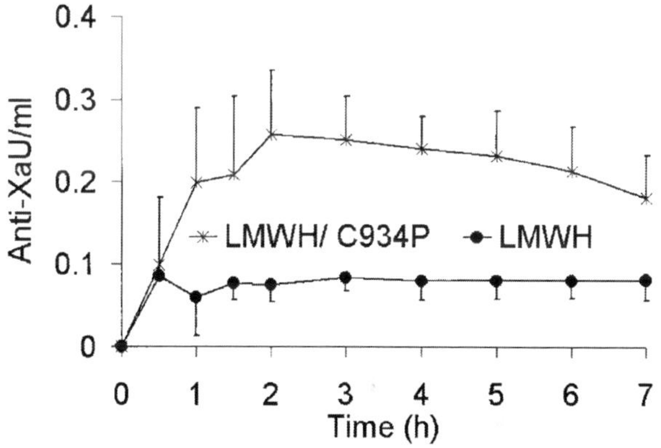

Fig. 6.3 Plasma anti-Xa levels after intraduodenal administration of low molecular weight heparin (5000 anti-Xa U/kg) with or without 1% (w/v) C934P in pigs (mean ± standard deviation; $n = 4$) (Thanou et al. 2001b)

6.3 Chitosan

6.3.1 Application, Mechanism and Safety Aspects

Chitosan (poly[β-(1-4)-2-amino-2-deoxy-D-glucopyranose]) is a cationic polysaccharide comprising copolymers of glucosamine and *N*-acetylglucosamine (Fig. 6.4). Nowadays, chitosan is available in different molecular weights (polymers 500,000–50,000 Da, oligomers 2000 Da), viscosity grades and degree of deacetylation (40–98%).

The bioadhesive properties were first described by Lehr et al. (1992b) demonstrating that chitosan in the swollen state is an excellent mucoadhesive polymer in the porcine intestinal mucosa and is also suitable for repeated adhesion. The authors also reported that chitosan undergoes minimal swelling in artificial

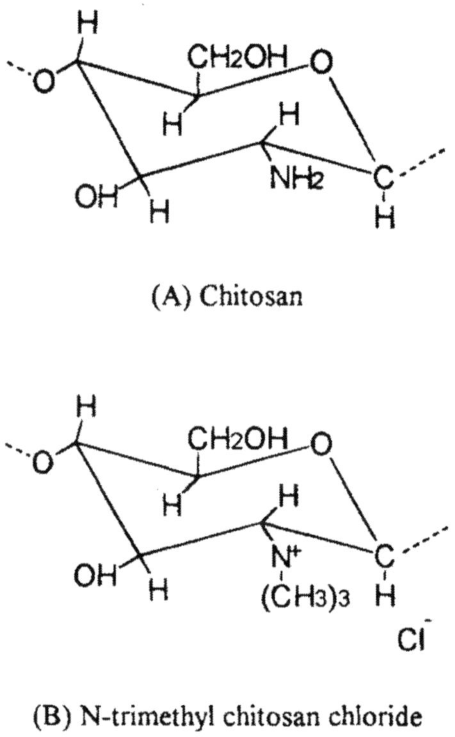

Fig. 6.4 Chemical structure of **(a)** chitosan and **(b)** N-trimethyl chitosan chloride (from Kotzé et al. 1998)

intestinal fluids due to its poor aqueous solubility at neutral pH values, proposing that substitution of the free $-NH_2$ groups with short alkyl chains would change the solubility and hence the mucoadhesion profile. The strong mucoadhesive properties of chitosan are due to the formation of hydrogen and ionic bonds between the positively charged amino groups of chitosan and the negatively charged sialic acid residues of mucin glycoproteins (Rossi et al. 2000).

Whereas for most low molecular weight absorption enhancers studied a strong cytotoxicity profile was evident in concentrations where they are able to act as penetration enhancers, chitosan gave contradictory results with respect to its safety profile (Carreno-Gomez and Duncan 1997). Dodane et al. (1999) investigated the effect of chitosan (degree of deacetylation 80%) solutions at pH 6.0–6.5 on the structure and function of Caco-2 cell monolayers. Using a series of microscopic techniques, the authors were able to show that chitosan had a transient effect on the tight junction's permeability and that viability of the cells was not affected. However, chitosan treatment slightly perturbed the plasma membrane, but this effect was reversible.

In a preliminary study Chae and coworkers (2005) investigated the molecular weight (MW)-dependent Caco-2 cell layer transport phenomena (in vitro) and the intestinal absorption patterns after oral administration (in rats in vivo) of water-soluble chitosans. The absorption of chitosans was significantly influenced by its MW. As the MW increases, the absorption decreases. The absorption both in vitro and in vivo of a chitosan with a MW of 3.8 kDA was about 25 times higher in comparison to a high MW chitosan (230 kDa). On the other side, the chitosans showed concentration- and MW-dependent cytotoxic effects: the chitosan oligosaccharides (MW $<$ 10 kDa) showed negligible cytotoxic effects on the Caco-2 cells whereas the high MW chitosans were more toxic in this experimental setting. However, the abundant use of chitosans in the food industry and the use of chitosan as excipient for peroral drug delivery systems prove also that chitosan with a high molecular weight can be regarded as safe.

6.3.2 Chitosan as Absorption Enhancer of Hydrophilic Macromolecular Drugs

Illum and coworkers (1994) reported at first that chitosan is able to promote the transmucosal absorption of small polar molecules as well as peptide and protein drugs across nasal epithelia. Immediately afterwards Artursson and collaborators (1994) reported that chitosan can increase the paracellular permeability of [^{14}C]mannitol (a marker for the paracellular route) across Caco-2 intestinal epithelia.

Chitosan gels were first tested in vivo for their ability to increase the intestinal absorption by Lueßen and coworkers (1996a). The absorption enhancement of the peptide analogue buserelin was studied after intraduodenal co-administration with chitosan (pH 6.7) in rats. Chitosan substantially increased the bioavailability of the peptide (5.1%) in comparison to control (no polymer) or Carbopol 934P containing formulations (see Fig. 6.1a and b). Borchard and his team (1996) investigated chitosan glutamate solutions at pH 7.4 for their effect in increasing the paracellular permeability of [^{14}C]mannitol and fluorescent-labelled dextran (MW 4400 Da) in vitro in Caco-2 cells. No effect on the permeability of the monolayer could be observed, indicating that at neutral pH value chitosan is not effective as absorption enhancer. The pH dependency of chitosan's effect on epithelial permeability was further investigated by Kotzé and coworkers (1998). Two chitosan salts (hydrochloride and glutamate) were evaluated for their ability to enhance the transport of [^{14}C]mannitol across Caco-2 cell monolayers at two pH values, 6.2. and 7.4. At low pH both chitosans showed a pronounced effect on the permeability of the marker, leading to 25-fold (glutamate salt) and 36-fold (hydrochloride salt) enhancement. However, at pH 7.4 both chitosans failed to increase the permeability due to their insolubilities in a more basic environment such as in the large intestine. These results made quite clear that chitosan (salts) cannot be used as absorption enhancers for in vivo studies when the drug should be released in the jejunum because of the insolubility and hence ineffective at pH values higher than 6.5.

Very recently a chitosan product could be manufactured by a special deacetylation process of chitin resulting in a chitosan which seems to be soluble also at pH values up to 7.4 (Personal communication of Jorg Thöming, UFT – Center for Environmental Research and Technology, Department of Production Engineering, Chemical Engineering – Recovery & Recycling, Leobener Str. UFT, D-28359 Bremen, Germany). This special chitosan may make the use of (quaternized) chitosan derivatives superfluous when proven that it is as effective in opening of the tight junctions as the quaternized chitosan derivatives.

6.3.3 N,N,N,-*Trimethyl Chitosan Hydrochloride (TMC)*

6.3.3.1 Synthesis and Characterization of TMC

Kotzé and collaborators (1998) based on the method of Domard et al. (1986) synthesized TMC with various degrees of quaternization. TMC is a partially quaternized derivative of chitosan which is prepared by reductive methylation of chitosan with methyl iodide in a strong basic environment at an elevated temperature (see Fig. 6.4). The degree of quaternization can be altered by increasing the number of reaction steps, by repeating them or by increasing the reaction time according to Sieval et al. (1998). TMC proved to be a derivative of chitosan with superior solubility and basicity, even at low degrees of quaternization, compared to chitosan salts. This quaternized chitosan shows much higher aqueous solubility than chitosan in a much broader pH and concentration range. The reason for this improved solubility is the substitution of the primary amine with methyl groups and the prevention of hydrogen bond formation between the amine and the hydroxylic groups of the chitosan backbone.

The absolute molecular weights, radius and polydispersity of a range of TMC polymers with different degrees of quaternization (22.1, 36.3, 48.0 and 59.2%) were determined with size exclusion chromatography and multi-angle laser light scattering (MALLS) (Kotzé et al. 1998). The absolute molecular weight of the TMC polymers decreased with an increase in the degree of quaternization. The respective molecular weights measured for each of the polymers were 2.02, 1.95, 1.66 and 1.43 g/mol $\times 10^5$. It should be noted that the molecular weight of the polymer chain increases during the reductive methylation process due to the addition of the methyl groups to the amino group of the repeating monomer. However, a net decrease in the absolute molecular weight is observed due to degradation of the polymer chain caused by exposure to the specific harsh reaction condition during the synthesis (Snyman et al. 2002). Polnok and coworkers (2004) investigated the influence of the methylation process on the degree of quaternization of *N*-trimethyl chitosan chloride. ^{1}H-Nuclear magnetic resonance spectra showed that the degree of quaternization was higher when using sodium hydroxide as base compared to dimethyl amino pyridine. The degrees of quaternization as well as O-methylation of TMC increased with the number of reaction steps.

The mucoadhesive properties of TMC with different degrees of quaternization, ranging between 22 and 49%, were investigated by the group of Snyman (2002). TMC was found to have a lower intrinsic mucoadhesivity compared to the chitosan salts, chitosan hydrochloride and chitosan glutamate, but if compared to the reference polymer, pectin, TMC possesses superior mucoadhesive properties. The decrease in the mucoadhesion of TMC compared to the chitosan salts was explained by a change in the conformation of the TMC polymer due to interaction between the fixed positive charges on the quaternary amino group, which possibly also decreases the flexibility of the polymer molecules. The interpenetration into the mucus layer by the polymer is influenced by a decrease in flexibility resulting in a subsequent decrease in mucoadhesivity (Snyman et al. 2003)

6.3.4 N-*Trimethyl Chitosan as Absorption Enhancer of Peptide Drugs*

TMC was firstly investigated for permeation-enhancing properties and toxicity (Kotzé et al. 1999), using the Caco-2 cells as a model for intestinal epithelium. Initially a trimethylated-chitosan having a degree of trimethylation of 12% (dimethylation 80%) was tested. This polymer (1.5–2.5%, w/v; pH 6.7) caused large increases in the transport rate of [^{14}C]mannitol (32–60-fold), fluorescent-labelled dextran 4400 (167–373-fold) and the peptide drug buserelin (28–73-fold). Confocal laser scanning microscopy (CLSM) confirmed that TMC opens the tight junctions of intestinal epithelial cells to allow increased transport of hydrophilic compounds along the paracellular transport pathway. No intracellular transport of the fluorescent marker could be observed (Kotzé et al. 1999).

Chitosan HCl and TMCs of different degrees of trimethylation were tested by Kotzé and collaborators (1999) for enhancing the permeability of [^{14}C]mannitol in Caco-2 intestinal epithelia at a pH value of 7.2. Chitosan HCl failed to increase the permeability of the marker substance across these monolayers and so did TMC with a degree of methylation of 12.8%. However, TMC with a degree of trimethylation of 60% increased significantly the [^{14}C]mannitol permeability across Caco-2 intestinal monolayers, indicating that a threshold value at the charge density of the polymer is necessary to trigger the opening of the tight junctions at neutral values.

TMC polymers were further investigated by Thanou et al. (see literature from 1999 and 2000a) to see if they provoke cell membrane damage on Caco-2 cell monolayers during enhancement of the transport of hydrophilic macromolecules. Using cell membrane impermeable fluorescent probes and CLSM, it was visualized that TMC polymers widen the paracellular pathways without cell damage. From such visualization studies it also appears that the mechanism of opening the tight junctions is similar to that of protonated chitosan (Thanou et al. 2001c). Because of the absence of significant toxicity, TMC polymers (particularly with a high degree of trimethylation) are expected to be safe absorption enhancers for improved transmucosal delivery of peptide drugs (Thanou et al. 1999).

The effects of TMC60 (degree of trimethylation 60%) polymers were subsequently studied in vivo in rats, using the peptide drug buserelin (pI = 6.8) and octreotide (pI = 8.2) (Thanou et al. 2000b, Thanou 2000). Octreotide formulations with or without TMC60 (pH 7.2) were compared with chitosan dispersions at neutral pH values after intraduodenal administration in rats. A remarkable increase in octreotide serum concentrations was observed after co-administration of the peptide with TMC60, whereas octreotide alone was poorly absorbed. In the presence of TMC60 octreotide was rapidly absorbed from the intestine having t_{max} at 40 min, whereas chitosan dispersions (at pH 7.2) showed a slight increase in octreotide absorption compared to the control. Chitosan did not manage to increase the octreotide concentrations to the levels achieved with TMC60. The absolute bioavailability of octreotide after co-administration with 1.0% TMC60 was 16.0%.

Octreotide was also administered to juvenile pigs with or without TMC60 at a pH of 7.4. The solutions were administered intrajejunally through an in-dwelling fistula that was inserted 1 week prior to the octreotide. Intrajejunal administration of 10 mg of octreotide, co-administered with 5 and 10% (w/v) TMC60, resulted in a 7.7-fold and 14.5-fold increase in octreotide absorption with absolute bioavailabilities of 13.9 ± 1.3 and 24.8 ± 1.8%, respectively. The results are presented in Fig. 6.5 (Thanou et al. 2001a).

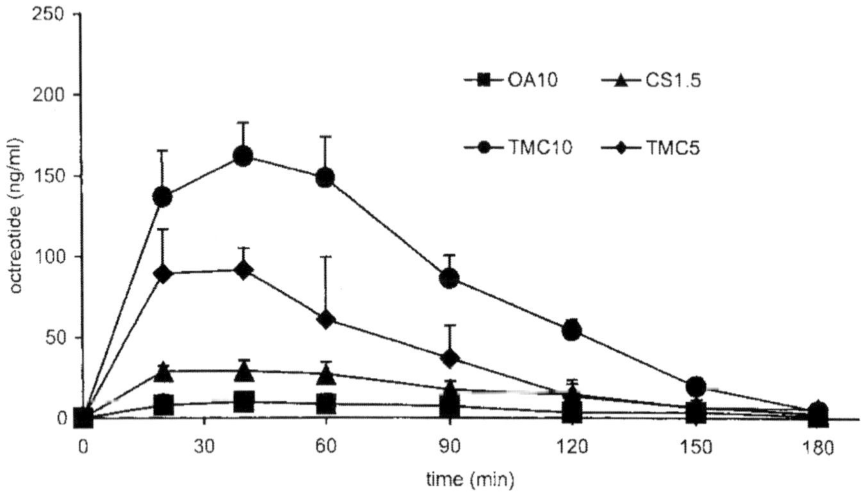

Fig. 6.5 Plasma octreotide concentration (mean ± SE) versus time curves after IJ administration of 10 mg/20 ml/pig with the polymers chitosan HCL [CS1.5: 1.5% (w/v); pH = 5.5; n = 6] and TMC [TMC10: 10% (w/v); pH = 7.4; n = 6 and TMC5: 5% (w/v); pH = 7.4; n = 3] or without any polymer [OA10: octreotide in 0.9% NaCl; pH = 7.4; n = 5] (Thanou et al. 2001a)

It is stated by the author that a gel was obtained with 10% (w/v) concentration of the polymer. This high concentration of the TMC60 polymer was chosen to counteract the dilution of the 20 ml administration volume by the luminal fluids and mucus of the intestinal tract and to ensure that substantial amounts of both peptide and enhancer could reach the absorptive site of the intestinal mucosa (Thanou 2001). Although the results show very high bioavailabilities (also taking into account the small absorptive area which is created by only widening of the tight junctions), the impracticality of administering such high concentrations in a solid dosage form cannot be overlooked as concentrations of 1–2 g of the polymer have to be administered in an attempt to obtain the same results (van der Merwe et al. 2004a).

In order to overcome these problems a completely new and different approach has been chosen by Dorkoosh and coworkers (2002). The platform of their delivery systems consists of superporous hydrogels (SPH) and superporous hydrogel composite (SPHC). These hydrogels can swell very rapidly and have the capacity to take up between 100 and 200 times of intestinal liquid of their original volume. Arriving in the intestine these SPHs swell quickly and bring the delivery systems (small tablet in which the drug is incorporated) which is attached to the outside of the SPH platform in direct contact with the absorbing surface. TMC at the outside of the small tablet will interfere at the interface between swollen SPH and intestinal wall as a polymeric penetration enhancer widening locally the tight junctions to allow for paracellular absorption of the peptide drug. In an in vivo study with pigs the achieved absolute bioavailabilities of octreotide were between $8.7 \pm 2.4\%$ and $16.1 \pm 3.3\%$ depending on the type of delivery system used. The value of $16.1 \pm 3.3\%$ was achieved with TMC60 as absorption enhancer (Fig. 6.6). After the peptide's release from the dosage form the SPH platforms get over-hydrated and are easily broken down by the peristaltic forces of the gut. Scintigraphic studies in human volunteers have shown good performance of these oral peptide drug delivery systems with prolonged residence times in the gut. Incorporating the SPH(C) delivery systems in enteric coated gelatin capsules of size 000 led to various stomach transit times (2–6 h in pigs and 0.5–2 h in humans). Capsules of smaller size and controlled food intake may reduce this variability in gastric transit times (Dorkoosh et al. 2004).

6.3.5 Mono-carboxymethyl Chitosan (MCC)

A usual approach to increase chitosan's solubility at neutral pH values is the substitution of the primary amine. Whereas N-substitution with alkyl groups (i.e. $-CH^3$ groups) can increase the aqueous solubility without affecting its cationic character, substitution with moieties bearing carboxyl groups can yield polymers with polyampholytic properties (Muzzarelli et al. 1982). Mono-carboxymethylated chitosan (MCC) was synthesized and further

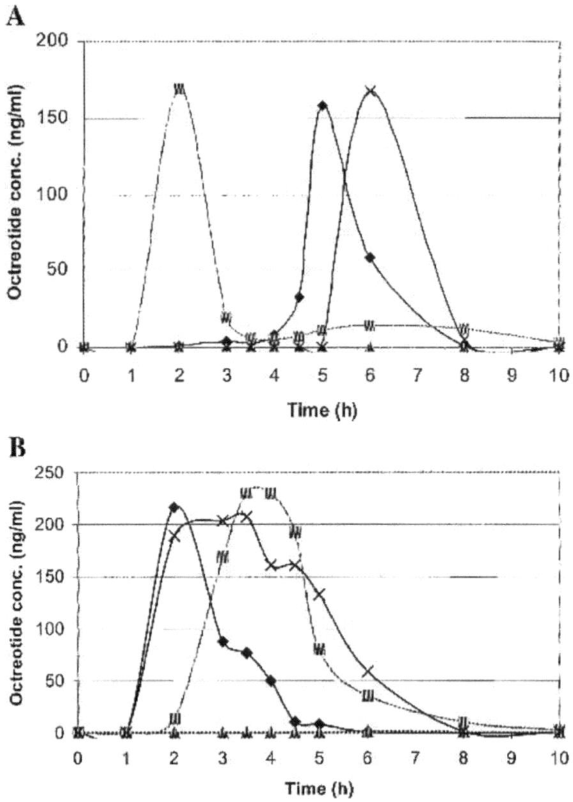

Fig. 6.6 Blood plasma profiles of octreotide after peroral administration of 15 mg/pig: A, Subject 2; B, Subject 6. Core (◆); core inside (■); octreotide without any polymer (▲); core outside with TMC (×) (Dorkoosh et al. 2002)

evaluated as potential absorption enhancer (Thanou et al. 2001d). This chitosan derivative (degree of substitution 87–90%) has a polyampholytic (zwitterionic) character, which allows the formation of clear gels or solutions (dependent on the concentration of the polymer) even in the presence of polyanionic compounds like heparins at neutral and alkaline pH values, whereas it aggregates at acidic pH. Chitosan and the quaternized derivative TMC form complexes with polyanions that precipitate out of the solution. In contrast MCC appeared to be compatible with polyanions.

Two viscosity-grade MCCs (high and low) were initially investigated to see if they were able to increase the permeation of low molecular weight heparin (4500 Da; LMWH) across Caco-2 intestinal cell monolayers. However, the MCC concentrations necessary to open the tight junctions were several times higher than that of TMC60 at neutral pH value. Low viscosity MCC induced higher transport of LMW when compared with the high viscosity derivative. Cell viability tests at the end of the experiments showed that this type of polymer had no damaging effect on cell membranes, whereas recovery of the transepithelial electrical resistance (TEER) values to initial levels indicated the functional integrity of the monolayer. The mechanism by which polyampholytic chitosans interacts with the tight junctions is not yet clear.

For in vivo studies, LMWH was administered intraduodenally with or without MCC to rats (Fig. 6.7); 3% (w/v) low viscosity MCC significantly increased the intestinal absorption of LMWH, reaching the therapeutic anticoagulant blood levels of LMWH for at least 5 h, determined by measuring Anti-Xa levels (Thanou et al. 2001d).

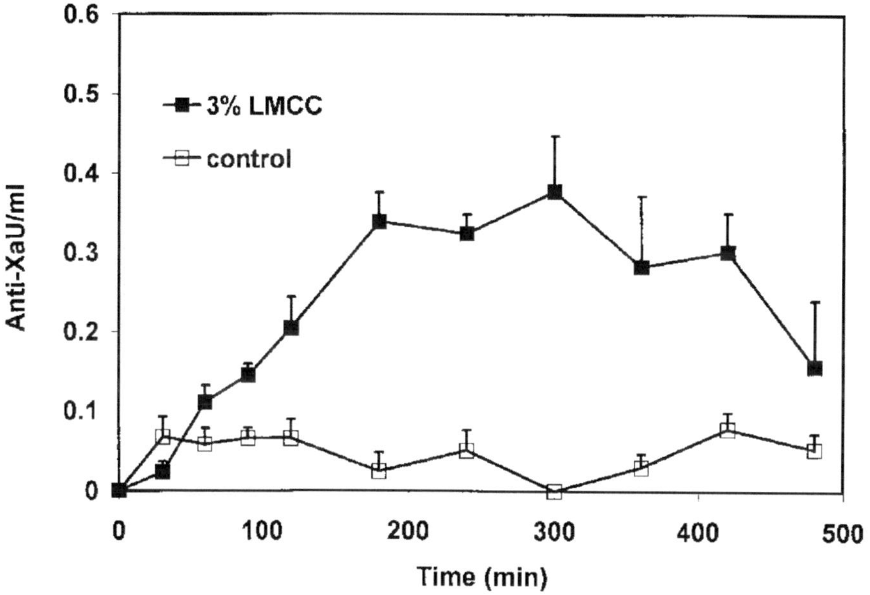

Fig. 6.7 Serum LMWH levels after intraduodenal administration without (control) and with 3% (w/v) LMCC in rats. LMWH was administered at 7200 anti-XaU/kg body weight (pH of administered formulation, 7.4; $n = 6$; mean ± SE) (Thanou et al. 2001d)

Just recently a series of other quaternized chitosan derivatives have been synthesized and characterized, namely, *N*,*N*-dimethyl, *N*-ethyl chitosan (DMEC) (Bayat et al. 2006), *N*-methyl, *N*,*N*-diethyl chitosan (DEMC) (Avadi et al. 2004) and *N*,*N*,*N*-triethyl chitosan (TEC) (Avadi et al. 2003). In a comprehensive study (Sadeghi et al. 2008a, b) the four quaternized derivatives of chitosan, trimethyl chitosan (TMC), diethylmethyl chitosan (DEMC), triethyl chitosan (TEC) and dimethylethyl chitosan (DMEC) with degree of substitution of approximately 50% were synthesized and their effect on the permeability of insulin across intestinal Caco-2 monolayers was studied and compared with chitosan both in free-soluble form and in nanoparticulate systems. Transepithelial electrical resistance (TEER) studies revealed that all four chitosan derivatives in free-soluble forms were able to decrease the TEER value in the following order TMC>DEMC>TEC=DMEC>chitosan, indicating their abilities to open the tight junctions. Recovery studies on the TEER showed that the effect of the polymers on Caco-2 cell monolayer is reversible and proves the viability of cells after incubation with all polymers. A similar rank order was also observed when measuring the zeta potentials of the various polymers. Transport studies of insulin together with soluble polymer across Caco-2 cell layers showed the following ranking: TMC>DMEC>DEMC>TEC>chitosan which is in agreement with the cationic charge of the polymer. In comparison to the free-soluble polymers, the nanoparticles prepared by ionic gelation of the chitosan and its quaternized derivatives had no significant effect on decreasing the TEER by opening of the tight junctions. In accordance with these results, the insulin-loaded nanoparticles showed much less permeation across the Caco-2 cell monolayer in comparison to the free-soluble polymers. Mass balance transport studies revealed that a substantial amount of the nanoparticles has been entrapped into the Caco-2 monolayer. It can thus be stated that while free-soluble polymers can reversibly open the tight junctions and increase the permeation of insulin, the nanoparticles had basically no effect on opening of the tight junction and the paracellular transport of insulin across the Caco-2 cell monolayer was minimal.

In a very recent study (Sadeghi et al. 2008a, b) two new derivatives of chitosan, C2–C6 trimethyl 6-amino-6-deoxy chitosan and C2–C6 triethyl 6-amino-6-deoxy chitosan were synthesized and characterized using ^{1}H-NMR and FTIR spectra. This means that besides the already existing *N*-trimethyl group or *N*-triethyl group on carbon atom 2 a second *N*-trimethyl group or *N*-triethyl group has been attached to carbon atom 6 with the aim to even increase the positive charge and hence the ability to open the tight junctions in a stronger way than the TMC or TEC molecules. The zeta potential and the antibacterial effect of these polymers were compared with chitosan and TMC. The results suggest that C2–C6 trimethyl 6-amino-6-deoxy chitosan and C2–C6 triethyl 6-amino-6-deoxy chitosan as highly water-soluble polymers have higher positive surface charge than both chitosan and TMC and they show higher antibacterial activity against gram-positive *Staphylococcus aureus* bacteria. However, first (yet unpublished) results show that the effect on TEER with the Caco-2 model was only moderately lower, e.g. much less than with TMC which was used as comparator substance.

6.4 Thiolated Polymers

6.4.1 Thiolated Polymers of Polyacrylates and Cellulose Derivatives

Thiolated polymers are synthesized by immobilizing thiol groups on polyacrylates or cellulose derivatives by the group of Bernkop-Schnürch and Steiniger (2000), Bernkop-Schnürch et al. (2000), Leitner et al. (2003a, b), Clausen et al. (2000), and Bernkop-Schnürch (2005). The main purpose of introducing free thiol groups into polymers which already have mucoadhesive properties is to further increase the strength of their mucoadhesiveness due to the chemical reaction of the thiol groups of the mucins and the thiol groups of the thiolated polymers by forming stable covalent disulphide bridges. With this elegant approach the mucoadhesivity of such polymers and additionally their cohesiveness could be strongly increased. Whereas the turnover time of mucus which has been estimated in the isolated intestinal loop of the rat by Lehr et al. (1991) to be in the order of 47–270 min, the mucus turnover in humans has been estimated to be in the range of 12–24 h (Forstner 1978; Allen et al. 1998). Hence, this longer possible residence time of these thiolated polymers and the delivery systems in the human gut makes them very interesting permeation enhancers for hydrophilic macromolecular drugs like peptides.

Bernkop-Schnürch and coworkers (2004) linked L-cysteine covalently to polycarbophil (PCP) and sodium carboxymethylcellulose (NaCMC), mediated by a carbodiimide (Fig. 6.8a and b). The resulting thiolated polymers displayed 100 ± 8 and 1,280 ± 84 μmol thiol groups per gram, respectively. In aqueous solutions these modified polymers were capable of forming inter and/or intramolecular disulphide bonds. Due to the formation of disulphide bonds within the thiol-containing polymers, the stability of matrix tablets could be strongly improved. Whereas tablets based on the corresponding unmodified polymer disintegrated within 2 h, the swollen carrier matrices of thiolated NaCMC and PCP remained stable for 6.2 h and for more than 48 h, respectively. With the model drug rifampicin controlled-release characteristics of these thiolated matrix tablets could be demonstrated. Tensile studies carried out with the unmodified and thiolated polymers at pH 3, 5 and 7, respectively, revealed that only if the polymer displays a pH value of 5, the total work of adhesion could be improved significantly due to the covalent attachment of thiol groups. The permeation-enhancing effect of thiolated polycarbophil on intestinal mucosa from guinea pigs showed weak enhancement ratios (1.1–1.5) in comparison to control tests.

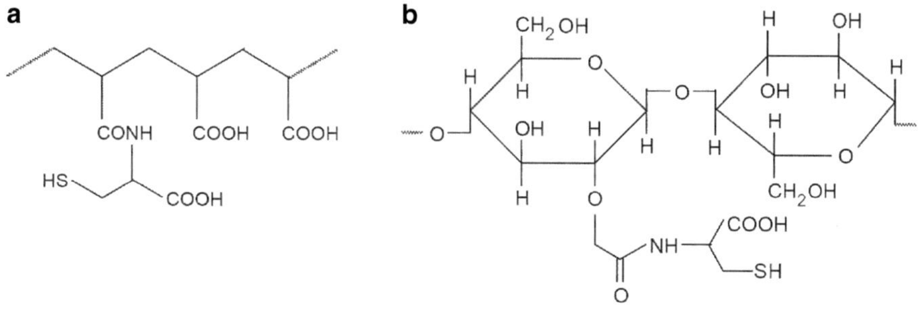

Fig. 6.8 **(a)** Poly(acrylic acid)-Cysteine (Bernkop-Schnürch 2005); **(b)** Carboxymethylcellulose-Cysteine (Bernkop-Schnürch 2005); **(c)** Chitosan-Thioglycolic acid (Bernkop-Schnürch 2005); **(d)** Chitosan-Cystein (Bernkop-Schnürch 2005); **(e)** Chitosan-Cystein (Bernkop-Schnürch 2005)

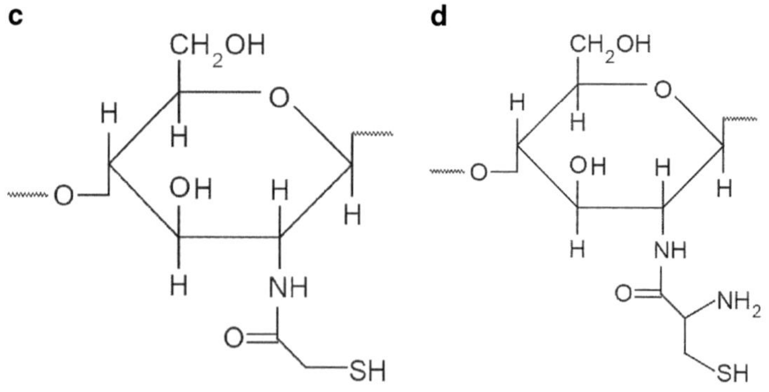

Fig. 6.8 (continued)

6.4.2 Thiolated Polymers of Chitosan

With the same aim as described in the previous paragraph chitosan has also been chemically modified by covalent binding of sulphur-containing moieties. To date, three different thiolated chitosan have been synthesized: chitosan–thioglycolic acid conjugates (Fig. 6.8c), chitosan–cysteine conjugates (Bernkop-Schnürch et al. 1999, 2001, Kast and Bernkop-Schnürch 2001; Hornof et al. 2003) (Fig. 6.8d) and chitosan–4-thio-butyl-amide (chitosan–TBA) conjugates (Bernkop-Schnürch et al. 2003) (Fig. 6.8e). These thiolated chitosans have numerous advantageous features in comparison to unmodified chitosan, such as significantly improved mucoadhesive properties and permeation-enhancing properties. The strong cohesive properties of thiolated chitosans make them highly suitable excipients for controlled drug release dosage forms (Bernkop-Schnürch et al. 2003; Kast et al. 2001). Moreover, solutions of thiolated chitosans display in situ gelling properties at physiological pH values which make them suitable for novel application systems to the eye (Bernkop-Schnürch et al. 2004)

The improved mucoadhesive properties of thiolated chitosans were explained by the formation of covalent bonds between thiol groups of the polymer and cysteine-rich subdomains of glycoproteins in the mucus layer (Leitner 2003b). These covalent bonds are supposed to be stronger than non-covalent bonds, such as ionic interactions of chitosan with ionic substructures as sialic acid moieties of the mucus layer. This theory was supported by the results of tensile studies with tablets of thiolated chitosan, which demonstrated a positive correlation between the degree of modification with thiol-bearing moieties and the adhesive properties of the polymer (Kast and Bernkop-Schnürch 2001; Roldo et al. 2004). These findings were confirmed by another in vitro mucoadhesion system, where the time of adhesion of tablets on intestinal mucosa was determined. The contact time of the thiolated chitosan derivatives increased with increasing amounts of immobilized thiol groups (Kast and Bernkop-Schnürch 2001; Bernkop-Schnürch et al. 2003). With chitosan–thioglycolic acid conjugates a 5–10-fold increase in mucoadhesion in comparison to unmodified chitosan was achieved.

6.5 Conclusions

With the advent of new biotechnological techniques endogenous compounds like insulin, buserelin or octreotide have become available at affordable prices. All of these substances still have to undergo needle application. Until today the development of alternative delivery systems for the nasal, buccal, peroral, rectal and pulmonary routes for the administration of those class III drugs according to the biopharmaceutics classification system (BCS) (Amidon et al. 1995) could not keep pace with this development of endogenous compounds or is not economic enough for the health care payers (e.g. insulin application via the pulmonary route).

Many multifunctional high molecular weight polymers as polyacrylates and chitosan with its multiple derivatives show promising properties as specific penetration enhancers for the paracellular absorption route of hydrophilic macromolecules with high enhancing potency of reversibly opening the tight junctions and practically without toxicity when applied in normal doses in vitro, the physical properties of these polymers, especially their high viscosity – and inherent with this their slow dissolution process – make the design of suitable delivery systems especially for the peroral route very difficult (van der Merwe et al. 2004b). Additionally the easy saturation of the mucoadhesive properties of the multifunctional polymers by soluble mucins in the intestinal liquids – a fact which also blocks their additional properties like opening of the tight-junctions and enzyme deactivation potency – makes it very difficult to develop suitable dosage forms for the peroral application. As a result of this such delivery systems should be able to quickly swell and expand in the (human) gut fluids and develop full mucoadhesive properties to reach the mucous linings in full activity. After adhesion to the gut mucus and widening of the tight junctions, the peptide drug should be released in the desired controlled way. Hence, the development of such dosage forms is still in its infancy, but there are promising perspectives (e.g. the systems as described by Dorkoosh et al. 2002 in a further developed state) that such delivery systems can be successfully developed in the near future.

References

Allen, A., Hutton, D.A., Pearson, J.P. and Sellers, L.A. et al. (1998) in M. O'Connor (Ed.): Mucus and Mucosa. Ciba Foundation Symposium, London, Vol. 109: p. 1984

Amidon, G.L., Lennernäs, H., Shah, V.P. and Crison, J.R. (1995) A theoretical basis for a biopharmaceutic drug classification: the correlation of in vitro drug product dissolution and in vivo bioavailability. *Pharm. Res.* 12: 413

Artursson, P., Lindmark, T., Davis, S.S. and Illum, L. (1994) Effect of chitosan on the permeability of monolayers of intestinal epithelial cells (Caco-2). *Pharm. Res.* 11: 1358–1361.

Avadi, M.R., Zohourian-Mehr, M.J., Younessi, P., Amini, M., Rafieeh-Tehrani, M., and Shafiee, A. (2003) Optimized synthesis and characterization of *N*-Triethyl Chitosan. *J. Bioact. Compat. Polym.* 18: 469–480.

Avadi, M.R., Sadeghi, A., Tahzibi, A., Bayati, K.H. Pouladzadeh, M. Zohourian-Mehr, M.J. and Rafiee-Tehrani, M. (2004) Diethylmethyl chitosan as antimicrobial agent: Synthesis, characterization and antibacterial effects. *Eur. J. Polym.* 40: 1355–1361.

Bayat, A., Sadeghi, M.M., Avadi, M.R., Amini, M, Rafiee-Tehrani, M., Shafiee, A. and Junginger, H.E. (2006) Synthesis of *N-N* dimethyl *N*-ethyl chitosan as carrier for oral delivery of peptide drugs. *J. Bioact. Compat Polym.* 21: 433–444.

Bernkop-Schnürch, A., Brandt, U.M., and Clausen, A.E. (1999) Synthesis and in vitro evaluation of chitosan-cysteine conjugates *Sci. Pharm.* 67: 196–206.

Bernkop-Schnürch, A. and Steiniger, S. (2000) Synthesis and characterization of mucoadhesive thiolated polymers. *Int. J. Pharm.* 194: 239–247.

Bernkop-Schnürch, A., Scholler, S. and Biebel, R.G. (2000) Development of controlled drug release systems based on thiolated polymers. *J. Control. Release* 66: 39–48.

Bernkop-Schnürch, A. and Hopf, T.E. (2001) Synthesis and in vitro evaluation of chitosan-thioglycolic acid conjugates. *Sci. Pharm.* 69: 109–118.

Bernkop-Schnürch, A., Hornof, M. and Zoidl, T. (2003) Thiolated polymers- thiomers: modification of chitosan with 2-iminothiolane. *Int. J. Pharm.* 260: 229–237.

Bernkop-Schnürch, A, Hornof, M. and Guggi, D. (2004) Thiolated chitosans. *Eur. J. Pharm. Biopharm.* 57: 9–17.

Bernkop-Schnürch (2005), Thiomers: A new generation of mucoadhesive polymers. *Adv. Drug Deliv. Rev.* 57: 1569–1582.

Borchard, G., Lueßen, H.L., de Boer, A.G., Verhoef, J.C., Lehr, C.-M. and Junginger, H.E. (1996) The potential of mucoadhesive polymers in enhancing intestinal peptide drug absorption. III: Effects of chitosan glutamate and carbomer on the epithelial tight junctions in vitro. *J. Control. Release* 39: 131–138.

Carreno-Gomez, B. and Duncan, R. (1997) Evaluation of the biological properties of soluble chitosan and chitosan microspheres. *Int. J. Pharm.* 148: 231–240.

Chae, S.Y., Jang, M.-K. and Nah, J.-W. (2005) Influence of molecular weight on oral absorption of water soluble chitosans. *J. Control. Release* 102: 383–394.

Clausen, A.E. and Bernkop-Schnürch, A. (2000) In vitro evaluation of permeation-enhancing effect of thiolated polycarbophil. *J. Pharm. Sci.* 89: 1253–1261

Dodane, V., Khan, A.M. and Merwin, J.R. (1999) Effect of chitosan on epithelial permeability and structure. *Int. J. Pharm.* 182: 21–32.

Dodou, D., Breedveld, P. and Wieringa, P.A. (2005) Mucoadhesives in the gastrointestinal tract: revisiting the literature for novel applications. *Eur. J. Pharm. Biopharm.* 60: 1–6.

Domard, A., Rinaudo, M. and Terrassin, C. (1986) New method for quaternization of chitosan. *Int. J. Biol. Macromol.* 8: 105–107.

Dorkoosh, F.A., Verhoef, J.C. Verheijden, J.H.M., Rafiee-Tehrani, M., Borchard, G. and Junginger, H.E. (2002) Peroral absorption of octreotide in pigs formulated in delivery systems on the basis of superporous hydrogel polymers. *Pharm. Res.* 19: 1532–1536.

Dorkoosh, F.A., Stokkel, M.P.M., Blok, D., Borchard, G., Rafiee-Tehrani, M., Verhoef, J.C. and Junginger, H.E. (2004) Feasibility study on the retention of superporous hydrogel (SPH) composite polymer in the intestinal tract of man using scintigraphy. *J. Control. Release* 99: 199–206.

Forstner, J.F. (1978) Intestinal mucins in health and disease. *Digestion* 17: 234–263

Hornof, M.D., Kast, C.E. and Bernkop-Schnürch, A. (2003) In vitro evaluation of the viscoelastic behavior of chitosan –thioglycolic acid conjugates. *Eur. J. Pharm. Biopharm.* 55: 185–190.

Illum. L., Farraj, N.F. and Davis, S.S. (1994) Chitosan as a novel nasal delivery system for peptide drugs. *Pharm. Res.* 11: 1186–1189.

Junginger, H.E. and Verhoef, J.C. (1998), Macromolecules as safe penetration enhancers for hydrophilic drugs – a fiction? *Pharm. Sci. Technol. Today* 1: 370–376.

Kast, C.E. and Bernkop-Schnürch, A. (2001) Thiolated polymers – thiomers: development and in vitro evaluation of chitosan-thioglycolic acid conjugates. *Biomaterials* 22: 2345–2352.

Kotzé, A.F., Lueßen, H.L., de Leeuw, B.J., de Boer, A.G., Verhoef, J.C. and Junginger, H.E. (1998) Comparison of the effect of different chitosan salts and *N*-trimethyl chitosan chloride on the permeability of intestinal epithelial cells (Caco-2). *J. Control. Release* 51: 35–46

Kotzé, A.F., Thanou, M.M., Lueßen, H.L., de Boer, A.G., Verhoef, J.C. and Junginger, H.E. (1999) Enhancement of paracellular drug transport with highly quaternized *N*-trimethyl chitosan chloride in neutral environment: in vitro evaluation in intestinal epithelial cells (Caco-2). *J. Pharm. Sci.* 88: 253–257.

Lehr, C.-M., Poelma, F.G.J., Junginger, H.E. and Tukker, J.J. (1991) An estimate of turnover time of intestinal mucus gel layer in the rat in situ loop. *Int. J. Pharm.* 70: 235–240.

Lehr, C.-M., Bouwstra, J.A., Kok, W., de Boer, A.G., Tukker, J.J., Verhoef, J.C., Breimer, D.D. and Junginger, H.E. (1992a) Effects of the mucoadhesive polymer polycarbophil on the intestinal absorption of a peptide drug in the rat. *J. Pharm. Pharmacol.* 44: 402–407.

Lehr, C.-M., Bouwstra, J.A., Schacht, E.H. and Junginger, H.E. (1992b) In vitro evaluation of mucoadhesive properties of chitosan and some other natural polymers. *Int. J. Pharm.* 78: 43–48.

Leitner, V.M., Marschütz, M.K. and Bernkop-Schnürch, A. (2003a) Mucoadhesive and cohesive properties of poly(acrylic acid)-cysteine conjugates with regard to their molecular mass. *Eur. J. Pharm. Sci.* 18: 89–96.

Leitner, .V.M., Walker, G.F. and Bernkop-Schnürch, A. (2003b) Thiolated polymers: evidence for the formation of disulphide bonds with mucus glycoproteins. *Eur. J. Pharm. Biopharm.* 56: 207–214

Lueßen, H.L., de Leeuw, B.J., Langemeijer, M.W.E., de Boer, A. G, Verhoef, J.C. and Junginger, H.E. (1996a) Mucoadhesive polymers in peroral drug delivery. VI. Carbomer and chitosan improve the intestinal absorption of the peptide drug buserelin in vivo. *Pharm. Res.* 13: 1668–1672

Lueßen, H.L., de Leeuw, B.J., Pérard, D., C.-M. Lehr, de Boer, A.G., Verhoef, J.C. and Junginger, H.E. (1996b) Mucoadhesive polymers in peroral drug delivery. I. Influence of mucoadhesive excipients on the proteolytic activity of intestinal enzymes. *Eur. J. Pharm. Sci.* 4: 117–128.

Lueßen, H.L., Rentel, C.-O., Kotzé, A.F., de Boer, A.G., Verhoef, J.C. and Junginger, H.E. (1997) Mucoadhesive polymers in peroral peptide drug delivery. IV. Polycarbophil and chitosan are potent enhancers of peptide transport across intestinal mucosae in vitro. *J. Control. Release* 45: 15–23.

Muzzarelli, R.A.A., Tanfani, F., Emmanueli, S. and Mariotti, S. (1982) *N*-(carboxymethylidene)-chitosans and *N*-(carboxymethyl)-chitosans: novel chelating polyampholytes obtained from chitosan glyoxylate. *Carbohydr. Res.* 107: 199–214.

Noach, A.B.J., Kurosaki, Y., Blom-Rosmalen, M.C.M., de Boer, A.G. and Breimer, D.D. (1993) Cell-polarity dependent effect of chelation on the paracellular permeability of confluent Caco-2 cell monolayers. *Int. J. Pharm.* 90: 229–237.

Polnok, A., Borchard, G., Verhoef, J.C., Sarisuta, N. and Junginger, H.E. (2004) Influence of methylation process on the degree of quaternization of *N*-trimethyl chitosan chloride. *Eur. J. Pharm. Biopharm.* 57: 77–83.

Roldo, M., Hornof, M., Caliceti, P. and Bernkop-Schnürch, A. (2004) Mucoadhesive thiolated chitosans as platforms for oral controlled drug delivery: synthesis and in vitro evaluation. *Eur. J. Pharm. Biopharm* 57: 115–121.

Rossi, S., Ferrari, F. Bonferoni, M.C. and Caramella, C. (2000) Characterization of chitosan hydrochloride-mucin interactions by means of viscosimetric and turbidimetric measurements. *Eur. J. Pharm. Sci.* 10: 251–257.

Sadeghi, A.M.M., Dorkoosh, F.A., Avadi, M.R., Bayat, A., Delie, F., Gurny, R., Rafieeh-Tehrani, M. and Junginger, H.E. (2008a) Permeation enhancer effect of chitosan and chitosan derivatives: comparison of formulations as soluble polymers and nanoparticulate systems on insulin absorption in Caco-2 cells. *Eur. J. Pharm. Biopharm.* 70: 270–278.

Sadeghi, A.M.M., Amin, A., Avadi, M.R. Siedi, F., Rafiee-Tehrani, M. and Junginger, H.E. (2008b) Synthesis, characterization and antibacterial effects of trimethylated and triethylated 6-NH2-6-deoxy-chitosan. *J. Bioact. Compat. Polym.* 23: 262–275.

Sieval, A.B., Thanou, M, Kotzé, A.F., Verhoef, J.C., Brussee, J. and Junginger, H.E. (1998) Preparation and NMR-characterization of highly substituted *N*-trimethyl chitosan hydrochloride. *Carbohydr. Polym* 36: 157–165.

Snyman, D., Hamman, J. H,. Kotzé, J.S., Rollings, J.E. and Kotzé, A.F. (2002) The relationship between the absolute molecular weight and the degree of quaternization of *N*-trimethyl chitosan chloride. *Carbohydr. Polym.* 50: 145–150.

Snyman, D., Hamman, J.H. and Kotzé, A.F. (2003) Evaluation of the mucoadhesive properties of *N*-trimethyl chitosan chloride. *Drug Dev. Ind. Pharm.* 29: 59–67.

Thanou, M. (2000) Chitosan Derivatives in Drug Delivery. Trimethylated and Carboxymethylated Chitosan as Safe Enhancers for the Intestinal Absorption of Hydrophilic Drugs, PhD Thesis, Leiden University, Leiden, pp. 91–108.

Thanou, M., Verhoef, J.C. Romeijn, S.G., Nagelkerke, J.F., Merkus, F.W.H.M. and Junginger, H.E. (1999) Effects of *N*-trimethyl chitosan chloride, a novel absorption enhancer, on Caco-2 intestinal epithelia and the ciliary beat frequency of chicken embryo trachea. *Int. J. Pharm.* 185: 73–82.

Thanou, M.M., Kotzé, A.F. Scharringhausen, T., Lueßen, H.L., de Boer, A.G., Verhoef, J.C. and Junginger, H.E. (2000a). Effect of degree of quaternization of *N*-trimethyl chitosan chloride for enhanced transport of hydrophilic compounds across intestinal Caco-2 cell monolayers. *J. Control. Release* 64: 15–25.

Thanou, M., Verhoef, J.C., Marbach, P. and Junginger, H.E. (2000b) *N*-trimethyl chitosan chloride (TMC) ameliorates the permeability and absorption properties of the somatostatin analogue in vitro and in vivo. *J. Pharm. Sci.* 89: 951–957.

Thanou, M., Verhoef, J.C., Verheijden, J.H.M. and Junginger, H.E. (2001a) Intestinal absorption of octreotide using trimethyl chitosan chloride: studies in pigs. *Pharm. Res.* 18: 823–828.

Thanou, M., Verhoef, J.C., Nihot, M.-T., Verheijden, J.H.M. and Junginger, H.E. (2001b) Enhancement of the intestinal absorption of low molecular weight heparin (LMWH) in rats and pigs using Carbopol 934P. *Pharm. Res.* 18: 1638–1641.

Thanou, M., Verhoef, J.C. and Junginger, H.E. (2001c) Oral drug absorption enhancement by chitosan and its derivatives. *Adv. Drug Deliv. Rev.* 52: 117–126.

Thanou, M., Nihot, M.T. , Jansen, M., Verhoef, J.C. and Junginger, H.E. (2001d) Mono--*N*-carboxymethyl chitosan (MCC), a polyampholytic chitosan derivative, enhances the intestinal absorption of low molecular weight heparin across intestinal epithelia in vitro and in vivo. *J. Pharm. Sci.* 90: 38–46.

Van der Merwe, S.M., Verhoef, J.C. Verheijden, J.H.M., Kotzé, A.F. and Junginger, H.E. (2004a) Trimethylated chitosan as polymeric absorption enhancer for improved peroral delivery of peptide drugs. *Eur. J. Pharm. Biopharm.* 58: 225–235.

Van der Merwe, Verhoef, J.C., Kotzé, A.F. and Junginger, H.E. (2004b) *N*-Trimethyl chitosan chloride as absorption enhancer in oral peptide drug delivery. Development and characterization of minitablet and granule formulations. *Eur. J. Pharm. Biopharm.* 57: 85–91.

Chapter 7
Strategies to Overcome Efflux Pumps

Florian Föger

Contents

Abstract Intestinal efflux pumps such as P-glycoprotein play a significant role in altering the absorption of a wide range of drugs. Anticancer agents, antibiotics, antivirals, calcium channel blockers, immunosuppressive agents, peptide drugs and several other therapeutic compounds have been reported to be substrates of one or more transmembrane efflux transporters.

Inhibition of these efflux pumps by various compounds can lead to enhanced absorption of several drugs across the intestine.

In this chapter, several efflux pump inhibitors such as low molecular mass inhibitors, polymeric inhibitors and advanced formulation approaches on how to overcome drug efflux are discussed.

7.1 Introduction

In order to achieve and sustain therapeutic blood levels by peroral administration of drugs, several hurdles have to be overcome. Besides well-known barriers such as enzymatic degradation, dissolution problems, mucus barrier and others,

F. Föger (✉)
Diabetes Research Unit, Oral Formulation Research,
Novo Nordisk A/S, Målφv, Denmark
e-mail: faf@novonordisk.com

A. Bernkop-Schnürch (ed.), *Oral Delivery of Macromolecular Drugs*,
DOI 10.1007/978-1-4419-0200-9_7, © Springer Science+Business Media, LLC 2009

it has been identified that efflux pumps play a significant role in altering the pharmacokinetics of various drugs (Werle 2008a).

Efflux transporters are expressed in many tissues including the surface of epithelial cells in the intestine where they play a significant role in absorption and disposition of not only endogenous substrates, toxins and therapeutic small molecules but also macromolecular drugs. Multidrug efflux transporters, such as P-glycoprotein (P-gp) and multidrug resistance protein (MRP), which are located in the apical membrane of enterocytes (Varma et al. 2006) limit the oral bioavailability of a lot of structurally diverse compounds. Anticancer agents (paclitaxel, doxorubicin), antibiotics (itraconazole, erythromycin), antivirals (saquinavir, ritonavir), calcium channel blockers (verapamil, diltiazem), immunosuppressive agents (cyclosporine, tacrolimus), macromolecular drugs (metkephamid, D-ala-leu-enkephalin, cyclosporine) and several other drugs (Werle 2008b) have been reported to be substrates of efflux pumps (Table 7.1 shows a summary of the most important efflux pump substrates). These transmembrane efflux proteins translocate substrates from the inner side of the membrane to the outer side. P-gp, which is often overexpressed in tumour cells, can lead to multiple drug resistance of such tumours (Hunter and Hirst 1997). Because of its expression in the intestinal tract, it limits the absorption of drugs from the intestine into systemic circulation (Hunter and Hirst 1997) (Fig. 7.1). As oral administration is one of the most convenient routes of drug administration, it is important to overcome this absorption hurdle.

It has been previously reported that inhibition of efflux pumps by various compounds can lead to enhanced absorption of drugs across the intestine (Kim 2002). P-gp can be inhibited by a range of substances that blocks its function either by acting as a high avidity substrate, like verapamil, diltiazem or cyclosporine (Gerrard et al. 2004), or by binding to it such as sulphydryl-substituted purines (Al-Shawi et al. 1994).

However, it is well known that most of these inhibitors themselves are pharmacologically active compounds, which have their own clinical indications. Furthermore, most of these drugs lead to undesired pharmacodynamic side effects caused by high concentrations necessary for sufficient gastrointestinal inhibition of P-gp. Therefore, much effort has been put into the development of inhibitors with less or no pharmacological activity combined with improved inhibitory properties. In contrast to low molecular weight P-gp inhibitors, polymeric inhibitors would offer the advantage of remaining more concentrated in the GI tract at the site of drug absorption and of not being absorbed per se thus avoiding systemic toxic side effects (Föger et al. 2006c).

7.2 Strategies to Overcome Efflux Pumps

7.2.1 Efflux Pump Inhibitors

There are several structural diverse compounds that are known to inhibit efflux pumps. These inhibitors have been used in cancer therapy in order to reduce

Table 7.1 Summary of various efflux pump substrates

Antimicrobial drugs	Chemotherapeutic drugs	HIV protease inhibitors	Immunosuppressive drugs	Cardiovascular drugs	Steroids	Macromolecular drugs (polypeptides)
Acyclovir	Actinomycin D	Amprenavir	Cyclosporine A	Digoxin	Aldosterone	Cyclosporine
Erythromycin	Daunorubicin	Indinavir	Tacrolimus	Quinidine	Cortisol	Metkephamid
Ivermectin	Doxorubicin	Nelfinavir		Verapamil	Corticosterone	Enkephalin
Itraconazole	Docetaxel	Ritonavir		Diltiazem	Dexamethasone	
Rifampin	Epirubicin	Saquinavir			Hydrocortisone	
	Etoposide					
	Imatinib					
	Irinotecan					
	Paclitaxel					
	Vinblastine					
	Vincristine					

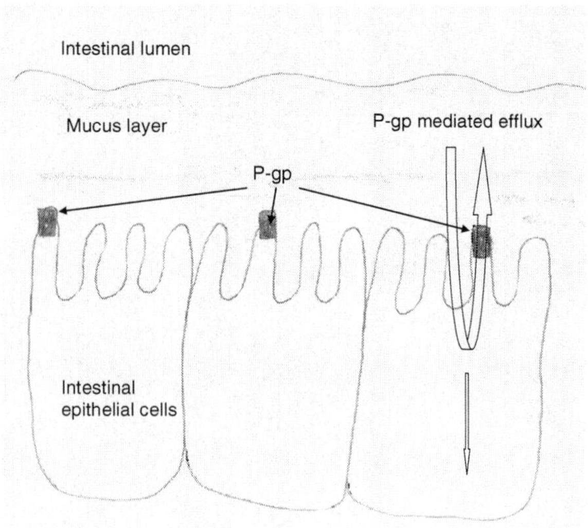

Fig. 7.1 P-glycoprotein (P-gp)-mediated efflux

multidrug resistance or just to improve the bioavailability of orally administered efflux pump substrates. There are three main mechanisms, how these inhibitors modulate or inhibit efflux. Varma et al. (2003) stated inhibition caused (I) by a blocking of drug binding sites either competitively or allosterically, (II) by interfering with ATP hydrolysis and (III) by altering the integrity of cell membrane lipids.

In this review efflux pump inhibitors are classified into two groups: low molecular mass inhibitors and polymeric inhibitors, because the high molecular mass of the polymeric excipients prevents absorption into systemic circulation after oral administration. In some cases, just a local inhibition of efflux transporters in the intestine is desired, whereas in other cases also an additional systemic modulation of efflux pumps can be of advantage. For chronical treatments, impact on the complex systemic efflux transporter system can result in severe complications. In this case, an enhanced intestinal absorption of efflux pump substrates can be achieved by using drug delivery systems based on polymeric inhibitors. On the other hand, in cancer therapy it would be of advantage to reduce efflux of anticancer compounds also in the systemic system because tumour tissues often overexpress these transporters. Then a low molecular mass efflux inhibitor could be useful.

7.2.1.1 Low Molecular Mass Inhibitors

Based on the specificity and affinity, low molecular mass efflux pump inhibitors are classified into three generations (Werle 2008a). First-generation P-gp inhibitors such as verapamil or cyclosporine are compounds that are in clinical use for other indications and exhibit additional inhibitory properties.

Because of their pharmacological activity and due to the high serum concentration that is required to inhibit efflux transporters, these first-generation inhibitors cannot be used as auxiliary agents in clinical applications to enhance oral bioavailability of efflux pump substrates. However, they have been used to demonstrate the feasibility of improving oral bioavailability mediated by efflux pump inhibition or generally to investigate efflux pump mechanisms. Coadministration of the antiarryhthmic drug quinidine increased the absorption of the P-gp substrate and anticancer drug etoposide in everted gut sacs prepared from rat jejunum and ileum (Leu and Huang 1995). Malingre et al. showed a strong enhancement of oral docetaxel bioavailability in the presence of cyclosporine (2001). In another study it has been shown that the oral bioavailability of tacrolimus (Floren et al. 2001) was doubled when coadministrating the inhibitor ketoconazole and in a further study digoxin bioavailability was improved in vivo, in presence with atorvastatin (Lennernas 2003). However, auxiliary agents in drug delivery systems should not only improve oral bioavailability but also be specific and safe (Werle 2008a).

Therefore much effort has been put into the development of second- and third- generation inhibitors.

Second-generation inhibitors display no or only low pharmacological activity in comparison to first-generation inhibitors. Moreover, they have enhanced inhibitory properties. Representatives of second-generation inhibitors are, for example, the cyclosporine analogue PSC833 (Twentyman and Bleehen 1991) or biricodar (Germann et al. 1997). Third-generation modulators are the most potent and selective inhibitors. KR30031, a verapamil analogue with fewer cardiovascular effects, improved paclitaxel bioavailability in rats 7.5-fold after oral dosing (Woo et al. 2003) (Fig. 7.2). In other studies, the oral bioavailability of paclitaxel has been improved by coadministering MS-209 (Kimura et al. 2002), GF120918 (Bardelmeijer et al. 2000) and SDZ PSC833 (van Asperen

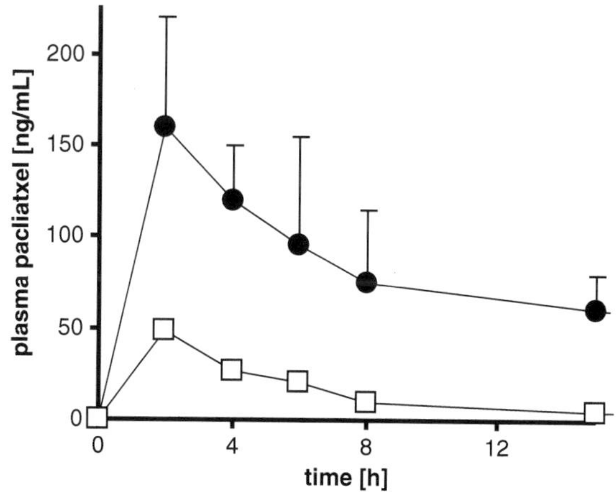

Fig. 7.2 Plasma concentration curves of paclitaxel in rats after oral administration of paclitaxel (25 mg/kg); -□- control, or in presence of 10 mg/kg -•- KR30031. Figure adapted from Woo et al. (2003)

et al. 1997). The third-generation inhibitor OC144-093 (ONT-093) has been reported to improve the oral uptake of docetaxel (Kuppens et al. 2005).

7.2.1.2 Polymeric Inhibitors and Surfactants

Besides low molecular mass inhibitors, it has also been shown that several polymeric compounds exhibit efflux pump inhibitory properties. In contrast to low molecular weight P-gp inhibitors, polymeric inhibitors would offer the advantage of remaining more concentrated in the GI tract at the site of drug absorption and of not being absorbed per se thus avoiding systemic toxic side effects (Föger et al. 2006c). It has to be taken into consideration that inhibition of efflux proteins can lead to drug–drug interactions, because P-glycoprotein, for example, is distributed within several tissues and organs implicated in the excretion and absorption of therapeutic agents as well as of xenobiotics (Balayssac et al. 2005). P-gp is, for example, linked to the integrity of blood–tissue barriers, such as the blood–brain barrier or placenta, and a partial blockage of P-gp could be responsible for a new drug distribution in the organism with possible increase of drug rates in organs behind these barriers. Therefore, concomitant administration of substrates and P-gp inhibitors would modify drug pharmacokinetics by increasing bioavailability and organ uptake, leading to more adverse drug reactions and toxicities (Balayssac et al. 2005). Consequently, the identification and comprehension of these drug–drug interactions remain important keys to risk assessment.

Limited local inhibition of these transporters by using polymeric inhibitors could reduce such systemic risks and side effects. Efflux pump modulating polymers and surfactants mentioned in this review have been distinguished between nonionic, ionic and thiolated polymers.

Nonionic Polymers

Polyethylene Glycol (PEG) and PEGylation

In a study by Johnson et al. it has been shown that concentrations of 1–20% of PEG 400 significantly decreased the basolateral to apical transport of digoxin through rat jejunal mucosa, indicating efflux pump inhibition (Johnson et al. 2002). In another study the effect of PEG 400, 2,000 and 20,000 on efflux pumps has been investigated (Shen et al. 2006). They showed in experiments with isolated rat intestine that the secretory transport of rhodamine 123 was inhibited by the addition of different concentrations (0.1–20% v/v or w/v) of PEGs, irrespective of their molecular weight. Additionally, they demonstrated in in situ closed loop studies that the absorption of rhodamine 123 was improved when formulated in solutions containing different concentrations of PEG 20,000. Hugger et al. demonstrated that the permeation of efflux pump substrates such as doxorubicin and paclitaxel through Caco-2 monolayers was improved in the presence of PEG 300. A concentration-dependent effect of

PEG 300 on paclitaxel transport was observed. It was concluded that the efflux pump inhibition mechanism of PEG 300 was mediated by changes in the microenvironment of Caco-2 cell membranes (Hugger et al. 2002). These changes are probably related to modifications in the fluidity of the polar head group regions of cell membranes.

Besides using PEG as an excipient, the covalent attachment of PEG (PEGylation) has been reported to improve the bioavailability of P-glycoprotein substrates. An increased uptake of PEGylated paclitaxel in comparison to unmodified paclitaxel after oral administration was observed. It was concluded that the water-soluble PEGylated prodrug was able to partly bypass P-glycoprotein efflux and CYP3A metabolism, which might explain the significantly improved absorption (Choi and Jo 2004).

D-Alpha-Tocopheryl Poly(Ethylene Glycol) Succinate 1000 (TPGS 1000)

Recently, efflux pump modulating effects of TPGS 1000 have been reported. Varma and Panchnagula for instance demonstrated that TPGS 1000 can improve the oral bioavailability of the P-gp substrate paclitaxel and that this effect is mediated by an improved drug solubility and P-glycoprotein inhibition. A formulation containing 25 mg/kg paclitaxel and 50 mg/kg TPGS 1000 improved the oral bioavailability in rats about 6-fold, leading to an oral bioavailability of about 30% (Varma and Panchagnula 2005). Collnot et al. investigated the influence of the length of the alkyl chain of various TPGS derivatives on their efflux pump inhibitory activity (Collnot et al. 2006). Results of ten different TPGS derivatives ranging from TPGS 200 to 6000 revealed that the commercially available derivative TPGS 1000 was the most potent efflux pump inhibitor.

Polysorbates (Tween)

The polysorbates used most regarding efflux pump inhibition are polyoxyethylene sorbitan monolaurates (Tween 20), polyoxyethylene sorbitan monopalmitates (Tween 40) and polyoxyethylene sorbitan monooleates (Tween 80). Various studies demonstrate the ability of polysorbates to inhibit efflux pumps. In transport experiments across intestinal mucosa, the efflux ratio (basolateral to apical drug transport/apical to basolateral drug transport) of rhodamine 123 was reduced in the presence of Tween 80 (Shono et al. 2004). In another study, Zhang et al. demonstrated enhanced absorption of the P-glycoprotein substrate digoxin in rats in the presence of Tween 80 (Zhang et al. 2003).

PEG-8 Glyceryl Caprylate/Caprate (Labrasol)

Recently an effect of PEG glyceryl fatty acid esters such as PEG-8 glyceryl caprylate/caprate (commercially known as Labrasol) on the transport of a P-glycoprotein substrate has been reported. In in situ absorption studies, Labrasol (0.1% (v/v)) significantly enhanced the intestinal absorption of rhodamine

123 in rats, although the absorption enhancing effect of Labrasol was much less than that of verapamil (Lin et al. 2007). These findings show that low concentrations of Labrasol enhance intestinal absorption and bioavailability of P-gp substrates.

POE Stearates (Myrj) and Alkyl-PEO Surfactants (Brij)

Some studies reported that polyoxyethylene (POE) stearates (under the brand name Myrj) and alkyl-polyethyleneoxide (PEO) surfactants (under the brand name Brij) can inhibit efflux pumps. The oral bioavailability of the P-gp substrate cyclosporine A administered in a solid dispersion of polyoxyethylene 40 stearate (Myrj 52) was in the same range as the oral bioavailability of the commercial product Sandimmune Neoral (Liu et al. 2006). In a study by Lo, it has been shown that apical to basolateral epirubicin transport across Caco-2 cells was enhanced in the presence of polyoxyethylene 40 stearate and the basolateral to apical transport was decreased. These results indicate that polyoxyethylene stearates effect efflux pumps (Lo 2003). Similar results were gained when using polyoxyethylene laurylether (Brij 30). In another study, tablets based on polyoxyethylene 40 stearate containing the P-gp substrate rhodamine 123 increased the oral bioavailability in rats by about 2.4-fold (Föger et al. 2006a).

Poloxamers (Pluronics)

Block copolymers of ethylene oxide (EO) and propylene oxide (PO) have been reported many times to inhibit efflux transporters. Johnson et al. for instance demonstrated significantly reduced digoxin flux in the presence of 0.1% (w/v) Pluronic P85. Poloxamers appear to exert their effect on P-glycoprotein via direct or indirect transporter inhibition, through effects on the membrane fluidity, adenosine triphosphate (ATP) depletion or through effects on the osmolarity (Johnson et al. 2002). In a further study, Batrakova reported that Pluronic P85 at concentrations below the critical micelle concentration enhanced the accumulation of rhodamine 123 in Caco-2 cells (Batrakova et al. 1998). However, Bogman et al. showed that Poloxamer 188 did not significantly alter talinolol absorption across Caco-2 cells or in vivo in healthy male volunteers (Bogman et al. 2005).

Ionic Polymers

Some naturally occurring polymers have been reported to exhibit efflux pump modulating properties. For example, a drug delivery system based on chitosan has been shown to nearly double the oral bioavailability of the P-glycoprotein substrate rhodamine 123 in vivo in rats in comparison with buffer control (Föger et al. 2006c).

In a patent by Carreno-Gomez and Duncan, the use of polysaccharides and dendrimers as efflux pump inhibitors for the oral delivery of antitumour, antineoplastic, antibiotic, antiviral, antifungal and antidepressant drugs has been claimed (Carreno-Gomez and Duncan 2002).

In experiments in everted gut sacs it has been reported that in the presence of xanthan gum the accumulation of the P-gp substrates vinblastin and doxorubicin in the gut cells was increased. An enhanced serosal transport of vinblastin but not of doxorubicin was observed (Carreno-Gomez and Duncan 2002). Also the potential efflux pump inhibitory activity of gellan gum was investigated. It has been demonstrated that at a concentration of 0.5 mg/ml gellan gum the serosal transport of vinblastin was improved, whereas the tissue level remained unchanged. At the same concentration, gellan gum improved the accumulation and the serosal transport of doxorubicin (Carreno-Gomez and Duncan 2002).

In another experiment, the alginate representatives flavicam and ascophyllum have been evaluated regarding their efflux pump modulation. In studies with everted gut sac cells, 0.5 mg/ml flavicam increased the accumulation of doxorubicin in the cells as well as the serosal transport of the drug, whereas no effect on vinblastin accumulation could be observed (Carreno-Gomez and Duncan 2002). Ascophyllum, at a concentration of 0.5 mg/ml, increased vinblastin and doxorubicin accumulation in everted gut sac cells. Furthermore the serosal transport of vinblastin was enhanced, whereas the serosal transport of doxorubicin could not be improved. Moreover, the effect of 250 mg/kg ascophyllum on the biodistribution of radioactive-labelled vinblastin after oral gavage in rats was investigated. The vinblastin blood level increased 1.7-fold in comparison to the control (Carreno-Gomez and Duncan 2002).

Thiolated Polymers

Recently, it was shown that thiolated chitosan significantly reduced secretion of P-gp substrate rhodamine 123 in guinea pig ileum (Werle and Hoffer 2006). Furthermore, it has been shown that a delivery system based on Ch-TBA enhanced rhodamine 123 absorption in vivo in rats (Föger et al. 2006c). It has been shown that oral administration of tablets based on a thiolated chitosan and reduced glutathione led to significantly improved plasma levels of the P-gp substrate rhodamine 123 in rats in comparison to tablets based on either poloxamer or Myrj 52 (Föger et al. 2006a). The results are shown in Fig. 7.3. Improved uptake of the efflux pump substrates, saquinavir or acyclovir, in the presence of a thiomer/GSH system has also already been demonstrated (Föger et al. 2006; Palmberger et al. 2008b). In another study an improved uptake of paclitaxel after oral administration in rats by using a thiolated polycarbophil-based delivery system has been demonstrated. Furthermore, it has been shown that this delivery concept reduced tumour growth in breast cancer-induced rats in comparison to vehicle control (Föger et al. 2008).

Moreover, it has been demonstrated that the transepithelial transport of sulphorhodamine representing an MRP2 substrate in the absorptive direction was improved up to about 4.5-fold in the presence of 0.5% (m/v) thiolated

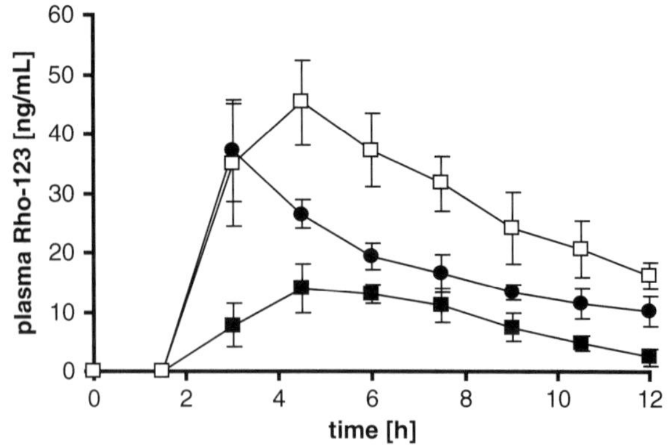

Fig. 7.3 Plasma curves of rhodamine 123 [ng/ml] after oral administration in Pluronic P85-tablets (■), Myrj 52-tablets (•) and Ch-TBA/GSH-tablets (□) to rats. Figure adapted from Föger et al. (2006a)

poly(acrylic acid) of various molecular mass (Grabovac and Bernkop-Schnürch 2006). These findings seem to indicate that thiomers might be useful tools to deliver various drugs which are affected by efflux transporters such as P-gp and MRP. One of the mechanisms responsible for P-gp inhibition by thiolated polymers could be due to their ability to covalently interact with cysteine residues of the gut mucosa which could lead to a disturbed lipid bilayer. Another explanation for their strong in vivo performance could be the prolonged residence time of these drug delivery systems in the upper part of the small intestine due to mucoadhesive properties of thiomers (Föger et al. 2006c).

7.2.2 Prodrug Modification

Another interesting strategy is to circumvent efflux pumps by prodrug modification. Jain et al. showed that prodrug derivatization of the HIV-protease inhibitor saquinavir can result in an increased permeability across P-gp overexpressing cells (Jain et al. 2007). In another study by using MDCKII-MDRI cell lines, they showed that prodrug derivatization of quinidine into val-quinidine can overcome P-gp-mediated efflux (Jain et al. 2004). Val-quinidine once bound to a peptide or amino acid transporter is probably not recognized and cannot be accessed by the P-gp efflux pump. Such a transporter-targeted prodrug derivatization seems to be a viable strategy for overcoming P-gp-mediated efflux.

This strategy is very interesting because prodrug modification would avoid the need of additionally administered efflux pump inhibitors.

7.2.3 Antisense Targeting of Efflux Pumps

Another approach to modulate drug efflux could be in the treatment of drug-resistant cells with MDR1-targeted siRNAs. Such a treatment resulted in reduction of P-glycoprotein (P-gp) expression, parallel reduction in MDR1 message levels, increased accumulation of the P-gp substrate rhodamine 123 and reduced resistance to antitumour drugs (Fisher et al. 2007).

7.2.4 Avoiding Exposure to Intestinal Efflux Pumps

7.2.4.1 Enhanced Paracellular Transport

Bypassing intestinal transmembrane transporters mainly by a paracellular absorption would avoid or limit exposure of the substrate to these efflux pumps. Improved paracellular uptake can be achieved by using fatty acids, calcium chelators such as EDTA, papain, bromelain, surfactants, chitosans, polyacrylic acid or thiolated polymers.

7.2.5 Absorption in the Upper Part of the Small Intestine

Because P-gp expression was reported to increase from the proximal to the distal regions of the small intestine (Yumoto et al. 1999), Lacombe et al. suggested gastroretentive or immediate release dosage forms for the delivery of P-gp substrates (Lacombe et al. 2004). It has been shown, for example, in everted gut sacs of rats that the transport of digoxin is about 2.5-fold higher in the proximal jejunum in comparison to the terminal ileum.

Therefore, increased bioavailability could be reached by using bioadhesive drug delivery systems, releasing the drug in the upper part of the small intestine where efflux pump activity is lower.

7.3 Conclusions

In summary it can be said that there are a lot of possibilities available to overcome the efflux pump-mediated absorption barrier in the intestinal tract. Further, more selective or more potent inhibitors will follow but it has to be carefully decided for each drug or therapy which type or class of inhibitor or efflux pump modulator might be best suited. Also drug delivery systems combining different efflux pump modulating properties have to be investigated in the future.

References

Al-Shawi M.K., Urbatsch I.L., Senior A.E. (1994) Covalent inhibitors of P-glycoprotein ATPase activity, *J Biol Chem* **269**(12):8986–8992.

Balayssac D., Authier N., Cayre A., Coudore F. (2005) Does inhibition of P-glycoprotein lead to drug-drug interactions? *Toxicol Lett*, **156**(3):319–29.

Bardelmeijer H.A., Beijnen J.H., Brouwer K.R., Rosing H., Nooijen W.J., Schellens J.H., van Tellingen O. (2000). Increased oral bioavailability of paclitaxel by GF120918 in mice through selective modulation of P-glycoprotein. *Clin Cancer Res,* **6**:4416–4421.

Batrakova E.V., Han H.Y., Alakhov V.Y., Miller D.W., Kabanov A.V. (1998) Effects of pluronic block copolymers on drug absorption in Caco-2 cell monolayers. *Pharm Res,* **15**(6):850–855.

Bogman K., Zysset Y., Degen L., Hopfgartner G., Gutmann H., Alsenz J., Drewe J. (2005) P-glycoprotein and surfactants: effect on intestinal talinolol absorption. *Clin Pharmacol Ther,* **77**(1):24–32.

Carreno-Gomez B. and Duncan R. Compositions with enhanced oral bioavailability. US patent, 2002.

Choi J.S. and Jo B.W. (2004) Enhanced paclitaxel bioavailability after oral administration of pegylated paclitaxel prodrug for oral delivery in rats. *Int J Pharm,* **280**:221–722.

Collnot E.M., Baldes C., Wempe M.F., Hyatt J., Navarro L., Edgar K.J., Schaefer U.F., Lehr C.M. (2006) Influence of vitamin E TPGS poly(ethlyene glycol) chain length on apical efflux transporters in Caco-2 cell monolayers. *J Control Release,* **111**:35–40.

Fisher M., Abramov M., Van Aerschot A., Xu D., Juliano R.L., Herdewijn P. (2007) Inhibition of MDR1 expression with altritol-modified siRNAs. Nucleic Acids Res, **35**(4):1064–74.

Floren L.C., Bekersky I., Benet L.Z., Mekki Q., Dressler D., Lee J.W., Roberts J.P., Hebert M.F. (2001) Tacrolimus oral bioavailability doubles with coadministration of ketoconazole. *Clin Pharmacol Ther,* **62**:41–49.

Föger F., Hoyer H., Kafedjiiski K., Thaurer M., Bernkop-Schnürch A. (2006a) In vivo comparison of various polymeric and low molecular mass inhibitors of intestinal P-glycoprotein. *Biomaterials*, **27**, (34), 5855–5860.

Föger F., Kafedjiiski K., Hoyer H., Loretz B., Bernkop-Schnürch A. (2006b) Enhanced transport of P-glycoprotein substrate saquinavir in presence of thiolated chitosan. *J Drug Targ*, **15**, (2), 132–139.

Föger, F., Schmitz T. and Bernkop-Schnürch A. (2006c) In vivo evaluation of an oral delivery system for P-gp substrates based on thiolated chitosan. *Biomaterials*, **27**, (23), 4250–4255.

Föger F., Malaivijitnond S., Wannaprasert T., Huck C., Bernkop-Schnürch A., Werle M. (2008) Effect of oral paclitaxel in presence of a thiolated polymer on absorption and tumor growth in rats. *J Drug Targeting,* **16**, (2), 149–55.

Germann U.A., Shlyakhter D., Mason V.S., Zelle R.E., Duffy J.P., Galullo V., Armistead D. M., Saunders J.O., Boger J., Harding M.W. (1997) Cellular and biochemical characterization of VX-710 as a chemosensitizer: Reversal of P-glycoprotein-mediated multidrug resistance in vitro. *Anticancer Drugs,* **8**:125–140.

Gerrard G., Payne E., Baker R.J., Jones D.T., Potter M., Prentice H.G. (2004) Clinical effects and P-glycoprotein inhibition in patients with acute myeloid leukaemia treated with zosuquidar trihydrochloride, daunorubicin and cytarabine, *Haematologica* **89**:782–790.

Grabovac V., Bernkop-Schnürch A. (2006) Thiolated polymers as effective inhibitors of intestinal Mrp2 efflux pump transporters. *Sci Pharm,* **74**.

Hugger E.D., Audus K.L., and Borchardt R.T. (2002) Effects of poly(ethylene glycol) on efflux transporter activity in Caco-2 cell monolayers. *J Pharm Sci,* **91**:1980–1990.

Hunter J., Hirst B.H. (1997) Intestinal secretion of drugs. The role of P-glycoprotein and related drug efflux systems in limiting oral drug absorption, *Adv Drug Deliv Rev* **25**:129–157.

Jain R., Majumdar S., Nashed Y., Pal D., Mitra A.K. (2004) Circumventing P-glycoprotein-mediated cellular efflux of quinidine by prodrug derivatization. *Mol Pharm,* **1**(4):290–9.

Jain R., Duvvuri S., Kansara V., Mandava N.K., Mitra A.K. (2007) Intestinal absorption of novel-dipeptide prodrugs of saquinavir in rats. *Int J Pharm,* **336**: 233–240.

Johnson B.M., Charman W.N., Porter C.J.H. (2002) An in vitro examination of the impact of polyehtylene glycol 400, pluronic P85 and vitamin E D-a-tocopheryl polyethylene glycol

1000 succinate on p-glycoprotein efflux and enterocyte-based metabolism in excised rat intestine. *AAPS PharmSci*, **4**: 1–13.

Kim R.B. (2002) Drugs as P-glycoprotein substrates, inhibitors, and inducers, *Drug Metab Rev*, **34**:47–54.

Kimura Y., Aoki J., Kohno M., Ooka H., Tsuruo T., Nakanishi O. (2002) P-glycoprotein inhibition by the multidrug resistance-reversing agent MS-209 enhances bioavailability and antitumor efficacy of orally administered paclitaxel. *Cancer Chemother Pharmacol*, **49**:322–328.

Kuppens I.E., Bosch T.M., van Maanen M.J., Rosing H., Fitzpatrick A., Beijnen J.H., Schellens J.H. (2005) Oral bioavailability of docetaxel in combination with OC144-093 (ONT-093). *Cancer Chemother Pharmacol*, **55**:72–78.

Lacombe O., Woodley J., Solleux C., Delbos J.M., Boursier-Neyret C., Houin G. (2004) Localisation of drug permeability along the rat small intestine, using markers of the paracellular, transcellular and some transporter routes. *Eur J Pharm Sci*, **23**:385–391.

Lennernas H. (2003) Clinical pharmacokinetics of atorvastatin. *Clin Pharmacokinet*, **42**:1141–1160.

Leu B.L., Huang J.D. (1995) Inhibition of intestinal Pglycoprotein and effects on etoposide absorption. *Cancer Chemother Pharmacol*, **35**:432–436.

Lin Y., Shen Q., Katsumi H., Okada N., Fujita T., Jiang X., Yamamoto A. (2007) Effects of Labrasol and Other Pharmaceutical Excipients on the Intestinal Transport and Absorption of Rhodamine123, a P-Glycoprotein Substrate, in Rats. *Biol Pharm Bull*, **30**(7) 1301–1307.

Liu C., Wu J., Shi B., Zhang Y., Gao T., Pei Y. (2006) Enhancing the bioavailability of cyclosporine a using solid dispersion containing polyoxyethylene (40) stearate. *Drug Dev Ind Pharm*, **32**:115–123.

Lo Y.L. (2003) Relationships between the hydrophilic–lipophilic balance values of pharmaceutical excipients and their multidrug resistance modulating effect in Caco-2 cells and rat intestines. *J Control Release*, **90**:37–48.

Malingre M.M., Ten Bokkel Huinink W.W., Mackay M., Schellens J.H., Beijnen J.H. (2001) Pharmacokinetics of oral cyclosporin A when co-administered to enhance the absorption of orally administered docetaxel. *Eur J Clin Pharmacol*, **57**:305–307.

Palmberger T.F., Hombach J., Bernkop-Schnürch A. (2008) Thiolated chitosan: Development and in vitro evaluation of an oral delivery system for acyclovir. *Int J Pharm*, **348**(1–2):79–85.

Shen Q., Lin Y., Handa T., Doi M., Sugie M., Wakayama K., Okada N., Fujita T., Yamamoto A. (2006) Modulation of intestinal P-glycoprotein function by polyethylene glycols and their derivatives by in vitro transport and in situ absorption studies. *Int J Pharm*, **313**:49–56.

Shono Y., Nishihara H., Matsuda Y., Furukawa S., Okada N., Fujita T., Yamamoto A. (2004) Modulation of intestinal P- glycoprotein function by cremophor EL and other surfactants by an in vitro diffusion chamber method using the isolated rat intestinal membranes. *J Pharm Sci*, **93**:877–885.

Twentyman P.R., Bleehen N.M. (1991) Resistance modification by PSC-833, a novel non-immunosuppressive cyclosporin. *Eur J Cancer*, **27**:1639–1642.

van Asperen J., van Tellingen O., Sparreboom A., Schinkel A.H., Borst P., Nooijen W.J., Beijnen J.H. (1997) Enhanced oral bioavailability of paclitaxel in mice treated with the P-glycoprotein blocker SDZ PSC 833. *Br J Cancer*, **76**:1181–1183.

Varma M.V., Ashokraj Y., Dey C.S., Panchagnula R. (2003) P-glycoprotein inhibitors and their screening: A perspective from bioavailability enhancement. *Pharmacol Res*, **48**:347–359.

Varma M.V., Panchagnula R. (2005) Enhanced oral paclitaxel absorption with vitamin E-TPGS: effect on solubility and permeability in vitro, in situ and in vivo. *Eur J Pharm Sci*, **25**:445–453.

Varma M.V., Perumal O.P., Panchagnula R. (2006) Functional role of P-glycoprotein in limiting peroral drug absorption: optimizing drug delivery. *Curr Opin Chem Biol,* **10**:367–373.

Werle M., Hoffer M. (2006) Glutathione and thiolated chitosan inhibit multidrug resistance P-glycoprotein activity in excised small intestine. *J Control Rel,* **111**, (1–2), 41–46.

Werle M. (2008a) Polymeric and low molecular mass efflux pump inhibitors for oral drug delivery. *J Pharm Sci,* **97**, (1), 60–70.

Werle M. (2008b) Natural and synthetic polymers as inhibitors of drug efflux pumps. *Pharm Res,* **25**, (3), 500–11.

Woo J.S., Lee C.H., Shim C.K., Hwang S.J. (2003) Enhanced oral bioavailability of paclitaxel by coadministration of the P-glycoprotein inhibitor KR30031. *Pharm Res,* **20**:24–30.

Yumoto R., Murakami T., Nakamoto Y., Hasegawa R., Nagai J., Takano M. (1999) Transport of rhodamine 123, a P-glycoprotein substrate, across rat intestine and Caco-2 cell monolayers in the presence of cytochrome P-450 3A-related compounds. *J Pharmacol Exp Ther,* **289**:149–155.

Zhang H., Yao M., Morrison R.A., Chong S. (2003) Commonly used surfactant, Tween 80, improves absorption of P-glycoprotein substrate, digoxin, in rats. *Arch Pharm Res,* **26**:768–772.

Chapter 8
Multifunctional Polymeric Excipients in Oral Macromolecular Drug Delivery

Claudia Vigl

Contents

Abstract More recently polymers exhibiting multifunctional properties such as mucoadhesive, enzyme inhibitory, permeation-enhancing properties and/or high buffer capacity turned out to be a powerful platform for oral delivery of macromolecular drugs. Several polymers are known to exhibit multifunctional properties such as chitosans, polyacrylates and cellulose derivatives. Chemical modification of these well-established polymers including the attachment of enzyme inhibitors, chelating agents or thiol moieties leads to further improvement or enlargement of their multifunctional profile. Delivery system based on multifunctional polymers can protect the incorporated drug from pH- and enzyme-dependent degradation down its way through the GI tract and can provide high drug concentrations at the target site and tight contact with the absorption membrane due to mucoadhesion. Furthermore, sustained or delayed release due to cohesive properties of the polymeric carrier system offers new possibilities for targeted macromolecular drug delivery.

C. Vigl (✉)
Thiomatrix Forschungs Beratungs GmbH, Trientlgasse 65, 6020 Innsbruck, Austria
e-mail: c.vigl@thiomatrix.com

A. Bernkop-Schnürch (ed.), *Oral Delivery of Macromolecular Drugs*,
DOI 10.1007/978-1-4419-0200-9_8,

8.1 Introduction

Most macromolecular drugs on the market have to be administered via the parenteral route, which is associated with pain, fear and inconvenience. Although alternative routes of application such as nasal or pulmonary can already offer some relief, the oral route is still favourable as it represents the easiest way of drug administration. Unfortunately, those drugs indicated for long-time treatment of chronic diseases such as diabetes or cancer exhibit poor oral bioavailability due to various barriers encountered with the gastrointestinal tract: the acidic and denaturizing environment in the stomach, the enzymatic barrier represented by proteases and peptidases (see Chapter 1), the mucus gel layer covering the absorptive tissues (see Chapter 2) and the absorption barrier itself (see Chapter 3). Strategies to overcome these barriers include the co-administration of auxiliary agents such as enzyme inhibitors, mucolytics, permeation enhancers or formulations being based on micro-, nanoparticulate or liposomal systems. More recently polymers exhibiting multifunctional properties such as mucoadhesiveness, enzyme inhibition, permeation enhancement release-controlling properties and/or high buffer capacity turned out to be useful excipients in oral macromolecular drug delivery. Multifunctional polymers are able to counteract most of the obstacles mentioned above at once, providing a powerful platform for oral delivery of macromolecular drugs. Several polymers are known to exhibit multifunctional properties such as chitosans, polyacrylates and cellulose derivatives. Chemical modification of these well-established polymers, for example, by the attachment of enzyme inhibitors, chelating compounds or thiol moieties, leads to further improvement or enlargement of their multifunctional profile. Improved mucoadhesion and permeation enhancement, for instance, can be achieved by covalent attachment of thiol group-bearing ligands to polymers or protection towards enzymatic degradation can be achieved or improved by conjugation of enzyme inhibitors. A summary of the most important multifunctional polymers and their

Table 8.1 Examples of multifunctional polymers and their properties

Multifunctional polymer	Properties	References
Chitosan	Mucoadhesive, permeation enhancing, high buffer capacity	Takeuchi et al. (1994) and Luessen et al. (1996a)
Polyacrylic acid	Mucoadhesive, permeation enhancing, enzyme inhibiting, high buffer capacity	Takeuchi et al. (1994), Bernkop-Schnürch and Gilge (2000), Luessen et al. (1996a) and Luessen et al. (1996b)
Thiomers	Mucoadhesive, permeation enhancing, enzyme inhibiting, efflux pump inhibiting	Bernkop-Schnürch et al. (2004), Werle (2008) and Palmberger et al. (2008)

Table 8.1 (continued)

Multifunctional polymer	Properties	References
Chitosan–inhibitor conjugates	Mucoadhesive, permeation enhancing, enzyme inhibiting	Guggi and Bernkop-Schnürch (2003), Guggi et al. (2003) and Bernkop-Schnürch (1999)
Polyacrylic acid–inhibitor conjugates	Mucoadhesive, permeation enhancing, enzyme inhibiting	Bernkop-Schnürch and Marschütz (1997)

properties is given in Table 8.1. This chapter is not intended to be comprehensive. Rather the focus is to address the likely most promising multifunctional polymers.

8.2 Multifunctional Polymers

8.2.1 Mucoadhesive Polymers

The mucus gel layer consists mainly of water (95–99%) and mucins (1–5%), the key components. Mucins are macromolecular glycoproteins with substructures linked by disulphide bonds susceptible to reductive disruption by thiols (Harding 2003). The mucus prevents peptides and other macromolecular drugs from getting into contact with the absorptive membrane lying underneath. Mucoadhesive polymers allow a close and prolonged contact of the delivery system with the mucosa and hence a high local drug concentration facilitating its permeation through the mucus and the absorption membrane due to a steep concentration gradient on the mucosa. Apart from increasing the active drug concentration at the site of absorption, the close contact of the dosage form with the mucosa can also contribute to protect the incorporated drug from luminally secreted degrading enzymes and to enhance drug uptake due to permeation-enhancing properties. In general, mucoadhesive polymers can

- increase residence time of dosage forms
- provide sustained drug release at target site
- provide high drug concentration gradient as driving force for absorption
- protect against enzymatic degradation
- interact with absorption membrane, thus enhancing permeation

Mucoadhesive polymers can be divided into non-covalent binding and covalent binding polymers. On the one hand the mechanism of mucoadhesion is based on hydrogen bonds, ionic interactions and van der Waal forces for non-covalent binding polymers, and on the other hand, on covalent bonds between the mucus and certain residues of the polymer. Moreover, physical interactions such as interpenetration of the polymer into the mucus gel layer entangle the polymer chains, which is strongly influenced by the swelling behaviour of the polymer. The faster the swelling is, the greater the interpenetration, but intense swelling will finally lead to loss in cohesion and shortened time of adhesion. Factors such as thickness of the mucosa, mucin types and mucus turnover also determine the efficiency of a mucoadhesive delivery system.

8.2.1.1 Non-covalent Binding Polymers

Anionic Mucoadhesive Polymers

The functional group of anionic polymers responsible for mucoadhesion is a carboxylic acid moiety. The interaction with the mucus takes place via the formation of hydrogen bonds with the mucus glycoproteins and polymers such as polyacrylates, sodium carboxymethylcellulose and alginate. The pH level of the surrounding medium strongly influences the swelling behaviour of those highly charged polymers as shown for polycarbophil in Fig. 8.1. At low pH swelling behaviour is low due to protonation of the carboxylic acid groups, whereas high pH can lead to overswelling and loss of mucoadhesive properties. Alginate is another anionic, naturally occurring, strong mucoadhesive polymer. Its pH sensitivity can be advantageous in order to minimize polymer swelling and thus drug release in the stomach. At low pH values, such as in the gastric fluid of the stomach, alginate shrinks and turns into a porous, insoluble alginic acid skin. Therefore, the encapsulated drug is not released. Passing along the gastrointestinal tract, the skin-like polymer becomes soluble and viscous, allowing a controlled release of the drug. Thiolation of alginate does not only improve mucoadhesion, but also swelling and cohesive properties, thus enhancing the stability of the network. Hydrophobic modification by introducing long alkyl chains leads to higher encapsulation rates, prevents fast dissolution and facilitates a sustained release. Cross-linking alginate or complexing it with

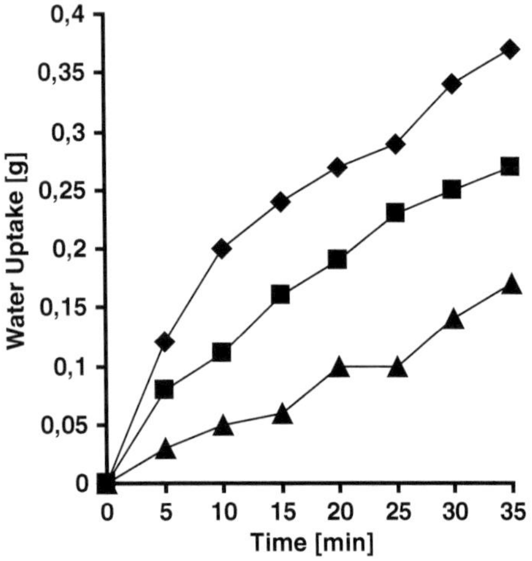

Fig. 8.1 Correlation between pH of anionic polymer polycarbophil and its swelling behaviour: pH 3 (▲), pH 5 (■), pH 7 (♦). Adapted from Bernkop-Schnürch and Steininger (2000)

other polymers, such as pectin, chitosan or Eudragits, increases the encapsulation efficiency by avoiding drug leaching out of the alginate hydrogel pores.

Cationic Mucoadhesive Polymers

Chitosan and its derivatives are the most widely used cationic polymeric excipients. Chitosan consists of β1→4 D-glucosamine units and is derived by the deacetylation of chitin from insects, crustaceans and fungi. It interacts ionically with the anionic substructures of sialic acid residues on the mucus layer. Chitosans are rapidly hydrated in a low pH environment like the gastric fluid and do not swell above pH levels of 6.5, exhibiting no more mucoadhesion.

Non-ionic Mucoadhesive Polymers

Non-ionic polymers are less dependent on parameters such as pH levels and electrolyte concentration of the surrounding fluids. The main mechanism of mucoadhesion seems to be just physical by interpenetration and subsequent chain entanglement. Some of the polymers such as polyethylene oxide can additionally form hydrogen bonds, but still play only a minor role in macromolecular drug delivery due to less pronounced mucoadhesive properties than the above-described charged polymers.

8.2.1.2 Covalent Binding Polymers

Among mucoadhesive polymers the class of thiolated polymers – so-called thiomers – have raised much attention as drug carrier systems. They are gained by the covalent attachment of thiol group-bearing ligands on well-established polymers such as chitosan and polyacrylates. Some examples of thiomers are given in Fig. 8.2. Like secreted mucus glycoproteins they are covalently anchored in the mucus layer via disulphide bonds formed with cysteine-rich domains of the mucus by thiol/disulphide exchange and/or oxidation reactions. Therefore, the mucoadhesion of thiomers is strongly improved in comparison to the corresponding non-modified polymers as can be seen in Fig. 8.3. Although they are readily hydrated they form highly cohesive gels without giving rise to the development of a liquid mucilage due to inter-/intramolecular disulphide bonds. This could be shown for various thiomers, for example, tablets of poly(acrylic acid)–cysteine showed prolonged mucoadhesion on porcine intestinal mucosa in vitro which significantly decreased after the addition of the disulphide bond breaker cysteine, suggesting cleavage of covalent bonds between thiomer and mucus, whereas adhesion of unmodified poly(acrylic acid) tablets was not influenced at all (Leitner et al. 2003). In order to guarantee sufficient mucoadhesion other factors have to be considered as well. Intra- and intermolecular disulphide bonds are also established within the polymer itself, thus leading to strengthened cohesiveness of the matrix. The cohesive properties of polymers play an important role as insufficient cohesion results in

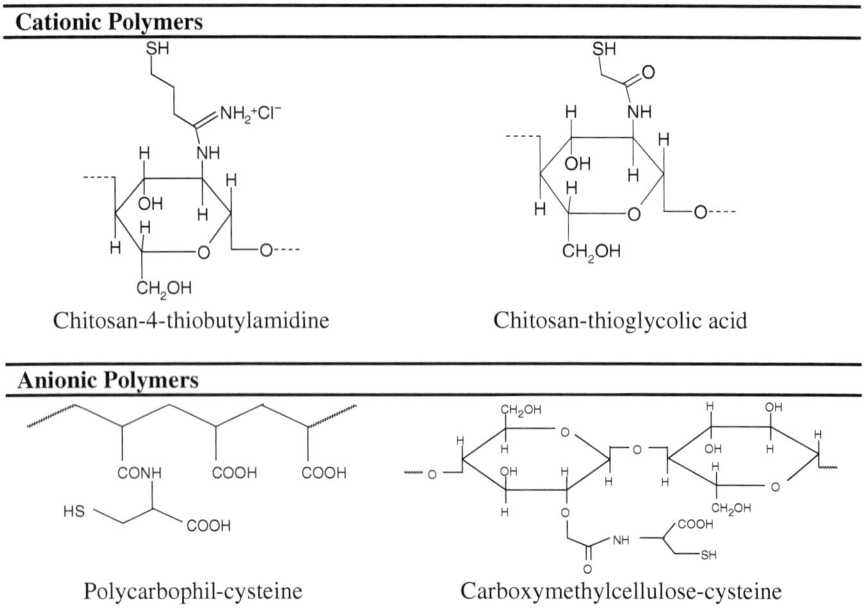

Fig. 8.2 Examples for thiolated polymers

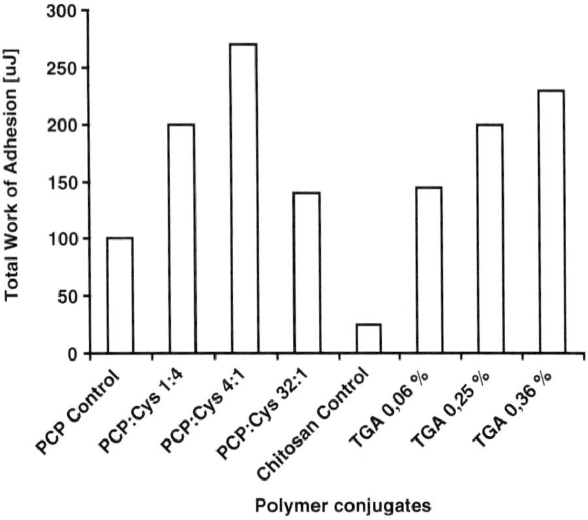

Fig. 8.3 Mucoadhesive properties of polycarbophil–cysteine (PCP:Cys) and chitosan–thioglycolic acid (TGA) in comparison to unmodified control (Bernkop-Schnürch et al. 1999; Kast and Bernkop-Schnürch 2001)

disintegration within the polymer network itself rather than between the polymer and the mucus layer. More recently the mucoadhesive properties of all polymeric excipients having been reported as mucoadhesive were evaluated via different methods (Grabovac et al. 2005). In Table 8.2 the ranking of the top ten

Table 8.2 Top ten most mucoadhesive polymers regarding their residence time on porcine mucosa; abbreviations: lyo = lyophilized polymer, pr = precipitated polymer (Grabovac et al. 2005)

Polymer	pH	Time of adhesion [h] mean ± SD (n = 3–5)
Chitosan–thiobutylamidine	pH 3 lyo.	161.2 ± 7.2
Chitosan–thiobutylamidine	pH 6.5 pr.	40.4 ± 2.1
Polycarbophil–cysteine	pH 3 lyo.	26.0 ± 0.9
Chitosan–thiobutylamidine	pH 6.5 lyo	20.4 ± 1.5
PAA450–cysteine	pH 3 lyo.	19.4 ± 0.8
Hydroxypropylcellulose	pH 7 pr.	15.2 ± 0.4
Carbopol 980	pH 7 pr	12.5 ± 0.9
Carbopol 974	pH 7 pr.	10.3 ± 0.9
Polycarbophil	pH 7 pr.	10.2 ± 0.8
Carbopol 980	pH 3 lyo	9.8 ± 0.2

most mucoadhesive polymers is listed demonstrating the advantage of thiomers over non-covalent binding mucoadhesive polymers.

8.2.2 Enzyme-Inhibiting Polymers

For macromolecular drugs being subject of an extensive presystemic metabolism the enzymatic barrier of the gastrointestinal tract is strongly limiting their oral bioavailability. Concepts to improve drug stability include formation of stable prodrugs due to chemical modification of the drug itself, co-administration of enzyme inhibitors and formulations providing protection towards an enzymatic attack including the use of multifunctional polymers. A list of enzyme-inhibiting multifunctional polymers is provided in Table 8.3. For

Table 8.3 Examples of various polymeric enzyme inhibitors

Enzyme	Polymeric inhibitor	References
Trypsin	Polyacrylic acid, thiomers, polymer–enzyme–inhibitor conjugates	Luessen et al. (1996b)
Chymotrypsin	Polymer–enzyme–inhibitor conjugates	Luessen et al. (1996b) and Bernkop-Schnürch and Thaler (2000)
Elastase	Polymer–enzyme–inhibitor conjugates	Guggi and Bernkop-Schnürch (2003) and Marschütz and Bernkop-Schnürch (2000)
Exopeptidases	Polyacrylic acid, thiomers, polymer–enzyme–inhibitor conjugates	Bernkop-Schnürch et al. (2006) and Luessen et al. (1996b)
Nucleases	Chitosan–aurintricarboxylic acid, chitosan–EDTA, thiomers	Loretz et al. (2006)

example, acrylic acid polymers inhibit a great number of luminal proteolytic enzymes. There is evidence that the mechanism of inhibition occurs by deprivation of metal ions required for protein structure and activity by the chelation of Zn^{2+} and Ca^{2+} ions. Due to conformational changes in the protein structure autolysis can be induced (Luessen et al. 1995). It has been shown that zinc-dependent enzymes such as carboxypeptidases A and B and aminopeptidase N can effectively be inhibited by chelating agents such as EDTA (Bernkop-Schnürch and Krajicek 1998). Additionally, it is assumed that polyacrylic acid polymers bind trypsin, thus leading to surface inactivation of the enzyme.

On the contrary, chitosan does not exhibit enzyme-inhibiting properties. In order to provide inhibitory activity chitosan can be conjugated with enzyme inhibitors such as Bowman-Birk inhibitor or elastinal or with complexing agents such as EDTA. The immobilization of EDTA on chitosan led to inactivation of Zn-dependent membrane-bound and secreted peptidases with stronger inhibitory effect on carboxypeptidase A and aminopeptidase N than acrylic acid polymers (Bernkop-Schnürch and Krajicek 1998). By the attachment of soybean trypsin inhibitor (Bownam-Birk) on a polyacrylate its trypsin- and chymotrypsin-inhibiting properties were strongly improved. However, a decrease in the mucoadhesive properties of these conjugates was shown (Bernkop-Schnürch 1999).

The use of thiomers for enzyme inhibition also proved to be a promising strategy. By the covalent attachment of cysteine to polycarbophil, the inhibitory effect of the polymer towards carboxypeptidase A, carboxypeptidase B and chymotrypsin was significantly improved (Bernkop-Schnürch and Thaler 2000). Thiolated polycarbophil had also a significantly greater inhibitory effect than unmodified polycarbophil on the activity of isolated aminopeptidase N and on aminopeptidase N present on intact intestinal mucosa (Bernkop-Schnürch et al. 2001). The strongly improved enzyme inhibitory properties of thiolated polycarbophil in comparison to unmodified polycarbophil can be explained by the inhibitory effect of L-cysteine itself towards carboxypeptidase A, carboxypeptidase B and aminopeptidase N due to the binding of the Zn^{2+} ion from the enzyme structure (Bernkop-Schnürch et al. 2001; Bernkop-Schnürch and Thaler 2000). In Fig. 8.4, the impact of a chitosan–inhibitor conjugate and thiolated chitosan on the decrease in the plasma calcium level is illustrated (Guggi et al. 2003).

Obviously, thiomers and polymer–inhibitor conjugates offer various advantages over traditional low molecular weight enzyme-inhibiting agents. They will not be absorbed from the gastrointestinal tract, as they are covalently bound to non-absorbable polymers and thus, they will not exhibit systemic toxicity. Due to mucoadhesive properties of the polymer the inhibitor will stay in contact with the absorption membrane for a prolonged period of time and the inhibitory effect will occur localized. In the case of enzyme inhibition via the complexation of essential metal ions, the polymer might act without coming into close contact with the target enzyme through diffusion mechanisms.

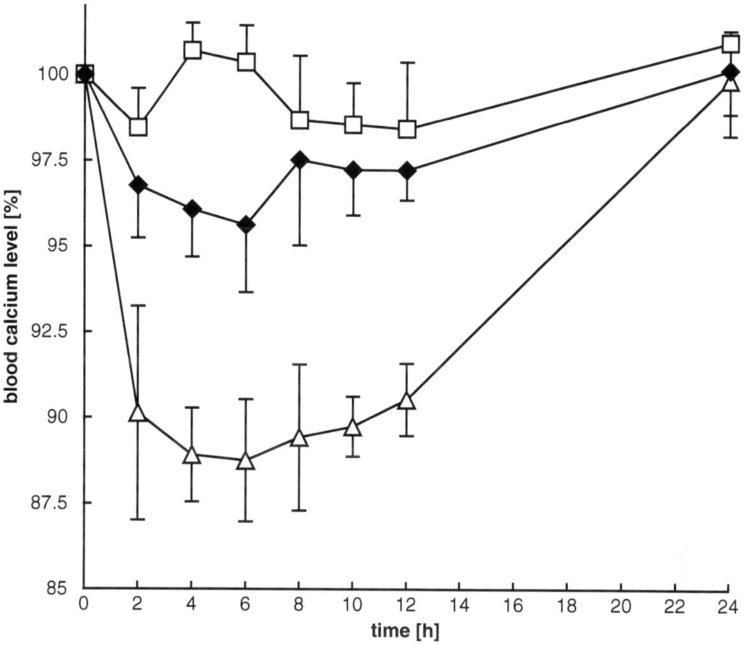

Fig. 8.4 Decrease in plasma calcium level as a biological response for the salmon calcitonin bioavailability in fasted rats after oral administration of chitosan minitablets (□), chitosan/chitosan–pepsin inhibitor conjugate minitablets (♦) and thiolated chitosan/chitosan–pepsin inhibitor conjugate minitablets (Δ), all containing 50 μg of the peptide drug. Indicated values are the mean results from five rats ± S.D. (Guggi et al. 2003)

Another feature of multifunctional polymers is their high buffer capacity. Generally anionic and cationic polymers can act as ion-exchange resins maintaining a stable pH level inside the polymeric network over a certain time period. This can contribute to the stability of the incorporated drug against pH-dependent denaturation and enzymatic degradation. Hydrated ionic polymers can provide a pH of approximately 7 even in the gastric fluid for several hours, thus protecting the embedded macromolecular drug from degradation. For example, tablets comprising a neutralized carbomer can buffer the pH inside the swollen carrier matrix for hours even in artificial gastric fluid at pH 2 (Bernkop-Schnürch and Gilge 2000). Protection against degradation in the intestine can be arranged by the use of polyacrylates in their acidic form. This pH prevents enzymes such as luminally secreted proteases which are only active above pH 4 from degrading the incorporated drug, as could be demonstrated for trypsin and chymotrypsin. Additionally, a significant blood glucose level reduction was observed after intraduodenal administration of insulin incorporated in the same polymer (Bai et al. 1996). Furthermore, aggregation of many enzymes can be reduced in low pH environment, as demonstrated for human calcitonin (Lu et al. 1999).

8.2.3 Permeation-Enhancing Polymers

The intestinal epithelium constitutes the major barrier for the absorption of orally administered peptides. The various uptake mechanisms are described in detail in Chapter 3. In order to reduce the barrier function of the mucosa the co-administration of permeation enhancers seems to be essential for oral delivery of most macromolecular drugs. In contrast to permeation enhancers such as surfactants, fatty acids, salicylates or chelating agents, multifunctional polymers offer the advantage of not being absorbed and remaining concentrated at the target site. Permeation studies across Caco-2 monolayers demonstrate a strong permeation-enhancing effect by chitosan and carbomer accompanied by a decrease in the transepithelial resistance, thus indicating loosening of the tight junctions (Borchard et al. 1996; Illum et al. 1994). Figure 8.5 shows a decrease in plasma glucose levels to 60% of the initial value after oral administration of tablets comprising PEGylated insulin and thiolated polyacrylate to diabetic mice. The underlying mechanism is not fully explained yet, but at least a mechanism being based on the depletion of Ca^{2+} ions can be excluded, as a chitosan–EDTA conjugate exhibiting a much higher Ca^{2+} ion binding capacity and affinity than polyacrylates did not show any permeation-enhancing properties at all (Bernkop-Schnürch and Krajicek 1998). Cationic polymers like chitosan might interact with negatively charged residues on the cell surface, resulting in conformational changes in the membrane structure and tight junction-associated proteins. In contrast, the attachment of thiol groups to mucoadhesive polymers can further improve their permeation-enhancing properties. Most likely, protein tyrosine phosphatase (PTP), which mediates the closing of tight junctions by dephosphorylating extracellular tyrosine groups, is inhibited

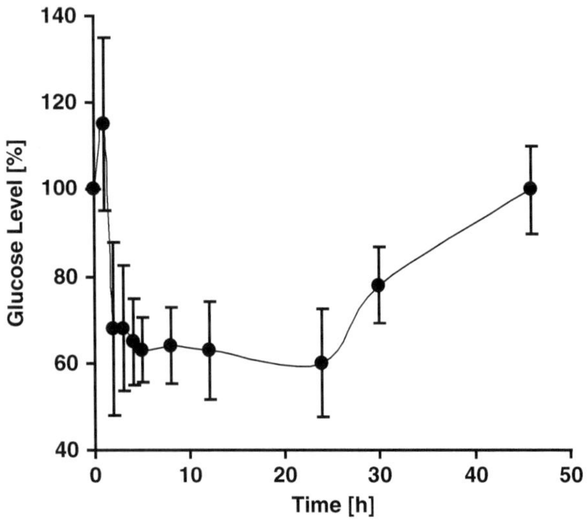

Fig. 8.5 Decrease in blood glucose level in diabetic mice after oral administration of thiolated polyacrylate minitablets comprising PEGylated insulin; indicated values are mean ± SD ($n = 10$) (Calceti et al. 2004)

by thiomers, as it is known that sulphydryl compounds like glutathione also inhibit this enzyme by covalent binding to cysteine residues (Clausen et al. 2002). Raising the cationic properties of chitosan such as trimethylation, however, did not further improve the permeation-enhancing properties of the polymer.

8.2.4 Efflux Pump-Inhibiting Polymers

Efflux pumps are considered to be another barrier to be dealt with in oral macromolecular drug delivery as most therapeutic peptides such as cyclosporine are substrate of efflux pumps. Their relevance and possibilities to overcome these restrictions are described in detail in Chapter 7. Among multifunctional polymers, thiomers exhibit the most pronounced efflux pump-inhibiting properties. This has been demonstrated in several studies in vitro and in vivo. A delivery system based on thiolated chitosan was evaluated both in vitro and in vivo using rhodamine-123 as representative P-gp substrate. The permeation-enhancing effect of chitosan-4 thiobutylamidine was found to be much higher than the effect of unmodified chitosan. This was also proven in vivo in rats, where tablets comprising the thiomer increased the area under the plasma concentration time curve of the model drug (Föger et al. 2006).

Further, the enhanced transport of the P-gp substrate acyclovir across excised rat intestinal mucosa and Caco-2 monolayers in the presence of thiolated chitosan was found to be due to efflux pump inhibition (Palmberger et al. 2008). Co-administration of paclitaxel and thiolated polycarbophil significantly improved paclitaxel plasma levels and led to a more constant pharmacokinetic profile and reduced tumour growth in mammary cancer-induced rats (Föger et al. 2008).

8.2.5 Polymers Providing Sustained/Delayed Release

8.2.5.1 Sustained Release

A sustained drug release is favourable for drugs with short elimination half-life. It can be controlled by hydration and diffusion mechanisms or ionic interactions between the drug and the polymeric carrier. In the case of diffusion control the stability of the carrier system is essential, as its disintegration leads to a burst release. Therefore, the cohesiveness of the polymer network plays a crucial role in order to control the release over several hours. Due to the formation of disulphide bonds within the network thiomers offer adequate cohesive stability. Almost zero-order release kinetics could be shown for insulin embedded in thiolated polycarbophil matrices (Clausen and Bernkop-Schnürch 2001). In the case of peptide and protein drugs release can be controlled via ionic interactions. An anionic or cationic polymer has to be chosen depending on the isoelectric point of the therapeutic agent. The polymer acts as an ion-exchange resin. Moreover, the release can be regulated by pH adjustment inside the polymeric carrier. A high buffer capacity of the system can ensure constant pH over a desired period of time. In most cases, the drug release is controlled by both hydration/diffusion and ionic mechanisms.

8.2.5.2 Delayed Release

According to several studies on site-specific delivery, it can be assumed that there are advantageous regions in the gastrointestinal tract for the absorption of macromolecular drugs. For example, leuprolide, an LH-RH analogue, is most favourably absorbed from ileum and colon in rats (Zheng et al. 1999), insulin preferably from ileum than from jejunum or colon in rats (Morishita et al. 1993). Therefore, a carrier system is required, which prevents drug release prior to reaching the target site. Many controlled-release systems are targeted to lower regions of the gastrointestinal tract where proteolytic activity is lower, thus circumventing the harsh acidic environment of the stomach and enzymatic degradation in the upper parts of the intestine. For this reason especially the colon has gained much interest. Strategies to address the colon include taking advantage of pH differences between the small and large intestine, enzymes of the colonic micro organisms, relatively short intestinal transit time and pressure increase due to strong colonic peristaltics. The small intestine can be reached through pH and time control. Systems acting via diffusion control can target the entire GI tract.

In order to overcome the drawback of multifunctional polymers of great swelling behaviour and subsequent disintegration the dosage form can be coated with a polymer insoluble at low pH. Commercially available Eudragit® products (methacrylic acid-methylmethacrylate co-polymers) guarantee dissolution at certain pH ranges depending on the type used. It was found that incorporation of Eudragit RL-100 into polycarbophil up to 20% led to reduced swelling of polycarbophil films, but improved mucoadhesion of Eudragit containing tablets (Tirosh et al. 1997). Polysaccharides such as chitosan have been shown to be an interesting tool for colonic delivery. Release of drug from such dosage forms takes place after degradation of polysaccharides due to cleavage by polysaccharidases found in the colon. As the primary amino groups of chitosan are rapidly hydrated under acidic conditions, it must be protected from the harsh environment in the stomach. The most frequent problem of polysaccharides – their relatively high water solubility – causing partial drug release in the upper GI tract can be overcome by the addition of cross-linking agents such as glutaraldehyde which renders them hydrophobic and less soluble. An obvious drawback of chemical modification of chitosan is the reduction or loss of its mucoadhesive properties and the incidence of toxicity for glutaraldehyde cross-linked chitosan (Carreno-Gomeza and Duncan 1997). Another class of enzymes which can be exploited for colon-targeting are azo reductases, which are present exclusively in the colon. So-called microbially degradable polymers or azo polymers can pass the hostile upper GI tract in intact form and liberate the drug upon reductive degradation by those enzymes (Chourasia and Jain 2003).

Another approach to reach delayed release from chitosan matrices are ionic interactions with polyanions. The formation of a polyelectrolyte complex membrane prevents early drug release. Via this method microcapsules can be formed between chitosan and alignate or even nanoparticles can be derived upon the complexation of chitosan with the polyanion tripolyphosphate.

8.3 Summary

Due to the constant emerge of new macromolecular therapeutics such as peptides, proteins and nucleic acids and the demand for simple administration there is a great need for new oral drug delivery systems. These drugs are usually indicated for chronic diseases and patients have to face a complicated and inconvenient treatment due to continuous injection of the therapeutic. Multifunctional matrices have gained much attention in this field, as they are based on polymers which exhibit mucoadhesive, enzyme inhibitory, permeation-enhancing and high buffering properties and can additionally provide controlled and/or targeted drug release. Their use to overcome the intestinal barriers is schematically summarized in Fig. 8.6. The interplay of those characteristics provides a good basis for the improvement of oral bioavailability.

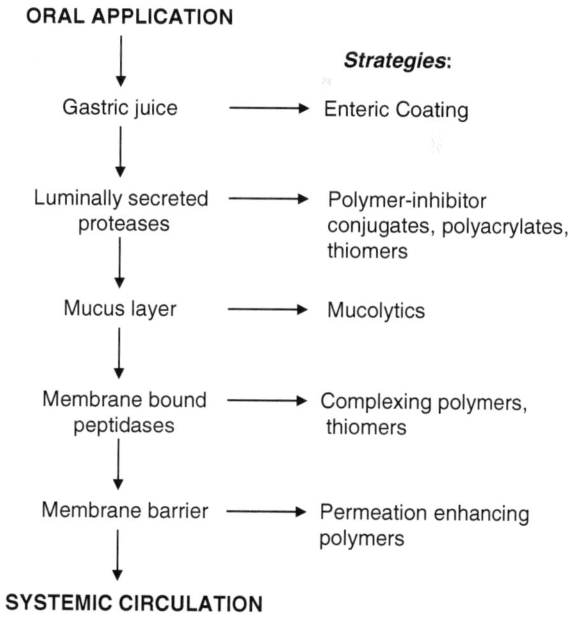

Fig. 8.6 Depiction of gastrointestinal barriers to oral peptide drug delivery and strategies to overcome them by the use of multifunctional matrices. Adapted from Bernkop-Schnürch and Walker (2001)

A delivery system based on multifunctional polymers can protect the incorporated drug from pH- and enzyme-dependent degradation down its way through the GI tract and provide high drug concentrations at the target site and tight contact with the absorption barrier due to mucoadhesion. Furthermore, sustained or delayed release due to cohesive properties of the polymeric carrier offers new possibilities for targeted macromolecular drug delivery.

References

Bai, J. P., Chang, L. L., and Guo, J. H. Effects of polyacrylic polymers on the degradation of insulin and peptide drugs by chymotrypsin and trypsin. J. Pharm. Pharmacol. 1996; 48, 1, 17–21.

Bernkop-Schnürch, A. Polymer-inhibitor conjugates: a promising strategy to overcome the enzymatic barrier to perorally administered (poly)peptide drugs? Pharm. Sci. 1999; 9, 78–87.

Bernkop-Schnürch, A. and Gilge, B. Anionic mucoadhesive polymers as auxiliary agents for the peroral administration of (poly)peptide drugs: influence of the gastric juice. Drug Dev. Ind. Pharm. 2000; 26, 2, 107–113.

Bernkop-Schnürch, A., Hoffer, M. H., and Kafedjiiski, K. Thiomers for oral delivery of hydrophilic macromolecular drugs. Expert. Opin. Drug Deliv. 2004; 1, 1, 87–98.

Bernkop-Schnürch, A. and Krajicek, M. E. Mucoadhesive polymers as platforms for peroral peptide delivery and absorption: synthesis and evaluation of different chitosan-EDTA conjugates. J. Control Release 1998; 50, 1–3, 215–223.

Bernkop-Schnürch, A. and Marschütz, M. K. Development and in vitro evaluation of systems to protect peptide drugs from aminopeptidase N. Pharm. Res. 1997; 14, 2, 181–185.

Bernkop-Schnürch, A., Schwarz, V., and Steininger, S. Polymers with thiol groups: a new generation of mucoadhesive polymers? Pharm. Res. 1999; 16, 6, 876–881.

Bernkop-Schnürch, A. and Steininger, S. Synthesis and characterisation of mucoadhesive thiolated polymers. Int J Pharm 2000; 194, 2, 239–247.

Bernkop-Schnürch, A. and Thaler, S. C. Polycarbophil-cysteine conjugates as platforms for oral polypeptide delivery systems. J. Pharm. Sci. 2000; 89, 7, 901–909.

Bernkop-Schnürch, A. and Walker, G. Multifunctional matrices for oral peptide delivery. Crit. Rev. Ther. Drug Carrier Syst. 2001; 18, 5, 459–501.

Bernkop-Schnürch, A., Weithaler, A., Albrecht, K., and Greimel, A. Thiomers: preparation and in vitro evaluation of a mucoadhesive nanoparticulate drug delivery system. Int. J. Pharm. 2006; 317, 1, 76–81.

Bernkop-Schnürch, A., Zarti, H., and Walker, G. F. Thiolation of polycarbophil enhances its inhibition of intestinal brush border membrane bound aminopeptidase N. J. Pharm. Sci. 2001; 90, 11, 1907–1914.

Borchard, G., Luessen, H. L., Verhoef, J. C., Lehr, C.-M., de Boer, A. G., and Junginger, H. E. The potential of mucoadhesie polymers in enhancing intestinal peptide drug absorption: III. Effects of chitosan-glutamate and carbomer on epithelial tight junctions in vitro. J. Control Release 1996; 39, 2–3, 131–138.

Calceti, P., Salmaso, S., Walker, G., and Bernkop-Schnürch, A. Development and in vivo evaluation of an oral insulin-PEG delivery system. Eur. J. Pharm. Sci. 2004; 22, 4, 315–323.

Carreno-Gomeza, B. and Duncan, R. Evaluation of the biological properties of soluble chitosan and chitosan microspheres. Int. J. Pharm. 1997; 148, 231–240.

Chourasia, M. K. and Jain, S. K. Pharmaceutical approaches to colon targeted drug delivery systems. J. Pharm. Pharm. Sci. 2003; 6, 1, 33–66.

Clausen, A. E. and Bernkop-Schnürch, A. Development an in vitro evaluation of a peptide drug delivery system based on thiolated polycarbophil. Pharm. Ind. 2001; 63, 312–

Clausen, A. E., Kast, C. E., and Bernkop-Schnürch, A. The role of glutathione in the permeation enhancing effect of thiolated polymers. Pharm. Res. 2002; 19, 5, 602–608.

Föger, F., Malaivijitnond, S., Wannaprasert, T., Huck, C., Bernkop-Schnürch, A., and Werle, M. Effect of a thiolated polymer on oral paclitaxel absorption and tumor growth in rats. J. Drug Target 2008; 16, 2, 149–155.

Föger, F., Schmitz, T., and Bernkop-Schnürch, A. In vivo evaluation of an oral delivery system for P-gp substrates based on thiolated chitosan. Biomaterials 2006; 27, 23, 4250–4255.

Grabovac, V., Guggi, D., and Bernkop-Schnürch, A. Comparison of the mucoadhesive properties of various polymers. Adv. Drug Deliv. Rev. 2005; 57, 11, 1713–1723.

Guggi, D. and Bernkop-Schnürch, A. In vitro evaluation of polymeric excipients protecting calcitonin against degradation by intestinal serine proteases. Int. J. Pharm. 2003; 252, 1–2, 187–196.

Guggi, D., Krauland, A. H., and Bernkop-Schnürch, A. Systemic peptide delivery via the stomach: in vivo evaluation of an oral dosage form for salmon calcitonin. J. Control Release 2003; 92, 1–2, 125–135.

Harding, S. E. Mucoadhesive interactions. Biochem. Soc. Trans. 2003; 31, Pt 5, 1036–1041.

Illum, L., Farraj, N. F., and Davis, S. S. Chitosan as a novel nasal delivery system for peptide drugs. Pharm Res. 1994; 11, 8, 1186–1189.

Kast, C. E. and Bernkop-Schnürch, A. Thiolated polymers – thiomers: development and in vitro evaluation of chitosan-thioglycolic acid conjugates. Biomaterials 2001; 22, 17, 2345–2352.

Leitner, V. M., Walker, G. F., and Bernkop-Schnürch, A. Thiolated polymers: evidence for the formation of disulphide bonds with mucus glycoproteins. Eur. J. Pharm. Biopharm. 2003; 56, 2, 207–214.

Loretz, B., Föger, F., Werle, M., and Bernkop-Schnürch, A. Oral gene delivery: Strategies to improve stability of pDNA towards intestinal digestion. J. Drug Target 2006; 14, 5, 311–319.

Lu, R. H., Kopeckova, P., and Kopecek, J. Degradation and aggregation of human calcitonin in vitro. Pharm. Res. 1999; 16, 3, 359–367.

Luessen, H. L., de Leeuw, B. J., Langemeyer, M. W., de Boer, A. B., Verhoef, J. C., and Junginger, H. E. Mucoadhesive polymers in peroral peptide drug delivery. VI. Carbomer and chitosan improve the intestinal absorption of the peptide drug buserelin in vivo. Pharm. Res. 1996a; 13, 11, 1668–1672.

Luessen, H. L., de Leeuw, B. J., Pérard, D., Lehr, C. M., de Boer, A. G., Verhoef, J. C., and Junginger H. E. Mucoadhesive polymers in peroral peptide drug delivery. I. Influence of mucoadhesive excipients on the proteolytic activity of intestinal enzymes. Eur. J. Pharm. Sci. 1996b; 4, 6, 117–128.

Luessen, H. L., Verhoef, J. C., Borchard, G., Lehr, C. M., de Boer, A. G., and Junginger, H. E. Mucoadhesive polymers in peroral peptide drug delivery. II. Carbomer and polycarbophil are potent inhibitors of the intestinal proteolytic enzyme trypsin. Pharm. Res. 1995; 12, 9, 1293–1298.

Marschütz, M. K. and Bernkop-Schnürch, A. Oral peptide drug delivery: polymer-inhibitor conjugates protecting insulin from enzymatic degradation in vitro. Biomaterials 2000; 21, 14, 1499–1507.

Morishita, M., Morishita, I., Takayama, K., Machida, Y., and Nagai, T. Site-dependent effect of aprotinin, sodium caprate, Na2EDTA and sodium glycocholate on intestinal absorption of insulin. Biol. Pharm. Bull. 1993; 16, 1, 68–72.

Palmberger, T. F., Hombach, J., and Bernkop-Schnürch, A. Thiolated chitosan: development and in vitro evaluation of an oral delivery system for acyclovir. Int. J. Pharm. 2008; 348, 1–2, 54–60.

Takeuchi, H., Yamamoto, H., Niwa, T., Hino, T., and Kawashima, Y. Mucoadhesion of polymer-coated liposomes to rat intestine in vitro. Chem. Pharm. Bull. (Tokyo) 1994; 42, 9, 1954–1956.

Tirosh, B., Baluom, M., Nassar, T., Friedman, M. and Rubinstein, A. The effect of Eudragit RL-100 on the mechanical and mucoadhesion properties of polycarbophil dosage forms. J. Control Release 1997; 45, 57–64.

Werle, M. Polymeric and low molecular mass efflux pump inhibitors for oral drug delivery. J Pharm Sci 2008; 97, 1, 60–70.

Zheng, Y., Qiu, Y., Lu, M. F., Hoffman, D., and Reiland, T. L. Permeability and absorption of leuprolide from various intestinal regions in rabbits and rats. Int. J. Pharm. 1999; 185, 1, 83–92.

Chapter 9
Nano- and Microparticles in Oral Delivery of Macromolecular Drugs

Gioconda Millotti and Andreas Bernkop-Schnürch

Contents

Abstract Oral delivery systems are highly demanded and preferred over parenteral systems as they offer an ease of administration and therefore a high patient compliance. A promising strategy in order to improve the oral uptake of macromolecular drugs is the use of micro- and nanoparticulate delivery systems. The most widely used materials and techniques applied in the development of micro- and nanocarrier systems are described as well as the strategies to modify the particle's surface in order to modulate their characteristics such as mucoadhesion, stability, protective effect, hydrophilicity or lipophilicity. Particles' properties such as membrane-passing properties, permeation-enhancing properties, mucoadhesive properties, and protective properties are extensively discussed. Furthermore, evidence for the potential of such systems is provided with examples from recent in vivo studies.

G. Millotti (✉)
Department of Pharmaceutical Technology, Leopold-Franzens-University Innsbruck, Innrain 52, A 6020 Innsbruck, Austria
e-mail: gioconda.millotti@uibk.ac.at

A. Bernkop-Schnürch (ed.), *Oral Delivery of Macromolecular Drugs*,
DOI 10.1007/978-1-4419-0200-9_9,

9.1 Introduction

Over the past years dramatic progress has been made in the fields of biotechnology, genomics, and proteomics, resulting in the capability to produce large numbers of potential therapeutic macromolecules in commercial quantities. These can be identified as high-purity macromolecular drugs such as pharmacologically active peptides and proteins, genes, products of recombinant DNA, and therapeutic RNAs. Moreover, these new techniques also allow for the production of peptide and protein vaccines based on antigens found on the surface of various infectious microorganisms and viruses. However, these drugs require the development of efficient systems that will allow their administration (Jung et al. 2000). In particular, oral delivery systems are highly on demand as they offer the greatest ease of administration and consequently the highest compliance. A promising strategy to deal with this problem is the development of micro- and nanoparticulate delivery systems. Nanocarriers are materials of nanoscale range (below 1 μm) made up of different materials like natural or synthetic polymers, lipids or phospholipids, and even organometallic compounds. They have a very high surface to volume ratio leading to increased dissolution rates (Rawat et al. 2006). Indeed, it is obvious that for poorly water-soluble compounds, the dissolution kinetics in the gastrointestinal tract is proportional to the specific surface area; therefore, the formulation of these compounds as submicroscopic colloidal systems may help to accelerate the dissolution process, thus increasing bioavailability (Uchegbu and Schätzlein 2006). Multiparticulate systems have definitely the advantage to distribute more uniformly in the GI tract compared to single-unit delivery systems such as tablets. Nano- and microparticles enable the protection of macromolecules, improving their stability and therefore increasing the duration of their therapeutic effect. The size of the systems was found to play an important role: whereas in the case of tablets and small patches bioadhesion may be hazardous and may lead to variable therapeutic responses (Eiamtrakarn et al. 2002), particulate systems of micrometer or nanometer size would present enormous advantages in terms of reproducibility (Habberfield et al. 1996), adhesion capability, residence time in the gut, and reducing the risk of irritation or ulceration.

The drug can be dissolved, entrapped, encapsulated, or attached to a nanoparticle matrix, and depending upon the method of preparation, nanoparticles, nanospheres, and nanocapsules can be obtained (Soppimath et al. 2001).

Polymeric nanoparticles have attracted a lot of attention in the last years. Polymeric materials exhibit several advantageous properties including biodegradability and ease of functionalization. They also allow for a greater control of pharmacokinetic behavior of the loaded drug leading to more steady levels of drugs (Rawat et al. 2006). Furthermore, they enable the modulation of the physicochemical properties of the surface such as Zeta potential and hydrophobicity/hydrophilicity. Many polymers used to develop nano- and microparticulate systems exhibit mucoadhesive properties enabling prolonged contact of the carried macromolecule with the absorption site. Polymeric materials used for the formulation of nanoparticles include polylactic acid (PLA) (Leo et al. 2004), polyglycolic acid (PGA), polylactic–glycolic acid (PLGA) (Fonseca et al. 2002; Dillen et al. 2004), polymethyl methacrylate (PMMA), poly(E-caprolactone) (PCL) (Barbault-Foucher et al. 2002), and poly(alkyl cyanoacrylates) (Chauvierre et al. 2003). Natural polymers have also been used such as chitosan (Galindo-Rodriguez et al. 2005), gelatin (Balthasar et al. 2005), and alginate (Johnson et al. 1997).

9.2 Preparation Methods

9.2.1 Different Methods

For both synthetic and natural polymers, different techniques can be used to prepare particles. In the following, the likely most promising techniques are described in more detail.

9.2.1.1 Dispersion of Preformed Polymers

This is a method to obtain colloidal drug delivery systems from preformed, well-defined macromolecular materials with known physicochemical and biological properties. Biodegradable nanoparticles from PLA, PLG, PLGA, and poly(E-caprolactone) have been prepared by dispersing the polymers (Vauthier et al. 1991; Couvreur et al. 1995).

9.2.1.2 Solvent Evaporation Method

The polymer is dissolved in an organic solvent such as dichloromethane, chloroform, or ethyl acetate. The drug is dissolved or dispersed into the preformed polymer solution, and this mixture is then emulsified into an aqueous solution to make an oil (O) in water (W) (O/W) emulsion by using a surfactant/emulsifying agent like gelatin, poly(vinyl alcohol), polysorbate 80, or poloxamer-188. After the formation of a stable emulsion, the organic solvent is evaporated either by increasing the temperature under pressure or by continuous stirring (Soppimath et al. 2001).

9.2.1.3 Polymerization Methods

Mostly poly(alkyl cyanoacrylates) (PACAs) are produced by this method. This technique was first introduced by Couvreur et al. to design nanoparticles with biodegradable polymers for the in vivo delivery of drugs (Couvreur et al. 1978,

1979). The group polymerized mechanically the dispersed methyl or ethyl cyanoacrylate in aqueous acidic medium in the presence of polysorbate-20 as a surfactant under vigorous mechanical stirring to polymerize alkyl cyanoacrylate. The polymerization follows an anionic mechanism since it is initiated in the presence of nucleophilic initiators like OH^- in an acidic medium (pH 1.0–3.5). The same group coated PACA nanoparticles with various polysaccharides introducing modifications in the method (Couvreur et al. 1978; Vauthier et al. 2003; Betrholon-Rajot et al. 2005).

9.2.1.4 Emulsion Cross-Linked Nanoparticles

This method is based on the reactive functional amino group of chitosan to crosslink it with aldehydes. In this method, a water-in-oil emulsion is prepared by emulsifying the chitosan aqueous solution in the oil phase. Aqueous droplets are stabilized using a suitable surfactant. The stable emulsion is cross-linked by using an appropriate cross-linking agent such as glutaraldehyde to harden the droplets (Agnihotri et al. 2004).

9.2.1.5 Coacervation/Precipitation

This method used the physicochemical properties of polymers like chitosan, which is insoluble in alkaline pH medium and therefore precipitates/coacervates when it comes in contact with alkaline solution. Particles are produced by blowing chitosan solution into an alkali solution like NaOH using a compressed air nozzle to form coacervate droplets (Agnihotri et al. 2004).

9.2.1.6 Ionic Gelation Method

The use of complexation between oppositely charged macromolecules to prepare nano- and microspheres is attractive due to the very mild and simple process. Furthermore, due to the electrostatic interaction, instead of chemical cross-linking, lower toxicity effects could be expected. Various polyanions have been used to form particles from positively charged polymers. For example, tripolyphosphate (TPP) is used to form particles with chitosan. In order to produce a high yield of stable and solid nanometric structures, the chitosan to TPP weight ratio should normally be within the range 3:1–6:1 (Janes et al. 2001). This method offers the following advantages: (i) particles are formed under extremely mild conditions; (ii) the size is adjustable; (iii) great capacity for the association with macromolecular drugs; (iv) possibility of modulation of the release of the carried drug depending on the composition of the particles (Janes et al. 2001).

9.2.2 Surface Modification

Colloidal carriers developed from biodegradable polymers, besides many advantages, have the main drawback to interact non-specifically with cells and proteins, leading to drug accumulation in non-target tissues. For a number of applications the surface of nanoparticles has to be highly hydrophilic and able to prevent protein absorption. Surface modifications are therefore helpful in order to improve the properties of nanoparticles. Surface modifications can be achieved either by surface coating with hydrophilic, stabilizing, mucoadhesive polymers/surfactants or by the development of biodegradable copolymers with hydrophilic segments (Soppimath et al. 2001). These modifications mainly change the Z potential of nanoparticles, hydrophobicity, stability, mucoadhesive properties, and protein adsorption at their surface. PEG has been used as a coating material for nanoparticles (Gref et al. 1994). Carbohydrates have also widely been used for the same purpose. Polysaccharides are involved in cell surface properties including tissue addressing and transport mechanism (Lemarchand et al. 2004). Furthermore several polysaccharides, such as chitosan, exhibit mucoadhesive properties. Among the most used polysaccharides are dextrans (Chouly et al. 1996; Passirani et al. 1999; Soma et al. 1999; Rouzes et al. 2000; Chauvierre et al. 2003) and chitosan (Yang et al. 2000; Vila et al. 2002; Chauvierre, et al. 2003). Chitosan and in particular thiolated chitosan-coated poly(isobutyl cyanoacrylate) nanoparticles showed by far stronger mucoadhesion in comparison to non-coated poly(isobutyl cyanoacrylate) particles, evidencing the advantageous aspect of coating (Bravo-Osuna, et al. 2007). Furthermore, the same particles proved to inhibit carboxypeptidase A in vitro while non-coated particles did not manifest such behavior (Bravo-Osuna et al. 2008). Different coating can provoke opposite results on a particulate system. Florence et al. (1995) reported that poloxamers appeared to block the uptake of 50-nm polystyrene in the small intestinal region of the gut, while tomato lectins bound to the surface of 500-nm polystyrene particles increased their uptake.

9.3 Properties of Micro- and Nanoparticles

9.3.1 Membrane-Passing Properties

Surface properties of particles and size have a great impact on particle uptake. In vitro studies performed by des Rieux et al. (2005) in co-cultures showed that the number of transcytosed nanoparticles having a diameter of 200 nm was seven times higher than that of 500 nm. Desai et al. (1996) studied the uptake by the Peyer's patch tissue of different-sized polylactic polyglycolic acid (50:50) copolymer nano- and microparticles concluding that the efficiency of uptake of 100-nm size particles by the intestinal tissue was 15–250-fold higher compared to larger-size microparticles (500 nm, 1 μm, 10 μm). Ten days daily dosing of 50–3000 nm sized unmodified polystyrene latex particles showed a clear size-dependence of the uptake process, favoring the smallest particles. Submicrometer particles accumulated after 10 days dosing to the extent of less than 1% in the spleen and from about 1 to 4% in the liver. However, smaller particles were detected in the spleen, blood, bone marrow, and kidney, while in the case of particles greater than 500 nm it was possible to detect them in the liver and the spleen (Florence et al. 1995). In general, the uptake of micro- and nanoparticles, however, seems to be too low to make use of this mechanism for systemic delivery of macromolecular drugs via the oral route.

9.3.2 Permeation-Enhancing Properties

As nanoparticles are poorly uptaken from the GI tract, the likely much more promising strategy is to generate nanoparticles releasing their drug load on the intestinal mucosa and providing an additional permeation-enhancing effect to improve drug absorption. Mi et al., for example, proposed a concept of the mechanism of orally given nanoparticles with excessive mucoadhesive *N*-trimethyl chitosan on their surface. The particles may adhere and infiltrate into the mucus of the intestinal tract and then mediate transiently opening the tight junctions between enterocytes. Consequently, the nanoparticles that infiltrated into the mucus must become unstable (swelling or disintegration); thus, their loaded insulin can be released and permeated through the paracellular pathway to the bloodstream (Mi et al. 2008).

9.3.3 Mucoadhesive Properties

After oral administration, the absorption efficacy of compounds can be improved by prolonging the residence time of the drug in front of the window of absorption. Intestinal transit results in reducing the drug residence time and therefore limits the drug's bioavailability. One possibility to increase the bioavailability of a substance which has low mucosal absorptive properties is to increase its residence time at mucosal or epithelial level, for example, by incorporating it in micro- and nanoparticles. When orally administered, micro- and nanoparticles can either transit through the GI tract without interacting with it and therefore undergo fecal elimination or interact with the GI tract and undergo mucoadhesion. When a suspension of micro- and nanoparticles is orally administered, they diffuse in the intestinal liquid medium and encounter the mucus at which they could adhere. Adhesion of particles to the mucosa allows a prolonged residence time of the carried drug at the site of absorption and the creation of a concentration gradient. The intestinal mucus is composed of high molecular weight glycoprotein covering the mucosa. The thickness of the mucin gel layer varies throughout the intestinal tract having a thickness of 50–500 μm in the stomach and 15–150 μm in the colon (Ponchel and Irache 1998). The particles can either adhere to the mucus in a non-specific manner like through Van der Waals forces or hydrophobic interactions or form specific interactions between complementary structures, such as the formation of disulfide bonds with glycoproteins. Non-specific mucoadhesion is based on physicochemical interactions. Some polymers have the ability to interact with the mucosa. Chitosan, due to its positive surface charges, is able to form ionic interactions with the negatively charged mucosal surfaces. Bernkop-Schnürch et al. proposed a method of specific interaction by covalently attaching thiol-bearing groups to the surface of various polymers. The thiol groups can therefore form disulfide bonds with thiol groups located on glycoproteins present in the mucus increasing the strength of mucoadhesion and therefore the residence time of these particles (Fig. 9.1) (Bernkop-Schnurch et al. 2006). A factor limiting the duration of mucoadhesion is the mucus turnover. Therefore, the release of the drug out of the carried system must take place during the mucoadhesion time.

There are various parameters that can have an influence on the extent of mucoadhesion, like the size of the carrier and the coating with polymeric materials. Lamprecht et al. described a size-dependent mucoadhesion. For 10-μm particles, only a fair deposition in the mucus was found. One-micrometer particles showed higher binding, while the highest binding was found for 0.1-μm particles (Lamprecht et al. 2001). Elcatonin-loaded chitosan-coated PLGA nanospheres had a greater efficiency in reducing the blood calcium level compared with the non-coated nanospheres probably because they offer a longer contact with the mucosa (Takeuchi et al. 2001) (Fig. 9.2)

The fate of nanoparticles has also been investigated in vivo. Albrecht et al. studied the in vivo mucoadhesive properties of thiomer formulations using magnetic resonance imaging and fluorescence detection (Albrecht et al. 2006). Following the hypothesis that unhydrated thiomers provide better mucoadhesion in vivo the group developed polycarbophil–cysteine microparticles loaded with fluoresceine diacetate (FDA) in Eutex capsules to maintain them in the dry

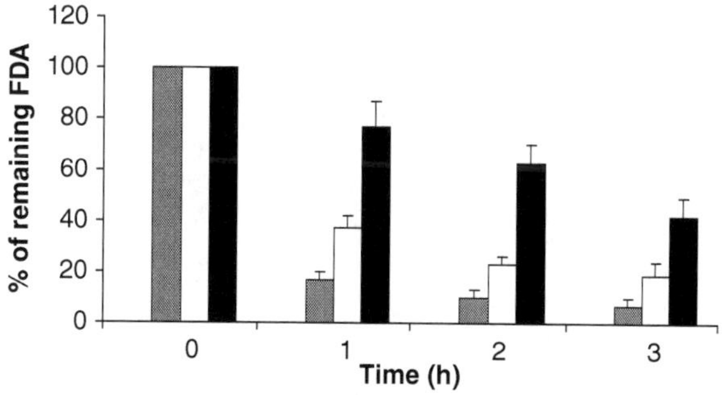

Fig. 9.1 Mucoadhesion studies: amount of FDA remaining on the intestinal mucosa when applying fluoresceine diacetate (FDA) alone (*grey bars*), incorporated into chitosan nanoparticles (*white bars*) and into thiolated chitosan nanoparticles (*black bars*). Adapted from Bernkop-Schnürch et al. (2006)

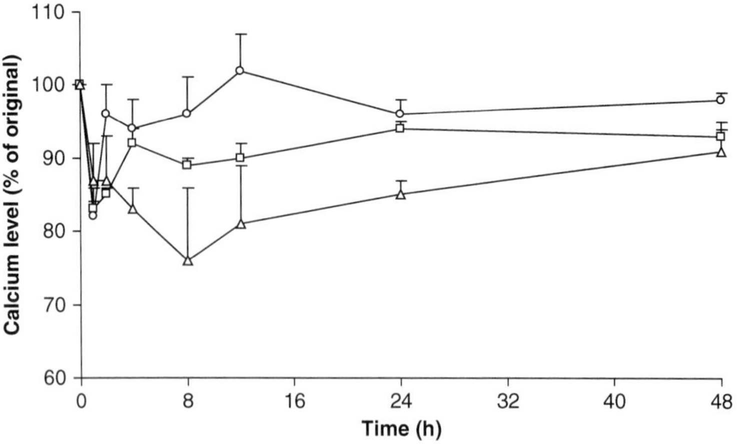

Fig. 9.2 Blood calcium level after intragastric administration of unmodified calcitonin-loaded PLGA nanospheres (*white squares*), chitosan-modified calcitonin-loaded PLGA nanospheres (*white triangles*), and calcitonin solution (*white circles*). Adapted from Takeuchi et al. (2001)

form. The results provided evidence for the validity of the hypothesis that unhydrated thiomers guarantee better mucoadhesive properties in vivo.

In other studies, after the peroral administration of radiolabeled poly(hexyl cyanoacrylate) nanoparticles to mice, the radiography showed that after 30 min, the particles were found exclusively in the stomach. After 4 h, a large quantity of radioactivity was found in the intestine (Kreuter et al. 1989; Kreuter 1991). The total radioactivity recovered in the intestine after 90 min amounted to about 60–80% of the administered dose. Then, it was rapidly eliminated up to 240 min. Radioactivity fell to trace values after 24 h, however, 0.04% of the 90-min radioactivity was still detectable after 6 days. Durrer et al (1994) also studied the intestinal mucoadhesive profile of poly(lactic acid) microparticles as a function of time after oral administration in rats. After intragastric administration of the microsphere suspension, the rats were sacrificed at different time-points. The intestine was removed and divided into five segments. They were rinsed with saline and afterward radioactivity of both the rinsed segments and fractions was measured in order to determine the amount of non-adhered particles. At 1 h after administration 98% of the particles remained in the GI tract. Distribution of particles, however, was found to move toward the lower part of the intestine over time and after 6 h all particles have passed through the small intestine. Ponchel and Irache (1998) proposed a description of the mechanism of particle mucoadhesion after oral administration. In a first step, the suspension of administered particles enters in contact with the mucosa. At a second step, as the particles are concentrated, adsorption of particles takes place at the mucus layer in an irreversible process. As third step, the suspension of particles will transit through the intestine with a simultaneous mucoadhesion. Finally, as fourth step, particles

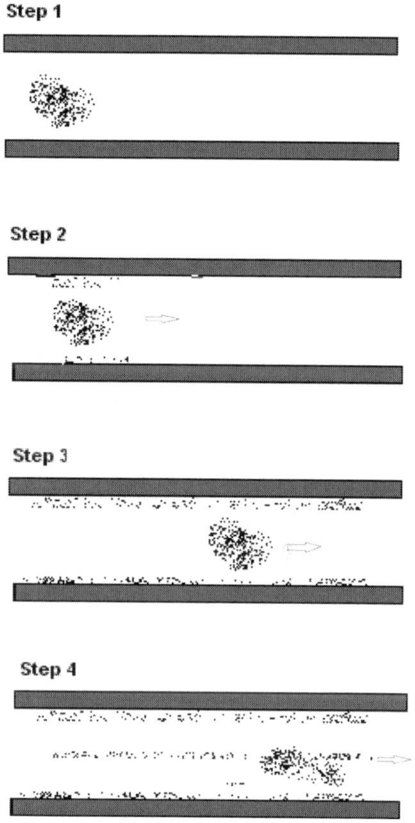

Fig. 9.3 The proposed four steps of the mechanism of particulate mucoadhesion. Step 1, administration. Step 2, initial adsorption of the particles. Step 3, mucoadhesion of particles and further transit in the lumen. Step 4, detachment of particles, further transit and further fecal elimination. Adapted from Ponchel et al. (1998)

start to detach from the proximal part of the intestine. The detached particles as well as the non-absorbed particles are eliminated in the feces (Fig. 9.3).

In vivo, however, various further aspects have to be taken into account such as variability in stomach emptying, intestinal transit, effect of the dilution of the particle suspension in the GI-tract fluids, and mixing with ingested food (Ponchel and Irache 1998).

9.3.4 Protective Properties

One of the biggest impediments of oral administration of macromolecules is their lack of stability in the gastrointestinal tract. Indeed, the majority of macromolecules such as peptides, proteins, oligonucleotides, and various types of RNA undergo enzymatic degradation before reaching their site of action or absorption. One of the strategies used to circumvent this problem is to incorporate such compounds into nano- or microcarriers. These carriers offer protection to the incorporated drug toward many of the enzymes encountered in the GI tract. Additionally, they can promote paracellular uptake of the drugs as well as intimate contact with the site of action or absorption. Many studies demonstrated a higher bioavailability when the drug is incorporated into nano- or microcarriers. Adsorption of insulin on the surface of nanoparticles did not protect against enzymatic degradation, whereas when insulin was incorporated into poly(isobutyl cyanoacrylate) nanocapsules, the peptide was protected and glycemia decreased by 50% in diabetic rats (Damge et al. 1988, 1990). It was, therefore, concluded that nanocapsules protect the peptide against enzymatic degradation. Alginate/chitosan nanoparticles for oral insulin delivery were also hypothesized to be able to stabilize and protect the incorporated insulin from degradation in the GI tract (Sarmento et al. 2007a, b). Sakuma et al. demonstrated that there was a good correlation between the in vitro stability of salmon calcitonin in the presence of nanoparticles and the ranking of the in vivo effectiveness of nanoparticles for enhancing the absorption of salmon calcitonin (Sakuma et al. 1997).

9.4 Proof of Concept

Of all the macromolecular drugs, special effort has been made to orally deliver insulin. Insulin-dependent patients are treated with daily subcutaneous injections, with low patient compliance. Many groups incorporated insulin into colloidal carriers with promising results. Pan et al. (2002), for instance, showed a decrease in plasma glucose levels induced by the oral administration of insulin-loaded chitosan nanoparticles, which was significantly more in comparison with the oral administration of insulin–chitosan solution. The authors explained this observation by the fact that proteins entrapped in nanoparticles are more stable and protected from degradation in the GI tract and by the mucoadhesive properties of chitosan providing also a transient opening of the tight junctions. Furthermore, insulin was orally administered loaded into alginate/chitosan nanoparticles produced by ionotropic pre-gelation/polyelectrolyte complexation (Sarmento et al. 2007a, b) (Fig. 9.4) The nanoparticles lowered serum glucose levels up to 55–59% of rat's basal glucose level depending on the administered dose. Sarmento et al. (2007a, b) incorporated insulin into polysaccharide nanoparticles. Insulin-loaded dextran sulfate/chitosan nanoparticles produced by polyelectrolyte complexation lowered serum glucose levels up to 64–67% of rat's basal glucose level depending on the administered dose. Poly(isobutyl cyanoacrylate) nanocapsules were shown to be able to incorporate insulin and to reduce glycemia after oral administration (Damge et al. 1988, 1990). Lin et al. (2007) also reported a reduction of blood glucose level after oral administration of a nanoparticulate system composed of chitosan and gamma-PGA (poly(gamma-glutamic acid)). Deutel et al. (2008) developed a novel poly(acrylic acid)–cysteine/PVP/insulin nanoparticulate

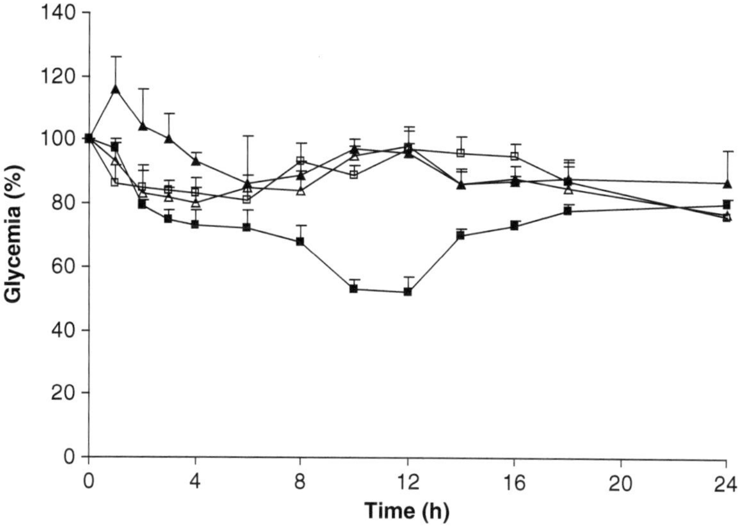

Fig. 9.4 Glycemia levels after oral administration of insulin-loaded nanoparticles 50 IU/kg (*black squares*), oral insulin solution (*white squares*), empty nanoparticles (*black triangles*) and physical mixture of empty nanoparticles, and insulin solution 50 IU/kg (*white triangles*). Adapted from Sarmento et al. (2007a, b)

drug delivery system for oral administration. The system proved to be efficient in order to increase insulin serum concentration and in reducing blood glucose level by 22%.

Calcitonin is another compound that was often incorporated into nanoparticles to enhance its oral absorption. Takeuchi et al. (2001) developed Elcatonin-loaded PLGA nanospheres coated with chitosan, observing a reduction of blood calcium level. Sakuma et al. (2002) hypothesizes that both mucoadhesion of nanoparticles incorporating salmon calcitonin into the GI mucosa (Sakuma et al. 1999, 2002) and increase in the stability of salmon calcitonin in the GI tract (Sakuma et al. 1997) result in the improvement of salmon calcitonin absorption. Moreover, chitosan–PEG nanocapsules increased the absorption of salmon calcitonin (Prego et al. 2006).

Nanoparticles offer also the new perspective of oral delivery of antigens. Ovalbumin was incorporated in PEGylated PLGA-based nanoparticles with the intention of oral vaccination (Garinot et al. 2007). The system was administered in mice and induced IgG response. The immunization with the nanoparticles induced both a humoral and a cellular immune response attesting that the antigen is efficiently presented to T lymphocytes by the antigen-presenting cells. Bovine serum albumin (BSA) incorporated into PLGA nanoparticles and orally administered gave a systemic IgG dose/response relationship.

Another very attractive application of nanoparticles is the oral gene delivery. For example, quaternized chitosan-trimethylated chitosan was shown to have the ability to encapsulate and protect pDNA and effectively transfer the green fluorescent protein (GFP) gene into cells both in vitro and in vivo (Zheng et al. 2007). In another study, DNA polyplexes composed of chitosan and Factor VIII DNA were developed and orally administered to hemophilia A mice (Bowman et al. 2008). Transgene DNA was detected in both local and systemic tissues. Despite the modest FVIII levels achieved in mice, detectable protein persisted for 1 month and phenotypic bleeding correction was observed in 65% of the mice given high or medium doses of chitosan–DNA nanoparticles (Bowman et al. 2008).

9.5 Conclusion

Due to the great progress in biotechnology, numerous macromolecules are being developed. The oral route remains the preferred way to administer them, due to the advantage of avoiding pain caused by injections. However, these macromolecules remain poorly bioavailable when administered orally, mainly due to their low mucosal permeability and very low stability in the harsh environment of the gastrointestinal tract. In recent years many efforts have been made to incorporate these macromolecules into very wide spectra of colloidal carriers in order to improve their oral availability. Indeed, nano- and microcarriers have shown to be able to improve the bioavailability of a number of macromolecules acting on various fronts: (i) by protecting the carried drug from enzymatic degradation; (ii) by increasing the residence time in the GI tract through a mucoadhesive mechanism; (iii) and possibly by promoting paracellular uptake. Nano- and microparticles represent therefore a promising tool for the oral delivery of macromolecules.

References

Agnihotri, S. A., N. N. Mallikarjuna, et al. (2004). Recent advances on chitosan-based micro- and nanoparticles in drug delivery. J Control Release 100(1): 5–28.

Albrecht, K., M. Greindl, et al. (2006). Comparative in vivo mucoadhesion studies of thiomer formulations using magnetic resonance imaging and fluorescence detection. J.Contol.Rel. 115: 78–84.

Balthasar, S., K. Michaelis, et al. (2005). Preparation and characterisation of antibody modified gelatin nanoparticles as drug carrier system for uptake in lymphocytes. Biomaterials 26(15): 2723–32.

Barbault-Foucher, S., R. Gref, et al. (2002). Design of poly-epsilon-caprolactone nanospheres coated with bioadhesive hyaluronic acid for ocular delivery. J Control Release 83(3): 365–75.

Bernkop-Schnurch, A., A. Weithaler, et al. (2006). Thiomers: preparation and in vitro evaluation of a mucoadhesive nanoparticulate drug delivery system. Int J Pharm 317(1): 76–81.

Betrholon-Rajot, I., D. Labarre, et al. (2005). Influence of the initiator system, cerium-polysaccharide, on the surface properties of poly(isobutylcyanoacrylate) nanoparticles. Polymer 46: 1407–1415.

Bowman, K., R. Sarkar, et al. (2008). Gene transfer to hemophilia A mice via oral delivery of FVIII-chitosan nanoparticles. J Control Release.

Bravo-Osuna, I., C. Vauthier, et al. (2007). Mucoadhesion mechanism of chitosan and thiolated chitosan-poly(isobutyl cyanoacrylate) core-shell nanoparticles. Biomaterials 28(13): 2233–43.

Bravo-Osuna, I., C. Vauthier, et al. (2008). Effect of chitosan and thiolated chitosan coating on the inhibition behaviour of PIBCA nanoparticles against intestinal metallopeptidases. J. Nanopart. Res DOI 10.1007/s11051-008-9364-5.

Chauvierre, C., D. Labarre, et al. (2003). Novel polysaccharide-decorated poly(isobutyl cyanoacrylate) nanoparticles. Pharm Res 20(11): 1786–93.

Chouly, C., D. Pouliquen, et al. (1996). Development of superparamagnetic nanoparticles for MRI: effect of particle size, charge and surface nature on biodistribution. J Microencapsul 13(3): 245–55.

Couvreur, P., C. Dubernet, et al. (1995). Controlled drug delivery with nanoparticles: Current possibilities and future trends. Eur. J. Pharm. Biopharm. 41: 2–13.

Couvreur, P., B. Kante, et al. (1978). [Perspective on the use of microdisperse forms as intracellular vehicles]. Pharm Acta Helv 53(12): 341–7.

Couvreur, P., B. Kante, et al. (1979). Polycyanoacrylate nanocapsules as potential lysosomotropic carriers: preparation, morphological and sorptive properties. J Pharm Pharmacol 31(5): 331–2.

Damge, C., C. Michel, et al. (1988). New approach for oral administration of insulin with polyalkylcyanoacrylate nanocapsules as drug carrier. Diabetes 37(2): 246–51.

Damge, C., C. Michel, et al. (1990). Nanocapsules as carriers for oral peptide delivery. J. Contol.Rel. 13: 233–239.

des Rieux, A., E. G. E. Ragnarsson, et al. (2005). Transport of nanoparticles across an in vitro model of the human intestinal follicle associated epithelium. Eur. J. Pharm. Sci 25: 455–465.

Desai, M. P., G. L. Labhasetwar, et al. (1996). Gastrointestinal uptake of biodegradable microparticles: effect of particle size. Pharm. Res. 13(12): 1838–1845.

Deutel, B., M. Greindl, et al. (2008). Novel insulin thiomer nanoparticles: in vivo evaluation of an oral drug delivery system. Biomacromolecules 9(1): 278–85.

Dillen, K., J. Vandervoort, et al. (2004). Factorial design, physicochemical characterisation and activity of ciprofloxacin-PLGA nanoparticles. Int J Pharm 275(1–2): 171–87.

Durrer, C., J. M. Irache, et al. (1994). Mucoadhesion of latexes II. Adsorption isotherms and desorption studies. Pharm Res 11(5): 680–683.

Eiamtrakarn, S., M. Itoh, et al. (2002). Gastrointestinal mucoadhesive patch system (GI-MAPS) for oral administration of G-CSF, a model protein. Biomat. 23: 145–152.

Florence, A. T., A. M. Hillery, et al. (1995). Nanoparticles as carriers for oral peptide absorption: studies on particle uptake and fate. J Control Release 36: 39–46.

Fonseca, C., S. Simoes, et al. (2002). Paclitaxel-loaded PLGA nanoparticles: preparation, physicochemical characterization and in vitro anti-tumoral activity. J Control Release 83(2): 273–286.

Galindo-Rodriguez, S. A., E. Allemann, et al. (2005). Polymeric nanoparticles for oral delivery of drugs and vaccines: a critical evaluation of in vivo studies. Crit Rev Ther Drug Carrier Syst 22(5): 419–64.

Garinot, M., V. Fievez, et al. (2007). PEGylated PLGA-based nanoparticles targeting M cells for oral vaccination. J Control Release 120(3): 195–204.

Gref, R., Y. Minamitake, et al. (1994). Biodegradable long-circulating polymeric nanospheres. Science 263(5153): 1600–3.

Habberfield, A., K. Jensen-Pippo, et al. (1996). Vitamin B12-mediated uptake of erythropoietin and granulocyte colony stimulating factor in vitro and in vivo. Int J Pharm 145(1–2): 1–8.

Janes, K. A., P. Calvo, et al. (2001). Polysaccharide colloidal particles as delivery systems for macromolecules. Adv Drug Deliv Rev 47(1): 83–97.

Johnson, F. A., D. Q. Craig, et al. (1997). Characterization of the block structure and molecular weight of sodium alginates. J Pharm Pharmacol 49(7): 639–43.

Jung, T., W. Kamm, et al. (2000). Biodegradable nanoparticles for oral delivery of peptides: is there a role for polymers to affect mucosal uptake? Eur J Pharm Biopharm 50(1): 147–60.

Kreuter, J. (1991). Peroral administration of nanoparticles. Adv Drug Deliv Rev 7: 71–86.

Kreuter, J., U. Müller, et al. (1989). Quantitative and microaudiographic study on mouse intestinal distribution of polycyanoacrylate nanoparticles. Int J Pharm 55: 39–45.

Lamprecht, A., U. Schäfer, et al. (2001). Size dependent bioadhesion of micro- and nanoparticulate carriers to the inflamed colonic mucosa. Pharm. Res. 18(6): 788–793.

Lemarchand, C., R. Gref, et al. (2004). Polysaccharide-decorated nanoparticles. Eur J Pharm Biopharm 58(2): 327–41.

Leo, E., B. Brina, et al. (2004). In vitro evaluation of PLA nanoparticles containing a lipophilic drug in water-soluble or insoluble form. Int J Pharm 278(1): 133–41.

Lin, Y. H., F. L. Mi, et al. (2007). Preparation and characterization of nanoparticles shelled with chitosan for oral insulin delivery. Biomacromolecules 8(1): 146–52.

Mi, F. L., Y. Y. Wu, et al. (2008). Oral delivery of peptide drugs using nanoparticles self-assembled by poly(gamma-glutamic acid) and a chitosan derivative functionalized by trimethylation. Bioconjug Chem 19(6): 1248–55.

Pan, Y., Y. J. Li, et al. (2002). Bioadhesive polysaccharide in protein delivery system: chitosan nanoparticles improve the intestinal absorption of insulin in vivo. Int J Pharm 249(1–2): 139–47.

Passirani, C., L. Ferrarini, et al. (1999). Preparation and characterization of nanoparticles bearing heparin or dextran covalently-linked to poly(methyl methacrylate). J Biomater Sci Polym Ed 10(1): 47–62.

Ponchel, G. and J. Irache (1998). Specific and non-specific bioadhesive particulate systems for oral delivery to the gastrointestinal tract. Adv Drug Deliv Rev 34(2–3): 191–219.

Prego, C., D. Torres, et al. (2006). Chitosan-PEG nanocapsules as new carriers for oral peptide delivery. Effect of chitosan pegylation degree. J Control Release 111(3): 299–308.

Rawat, M., D. Singh, et al. (2006). Nanocarriers: promising vehicle for bioactive drugs. Biol Pharm Bull 29(9): 1790–8.

Rouzes, C., R. Gref, et al. (2000). Surface modification of poly(lactic acid) nanospheres using hydrophobically modified dextrans as stabilizers in an o/w emulsion/evaporation technique. J Biomed Mater Res 50(4): 557–65.

Sakuma, S., Y. Ishida, et al. (1997). Stabilization of salmon calcitonin by polystyrene nanoparticles having surface hydrophilic polymeric chains, against enzymatic degradation. Int J Pharm 159: 181–189.

Sakuma, S., R. Sudo, et al. (1999). Mucoadhesion of polystyrene nanoparticles having surface hydrophilic polymeric chains in the gastrointestinal tract. Int J Pharm 177(2): 161–72.

Sakuma, S., R. Sudo, et al. (2002). Behavior of mucoadhesive nanoparticles having hydrophilic polymeric chains in the intestine. J Control Release 81(3): 281–90.

Sarmento, B., A. Ribeiro, et al. (2007a). Oral bioavailability of insulin contained in polysaccharide nanoparticles. Biomacromolecules 8(10): 3054–60.

Sarmento, B., A. Ribeiro, et al. (2007b). Alginate/chitosan nanoparticles are effective for oral insulin delivery. Pharm Res 24(12): 2198–206.

Soma, C. E., C. Dubernet, et al. (1999). Ability of doxorubicin-loaded nanoparticles to overcome multidrug resistance of tumor cells after their capture by macrophages. Pharm Res 16(11): 1710–6.
Soppimath, K. S., T. M. Aminabhavi, et al. (2001). Biodegradable polymeric nanoparticles as drug delivery devices. J Control Release 70(1–2): 1–20.
Takeuchi, H., H. Yamamoto, et al. (2001). Mucoadhesive nanoparticulate systems for peptide drug delivery. Adv Drug Deliv Rev 47(1): 39–54.
Uchegbu, I. F. and A. G. Schätzlein (2006). Polymers in drug delivery. Boca Raton, London, New York, Taylor & Francis Group LLC.
Vauthier, C., S. Beanabbou, et al. (1991). Methodology of ultradispersed polymer system. S. T.P. Pharm. Sci. 1: 109–116.
Vauthier, C., C. Dubernet, et al. (2003). Drug delivery to resistant tumors: the potential of poly(alkyl cyanoacrylate) nanoparticles. J Control Release 93(2): 151–60.
Vila, A., A. Sanchez, et al. (2002). Design of biodegradable particles for protein delivery. J Control Release 78(1–3): 15–24.
Yang, S. C., H. X. Ge, et al. (2000). Formation of positively charged poly(butyl cyanoacrylate) nanoparticles stabilized with chitosan. Colloid Polym. Sci. 278: 285–292.
Zheng, F., X.-W. Shi, et al. (2007). Chitosan nanoparticle as gene therapy vector via gastrointestinal mucosa administration: Results of an in vitro and in vivo study. Life Sci. 80: 388–396.

Chapter 10
Liposome-Based Mucoadhesive Formulations for Oral Delivery of Macromolecules

Pornsak Sriamornsak, Jringjai Thongborisute, and Hirofumi Takeuchi

Contents

Abstract The design and evaluation of liposome-based mucoadhesive dosage form containing mucoadhesive polymers (e.g., chitosan, pectin, carbopol, and PVA-R) have been described. Liposomes with different compositions were prepared and surface-modified by various polymers. The surface coating or complexation of liposomes was confirmed by monitoring the changes in surface charge before and after surface modification. The mucoadhesion of the liposome-based formulations has been examined both in vitro and in vivo. Using CLSM, the penetration of polymer-coated liposomes or polymer–liposome complexes into a layer of mucus in the rats after oral administration was confirmed. The pharmacological action of the macromolecular drugs (e.g., insulin and calcitonin) was observed after oral administration. In most cases, the polymer-coated liposomes and/or polymer–liposome complexes showed superior pharmacological action compared to other systems without mucoadhesive polymers.

H. Takeuchi (✉)
Laboratory of Pharmaceutical Engineering, Gifu Pharmaceutical University, Gifu 502-8585, Japan
e-mail: takeuchi@gifu-pu.ac.jp

A. Bernkop-Schnürch (ed.), *Oral Delivery of Macromolecular Drugs*,
DOI 10.1007/978-1-4419-0200-9_10,

10.1 Introduction

Drug delivery systems (DDS) are used as a medium or carrier for the controlled delivery of drugs to a patient. DDS allow a suitable amount of drug to be delivered to the appropriate region of the body at the appropriate time. In considering the concept of DDS as an ideal formulation of drug (i.e., the broad view of DDS), it is necessary to focus on the delivery of the drug into the body by selecting the most suitable administration route. Because several new classes of drugs, including peptide drugs, have been developed in recent years, it is important to identify new administration methods in order to maximize the efficacy of drugs. Although some drugs are administered as injections, it is preferable from the point of view of the patient to have a less invasive route, such as oral administration. Such a route also leads to better patient compliance. Not surprisingly, the demand for orally administered drugs has increased.

Drug administration via absorption through a mucus membrane, such as the gastrointestinal (GI) mucosa, nasal mucosa, or lung mucosa, has received much attention as an alternative method of injection. In comparison with the absorption from transdermal system, the absorption from the mucosa is very efficient, because the surface area of the mucosa is about 200 times larger than that of the epidermis. However, some drugs cannot be absorbed from the mucosa due to their particular characteristics, e.g., high molecular weight, high hydrophilicity, or high hydrophobicity. These disadvantageous characteristics are especially typical of the new peptide drugs, which are therefore good candidates for new dosage forms. In addition, these drugs are susceptible to enzymatic degradation on the mucosa when administered by other routes. Fine particles are therefore recommended as a carrier to transport this type of drug, and the design, size, and structure of these particles must be suitable for effective drug delivery.

Many researchers have reported on the problems of oral administration of peptide drugs, i.e., the physical and chemical instability of peptide drugs in the GI tract or biological environment, the very low absorption at the site of administration, the degradation by enzymes, and so on (Iwanaga et al. 1997; Kotze et al. 1997; Takeuchi et al. 1996, 2003). Many strategies have been adopted to solve these problems, such as the use of mucoadhesive polymers in tablet dosage forms (Bernkop-Schnurch et al. 1998), the synthesis and modification of mucoadhesive polymers to improve drug efficiency (van der Merwe et al. 2004a, b), the development of special dosage forms by utilizing micro- or nanoparticles or nanocapsules as a drug carrier (Aboubakar et al. 1999; Agnihotri et al. 2004; Hu et al. 2004; Sakuma et al. 2002), the use of enzymatically controlled delivery systems targeting the oral dosage form to the colon (Mackay et al. 1997; Sriamornsak 1998, 1999), and the use of pH-sensitive hydrogel nanospheres and a copolymer network as carriers to the targeting site (Ichikawa and Peppas 2003; Torres-Lugo et al. 2002).

Liposomal formulations are one of the most promising particulate carriers and have been used for several years for oral delivery of peptide drugs such as insulin. There have been several reports demonstrating the effectiveness of oral administration of insulin using liposomes (e.g., Patel and Ryman 1981). However, the results of these studies indicate that the influences of the liposomal formulations on drug absorption are not predictable or reproducible, and in fact some of them have reported a marked hypoglycemic response. To overcome this discrepancy, Takeuchi and coworkers have tried to develop more effective liposomal formulations, including polymer-coated liposomes, as drug delivery carriers (Takeuchi et al. 1996, 2001, 2003, 2005a). There have also been several reports concerning modified liposomal formulations (Gutierrez de Rubalcava et al. 2000; Katayama et al. 2003; Kim et al. 1999; Li et al. 2003). Charged polymers can form complexes with oppositely charged liposomes to produce so-called polymer–liposome complexes. These complexes have shown good mucoadhesive properties and pharmacological action (Sriamornsak et al. 2008; Thirawong et al. 2008b).

In this chapter, the preparation methods and efficacy of novel lipid particles such as polymer-coated liposomes and polymer–liposome complexes for peptide drug delivery are described. Both systems have common characteristics of mucoadhesion.

10.2 Oral Administration of Peptide Drugs with Mucoadhesive Liposomal Formulations

10.2.1 Mucoadhesive Dosage Forms

Mucoadhesive dosage forms are an attractive method to improve the bioavailability of peptide drugs. Since mucoadhesion can prolong the residence time of drug carriers at the absorption sites, improved drug absorption is expected with a combination of mucoadhesiveness and controlled drug release from the devices. Longer and colleagues (Longer et al. 1985) first showed that delayed GI transit induced by bioadhesive polymers could lead to increased oral bioavailability of a drug. Other researchers have prepared a multiunit bioadhesive system by coating microspheres of poly-hydroxyethyl-methacrylate with mucoadhesive polymers using laboratory-scale equipment (Lehr et al. 1990, 1992). Akiyama and coworkers (Akiyama et al. 1995) prepared a polyglycerol ester of fatty acid-based microspheres coated with Carbopol934P (CP) and CP-dispersing microspheres to evaluate their mucoadhesive properties. In developing colloidal drug delivery systems, Lenaerts and colleagues (1990) confirmed that the bioavailability of vincamine was improved with the nanoparticulate systems having mucoadhesive properties.

So far, a wide range of polymeric materials, particularly hydrophilic polymers containing numerous hydrogen bond (H-bond)-forming groups, have been tested for their mucoadhesive properties (Grabovac et al. 2005). It has been proposed that the interaction between the mucus and mucoadhesive polymers is a result of physical entanglement and secondary bonding, mainly H-bonding and van der Waals attraction. These forces are related to the chemical structure of the polymers (Duchene et al. 1988). The types of surface chemical groups of mucoadhesive polymers (e.g., Fig. 10.1) that contribute to mucoadhesion include hydroxyl, carboxyl, amine, and amide groups (Gurny et al. 1984). Peppas and Buri (1985) have suggested that the polymer characteristics that are necessary for mucoadhesion are (i) strong H-bonding groups, (ii) strong anionic charges, (iii) high molecular weight, (iv) sufficient chain flexibility, and (v) surface energy properties that readily allow the material to spread across mucus.

Park and Robinson (1984) found that cationic and anionic polymers bound more effectively than neutral polymers, polyanions were better than polycations in terms of binding/potential toxicity, and water-insoluble polymers allowed greater flexibility in dosage form design compared to rapidly or slowly dissolving water-soluble polymers. Anionic polymers with sulfate groups bound more effectively than those with carboxylic groups. The degree of binding was proportional to the charge density on the polymer, and highly binding polymers included carboxymethyl cellulose, gelatin, hyaluronic acid, carbopol, and polycarbophil. Rao and Buri (1989) showed that polycarbophil and sodium carboxymethylcellulose (NaCMC) adhered more strongly to mucus than hydroxypropylmethylcellulose (HPMC), methylcellulose, or pectin. Better adhesion occurred in the stomach than in the intestine.

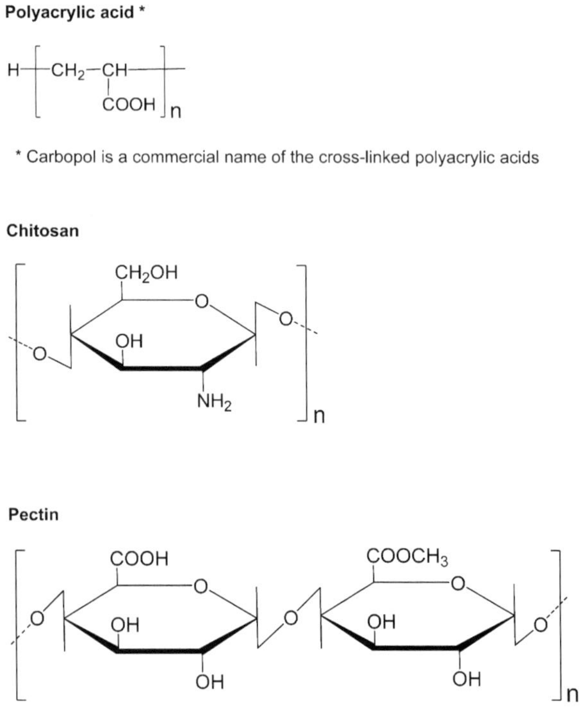

Fig. 10.1 Structure of some mucoadhesive polymers

The incorporation of protein into the matrices of mucoadhesive polymers (e.g., alginate, chitosan, and pectin) can be done under relatively mild environment and hence the chances of protein denaturation are minimal. The limitations of these polymers, like drug leaching during preparation can be overcome by different techniques which increase their encapsulation efficiency. Cross-linked alginate or pectin has more capacity to retain the entrapped drugs and mixing them with other polymers such as neutral gums, chitosan, and Eudragit® has been found to solve the problem of drug leaching (George and Abraham 2006). The favorable properties like biocompatibility, biodegradability, pH sensitiveness, and mucoadhesiveness have enabled these polymers to become the choice of the pharmacologists as oral delivery matrices for protein or peptide drugs.

Several properties of chitosan make it a good candidate for a mucoadhesive polymer coating, i.e., its non-toxicity, biocompatibility, and biodegradability (Galovic Rengel et al. 2002; Prabaharan and Mano 2004). Moreover, several studies have highlighted the potential use of chitosan to enhance the absorption of drugs into intestinal epithelial cells and to rearrange tight junction proteins (e.g., Pan et al. 2002). Recently, some research groups have found that differences in the properties of chitosan lead to differences in the effectiveness of the drug delivery system. For example, Luangtana-anan et al. (2005) reported that the physicochemical properties of nanoparticles were dependent on the types and molecular weights of chitosan and that the salt form and molecular weight of chitosan played a crucial role for optimization of protein-loaded nanoparticles, leading to a broad applicability to pharmaceutical applications. Chae et al. (2005) concluded that the absorption of chitosan was significantly influenced by its molecular weight, with higher molecular weight leading to reduced absorption.

Pectin is regarded as safe for human consumption and has been used successfully for many years in food and pharmaceutical industries. Because it is rich in carboxylic groups and can interact with functional groups in the mucus layer, it has been used as a mucoadhesive polymer for controlled drug delivery (e.g., Liu et al. 2005; Schmidgall and Hensel 2002; Thirawong et al. 2007, 2008a). Liu et al. (2005) reported that pectin with higher net electrical charges showed a higher mucoadhesion with porcine colonic tissues than the less charged ones. Recently, Thirawong et al. (2007, 2008a) reported the mucoadhesive performance of various pectins on the gastrointestinal tract, investigating by texture analysis and viscometric study.

10.2.2 Mucoadhesive Liposomes

Mucoadhesive liposomes are a new type of particulate drug carrier for oral administration of drugs. They are easily prepared by mixing a liposomal suspension with mucoadhesive polymers such as chitosan and carbopol. The polymer-coated liposomes were then formed consecutively. The basic mechanism in the preparation of polymer-coated liposomes is the formation of ion-complexes on the surface of liposomes. In the case of positively charged polymers (e.g., chitosan), negatively charged liposomes are prepared and mixed with the chitosan solution. Takeuchi and coworkers developed mucoadhesive liposomes by coating the anionic liposomal surface with a cationic mucoadhesive polymer, i.e., chitosan, and observed their mucoadhesive properties both in vitro and in vivo (Takeuchi et al. 1994, 1996). The typical anionic liposomal formulations contain L-α-dipalmitoylphosphatidylcholine (DPPC) and dicetyl phosphate (DCP) in a ratio of 8:2 or L-α-distearoylphosphatidylcholine (DSPC), DCP, and cholesterol in a ratio of 8:2:1. In the case of the preparation of carbopol-coated liposomes, stearyl amine (SA) has been used to confer the positive charge to the formulated liposomes (Takeuchi et al. 2003). However, the amount of SA used was less than one-tenth the amount of DCP, because a higher amount of SA in the formulation caused the carbopol-coated liposomes to aggregate.

In a typical procedure, 161 mg (204 μmol) of DSPC, 27.9 mg (51 μmol) of DCP, and 9.7 mg (25 μmol) of cholesterol were dissolved in a small amount of chloroform, and the solution was rotary evaporated at 40°C to obtain a thin lipid film. The thin lipid film was dried in a vacuum oven overnight to ensure complete removal of the solvent, and hydration was carried out with 5 mL of acetate buffer solution (pH 4.4, 100 mM) by vortexing, followed by incubation at 10°C for 30 min. The size of multilamellar liposomes can be controlled by applying sonication or by use of an extruder. The particle size of liposomes can be adjusted by controlling the sonication time or by using an extruder filter with the appropriate pore size. For example, submicron-sized liposomes (ssLip) of ca. 200 nm were prepared by sonicating three times for 3 min each. The particle size of the resultant liposomes was confirmed by dynamic light scattering analysis. For preparation of chitosan-coated liposomes (CS-Lip), an aliquot of the liposomal suspension was mixed with the same volume of acetate buffer solution (pH 4.4) of chitosan (0.6%), followed by incubation at 10°C for 1 h. The submicron-sized chitosan-coated liposomes (ssCS-Lip) were prepared in the same manner.

The formation of a coating layer on the surface of liposomes was confirmed and detected by measuring the zeta potential. As shown in Fig. 10.2, the zeta potential of liposomes was changed by increasing the concentration of polymers, because the surface charge of liposomes was neutralized by the opposite charge of the coating polymer.

These polymer-coated liposomes showed high potency in oral delivery of peptide drugs such as insulin and calcitonin, mainly because of the mucoadhesion of the chitosan-coated liposomes to the intestinal tract (Takeuchi et al. 1996, 2001, 2003, 2005a). Similar trials have been reported by Guo et al. (2003), who investigated the effect of chitosan concentration and lipid type on the characteristics of chitosan-coated liposomes and their interactions with leuprolide. They found that a thicker adsorptive layer could be realized by using low

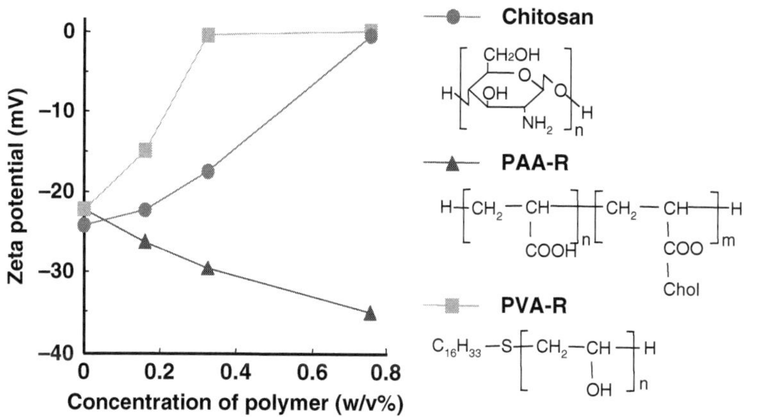

Fig. 10.2 Zeta potential of mucoadhesive liposomes in a phosphate buffer solution (pH 7.4) (taken from Takeuchi et al. 1996). Lipid composition: DPPC:DCP = 8:2

purity lipids resulting from a stronger electrostatic attraction with chitosan. As to particles from high purity lipids, polymer bridging caused flocculation at low polymer concentration, while at high concentration, the adsorbed chitosan molecule led to steric stabilization. Drug entrapment efficiency decreased as chitosan was added to liposomes, showing the disturbance of bilayers. The interaction between chitosan and the polar head groups on the surface of phospholipid bilayers may interfere with leuprolide entrapped in liposomes and result in the leakage of leuprolide.

In addition to complex formation, another mechanism of polymer coating on the surface of liposomes is anchoring of the hydrophobic part of the polymer molecules (Fig. 10.3). Takeuchi and coworkers have demonstrated the formation of a thick coating layer by using polyvinyl alcohol molecules ending in long alkyl chains (PVA-R) (Takeuchi et al. 2000). Thongborisute et al. (2006b) prepared hydrophobically modified chitosan, i.e., dodecylated chitosan. The zeta potential of dodecylated chitosan-coated liposomes showed positive values in both liposomal formulations, i.e., negatively charged and neutral-charged liposomes. These results indicated that dodecylated chitosan could be considered a more suitable polymer for coating neutral-charged liposomes than chitosan because the hydrophobic side chain of dodecylated chitosan inserts itself into the lipid bilayer of liposomes. Moreover, chitosan seemed to be less effective for coating a neutral-charged liposome because of the low positive values of its zeta potential. Chitosan provided only an electrostatic force when used for coating liposomes, while dodecylated chitosan provided both an electrostatic and a hydrophobic force due to the long alkyl chain in its backbone.

In designing polymer-coated liposomes containing peptide drugs, attention should also be paid to their structure. Drugs can be loaded into a liposomal

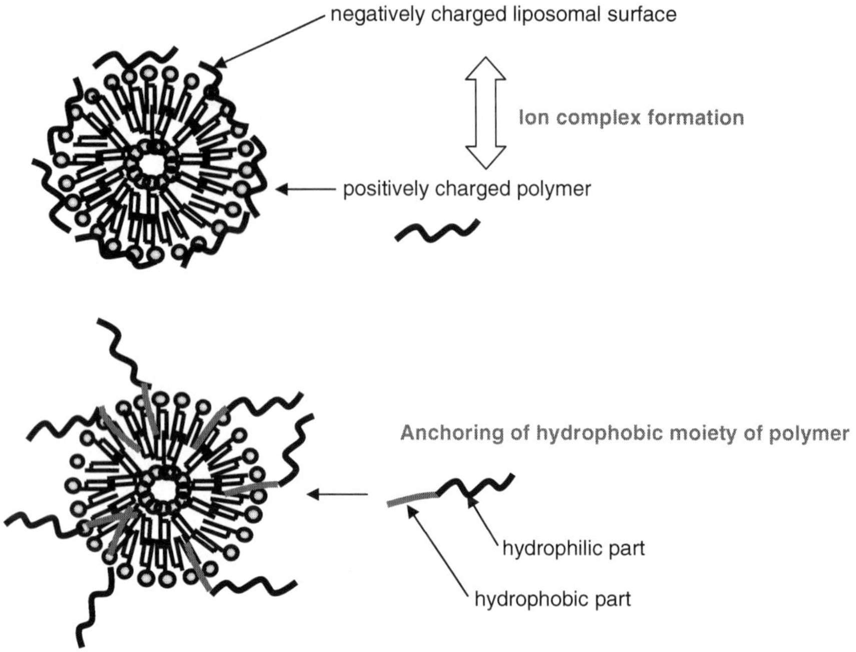

Fig. 10.3 Schematic drawings of polymer-coated liposomes

system in one or both of two ways. They can be encapsulated within liposome particles or they can be entrapped on the outside of liposome particles by adsorption onto their surfaces. In preparing the liposomes with the hydration method, at least a part of drug can be encapsulated into the liposomal particles, while all of the drug molecules will exist on the outside of liposomal particles when the preformed liposomal suspension is mixed with a drug solution. Although the difference in the drug-loading method is important, few papers have reported on the effects of the resulting pharmaceutical function of these two loading systems (Thongborisute et al. 2006c).

It is not easy to evaluate the mucoadhesive properties of fine particulate drug carriers such as liposomes. In developing colloidal drug delivery systems, Lenaerts et al. (1990) demonstrated the mucoadhesive properties of polyalkylcyanoacrylate nanoparticles with autoradiographic studies. Pimienta et al. (1992) investigated the bioadhesion of hydroxypropylmethacrylate nanoparticles or isohexylcyanoacrylate nanocapsules coated with poloxamers and poloxamine on rat ileal segments in vitro using a labeled compound.

Takeuchi et al. (1996) have confirmed the mucoadhesive properties of polymer-coated liposomes with a simple particle counting method in vitro. This mucoadhesion test was carried out using intestines isolated from male Wistar rats. The intestines were washed with saline solution, filled with a liposomal suspension that was diluted 100-fold with a buffer solution (pH 7.4, 6.5, 5.6 or 1.2) or saline solution (pH 6.0), and sealed with closures. The intestines were incubated in saline solution at 37C and the number of liposome particles was measured with a Coulter counter both before and after incubation. The adhesive% was calculated using the following equation:

$$Adhesive\ \% = \frac{N_o - N_s}{N_o} \times 100,$$

where N_o and N_s are the number of liposomes before and after incubation, respectively. As shown in Fig. 10.4, the adhesive% of polymer-coated liposomes was significantly higher than that of non-coated liposomes. When the concentration of chitosan used for coating was decreased, the adhesive% of chitosan-coated liposomes decreased. These results confirmed that the mucoadhesive properties of the polymer-coated liposomes were conferred by the polymer layer fixed on the surface of the liposomes.

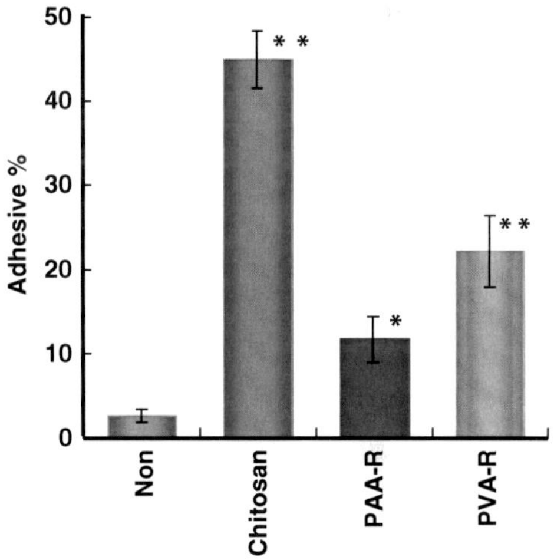

Fig. 10.4 Adhesive% of polymer-coated liposomes in rat intestine (taken from Takeuchi et al. 1996). Lipid composition: DPPC:DCP=8:2. Dispersion medium: phosphate buffer solution

The mucoadhesive properties was also confirmed by using a confocal laser scanning microscope (CLSM) and an in vivo test in rats (Takeuchi et al. 2005a, b). The protocol is shown in Fig. 10.5. This CLSM observation confirmed the retention of liposomes in the intestinal tract after oral administration. Chitosan coating of the liposomal surface could lead to an increase in the retention time in the intestinal tract due to the resulting mucoadhesion. It was also found that the retention profiles were dramatically changed by reducing the particle size of liposomes. As shown in Fig. 10.6, the submicron-sized chitosan-coated liposomes (ssCS-Lip) were shown to penetrate into the mucosal layer deeply. Because large multilamellar liposomes did not exhibit this behavior, it was confirmed that the particle size was a very important factor for controlling the behavior of the drug carriers in the intestinal tract.

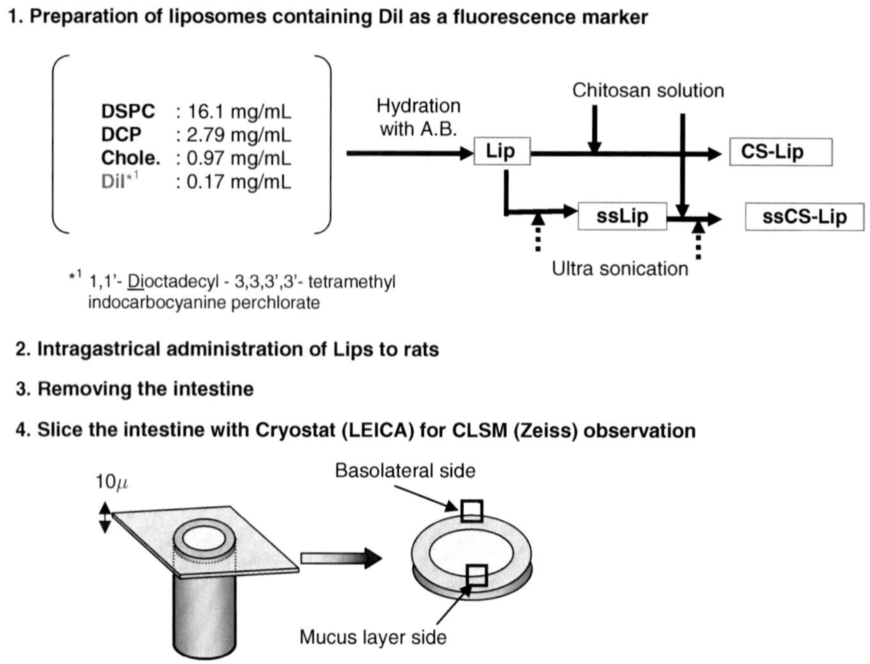

Fig. 10.5 Protocol for measurement of the mucoadhesive behavior of polymer-coated liposomes with confocal laser scanning microscopy (CLSM)

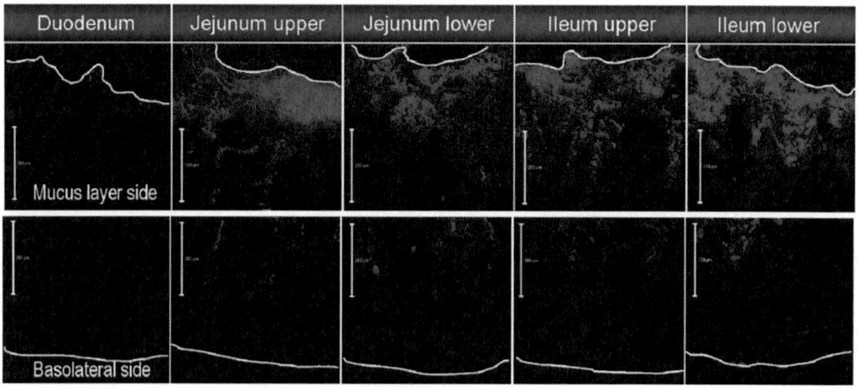

Fig. 10.6 Confocal laser scanning microscopy photographs of various parts of the intestinal tract of the rat at 2 h after intragastric administration of ssCS-Lips (taken from Takeuchi et al. 2005a). The measured mean diameter is 281.2 nm. The formulation of liposome is DSPC:DCP:cholesterol = 8:2:1

In order to observe the behavior of chitosan-coated liposomes more precisely, chitosan was labeled with fluorescein isothiocyanate (FITC) via chemical reaction at the isothiocyanate group of FITC and the primary amino group of chitosan; the liposomes (Lips) were marked by incorporation of DiI into the liposomal formulation. FITC-labeled chitosan (FITC-CS), non-coated liposomes, and FITC-labeled chitosan-coated liposomes (FITC-CS-Lips) were intragastrically administered into male Wistar rats, and then the behavior of the molecules was visualized by CLSM (Thongborisute et al. 2006a). The results demonstrated that the chitosan molecules themselves, as well as the liposomes, could penetrate across the intestinal mucosa. Moreover, the CLSM images demonstrated a lack of separation of the chitosan molecules from the surface of the liposomes after the administration of chitosan-coated liposomes.

10.2.3 Mucoadhesive Polymer–Liposome Complexes

Adsorption of polyions onto oppositely charged particles or liposomes has attracted considerable attention in the last decades. This phenomenon, which plays an important role in many aqueous solutions of biological samples, is of great interest in gene therapy, where the lipoplexes, i.e., complexes formed between deoxyribonucleic acid (DNA) and positively charged liposomes (or lipids), have been intensively investigated. Recently, the interactions between charged lipids and oppositely charged biopolymers have been studied (Antonietti and Wenzel 1998; Raviv et al. 2005). The highly ordered structures formed by polyelectrolyte–lipid complexes are of great interest in material sciences as templates and building blocks for hierarchiral supramolecular assembly (Antonietti and Wenzel 1998).

The inclusion of polymers in liposomes has been studied as a means to increase the stability of liposomes and to modify their functional properties. Grafting hydrophilic polymers onto the phospholipid head groups has been shown to increase the circulation time as well as to inhibit the liposome fusion (Cho et al. 2007). Conformational changes in the adsorbed polymers can cause a partial or total rearrangement in the structure of liposomes. Seki and Tirrell (1984) were the first to report that the structure of liposomes is affected by the addition of amphiphilic polyelectrolytes. Henriksen et al. (1994) reported the formation of chitosomes, i.e., liposome–chitosan complexes. The complexes were formed by an initial adsorption of chitosan polymer onto the liposomal surface and then an alignment of these positive surface patches with negative patches on approaching liposomes. This phenomenon occurred when small quantities of chitosan were used. With an excess amount of chitosan, the restabilization occurred due to a charge reversal of the polymer-coated liposomes. The ionic strength, chitosan quality, lipid/polymer ratio, and pH had a significant effect on the resulting aggregate size (Henriksen et al. 1997).

Recently, self-assembling nanocomplexes between pectin and cationic liposomes, i.e., pectin–liposome nanocomplexes (PLNs), have been designed using a method similar to that for the ion-complex formation of carbopol with positively charged ions on the surface of the liposomes (Sriamornsak et al. 2008). Cationic liposomes containing SA were prepared, since the SA led to the complexation between liposome particles and pectin molecules when mixing the liposome samples with the pectin dispersion. The electrostatic interactions between the positively charged SA in liposomes and the negatively charged pectin chain were confirmed by FTIR studies (Sriamornsak et al. 2008). The high resolution of the AFM was used to characterize the structure of pectin, liposomes, and PLNs. Chain-like structures with a small number of branches were imaged for the pectin sample while spherical structures of the small unilamellar vesicles of liposomes were observed (Fig. 10.7). The attachment or association of cationic liposomes on the pectin chain forming self-assembling

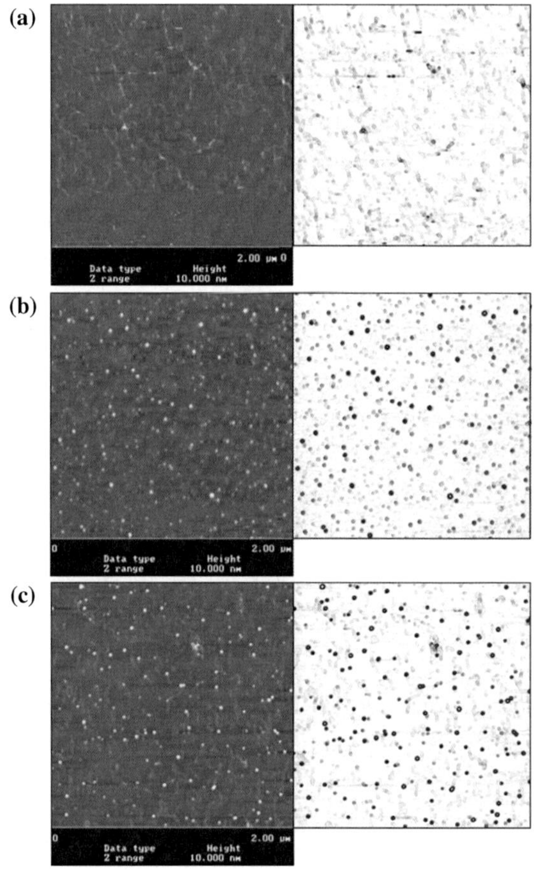

Fig. 10.7 Topographical (*left*) and equivalent processed (*right*) images from atomic force microscopy (AFM) of (**a**) pectin dispersion, (**b**) cationic liposomes, and (**c**) pectin–liposome nanocomplexes (PLNs) (modified from Sriamornsak et al. 2008)

supramolecular complexes (i.e., PLNs) was also visualized (Fig. 10.7). Figure 10.8 demonstrates the proposed model of the interaction between pectin and cationic liposomes to form PLNs, derived from the AFM imaging results. The PLN structure is different from the structure of lipoplexes for which the AFM images exhibited a large variety of different structures, such as loops of DNA extending from a central core of liposomes, globular structures, and so on (Wangerek et al. 2001).

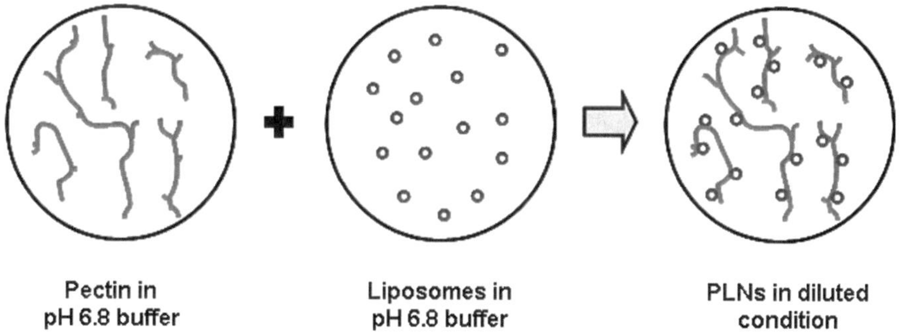

Fig. 10.8 Schematic representation of the formation of pectin–liposome nanocomplexes (PLNs) (taken from Sriamornsak et al. 2008)

In order to measure the mucoadhesive properties of PLNs, fluorescein isothiocyanate-dextran, which has a molecular weight of 4300 Da (FD4) was used as a fluorescent marker for CLSM observation (Thirawong et al. 2008b). The mucoadhesive behavior of PLNs in the GI mucosa was evaluated by observing the residual FD4 with CLSM. High FD4 intensity after intragastric administration of FD4 solution into male Wistar rats was rarely found in the stomach and colon, but a very small amount of FD4 could be observed in the small intestine at 1 h after intragastric administration (Fig. 10.9). This result implied that the FD4 solution could not be retained in the GI tract after intragastric administration, but rather was eliminated or washed out from the GI tract during the first hour. After administration of the mixture of pectin and FD4, a high intensity of FD4 was seen in the duodenum and jejunum after the first hour (Fig. 10.9), and this intensity decreased after the second and sixth hour. The explanation for the mucoadhesive effect of pectin was that pectin adhered at the mucosal surface, particularly at the duodenum and jejunum, but it could not reach the colon during the in vivo test. Subsequently, FD4 in the mixture could penetrate through epithelial cells. This result agreed with the in vitro study by Schmidgall and Hensel (2002), who reported that the

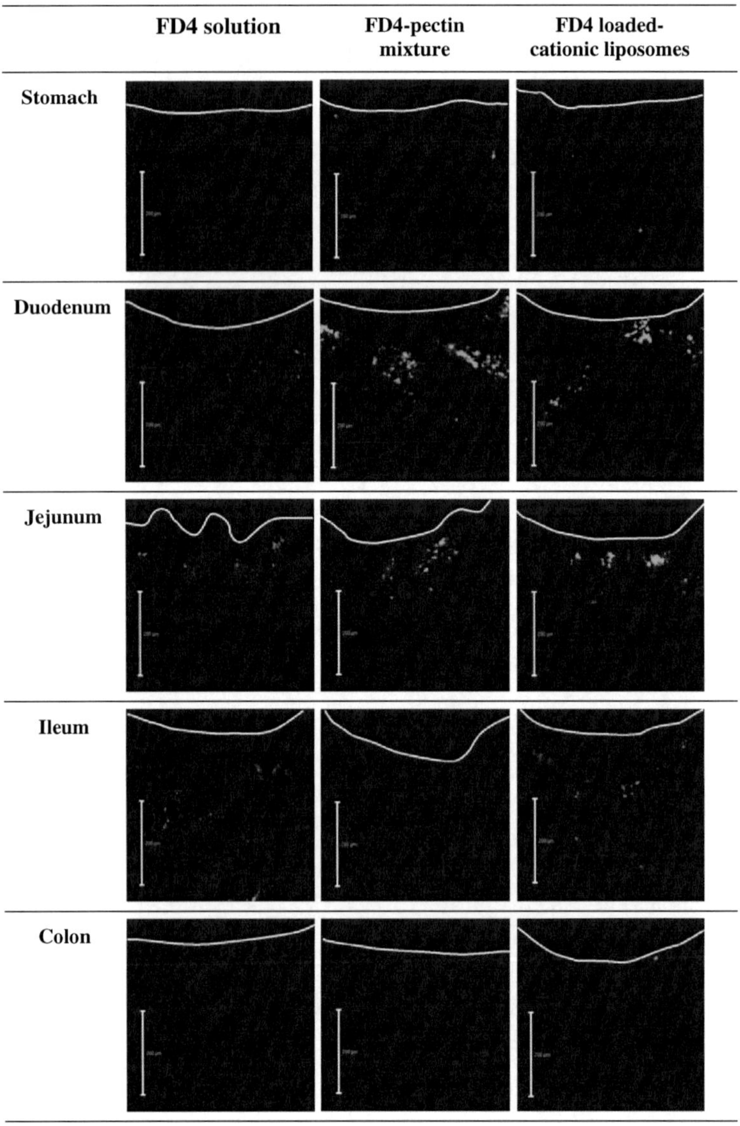

Fig. 10.9 CLSM images of FD4 intensities in the stomach, small intestine, and colon at 1 h after intragastric administration of FD4 solution, FD4–pectin mixture, and FD4-loaded cationic liposomes (taken from Thirawong et al. 2008b)

polygalacturonides derived from pectin could be adsorbed onto the model surface (i.e., colonic mucosa) after incubating the mucosa with FITC-abeled sugar beet pectin, indicating a specific binding to surface structure of the mucosa. It is well known that the mucosa restrict the penetration of large molecules and possibly of particles as well. It has been generally believed that macromolecules hardly pass through the absorptive enterocytes of the mucosa. However, as shown in the CLSM micrographs in Fig. 10.9, the intensity of FD4 was greater, especially in the duodenal part of the small intestine, when a mixture of pectin and FD4 was administered than when a solution of FD4 alone was used, suggesting the penetration of FD4 into the mucosa. It is probable that pectin behaved as a mucopenetrative vehicle through the rat intestinal mucosa. This is similar to the result using chitosan-coated liposomes as observed by Takeuchi et al. (2005a). CLSM images of FD4 intensities in the stomach, small intestine, and colon at 1 h after intragastric administration of FD4-loaded cationic liposomes are also shown in Fig. 10.9. A small amount of FD4 was observed in the epithelium of the stomach and colon, while a larger amount was observed in the epithelium of the small intestine after administration of FD4-loaded cationic liposomes. This observation may have resulted from the electrostatic interaction between positively charged cationic liposomes and negatively charged sialic acid residue of mucin, which facilitated the adhesion to mucus membrane.

After administration of PLNs, a small amount of FD4 was detected in the stomach and colon, but a very large amount of FD4 was detected in the duodenum, jejunum, and ileum from the first hour until the sixth hour (Fig. 10.10). FD4 from PLNs most likely remained in the duodenum, jejunum, and ileum, which suggested their excellent intestinal mucoadhesive properties. PLNs showed a stronger mucoadhesion than cationic liposomes. The viscosity of the pectin in PLNs may have partly contributed to the prolongation of the residence time of the liposomes in the intestine. The results also suggested that pectin could prolong the adherence of liposomes to the GI mucosa. This implied that the number of carboxyl groups was a predominant factor for the in vivo mucoadhesive properties of pectin.

The mucoadhesive effect of pectin involved the H-bonding and, perhaps, the hydrophobic interaction between the pectin chain and mucin in the mucus layer rather than the electrostatic attraction (Thirawong et al. 2007). However, most of the carboxyl groups in pectin and sialic acid in mucin can be ionized at physiological pH which is higher than their p*K*a. Then, the electrostatic repulsion between pectin and mucin may play a more important role than H-bonding, leading to a coil expansion which facilitates molecular entanglements (Liu et al. 2005). The other possible explanation is that a polymer with a longer chain length may constitute a barrier to the absorption of drugs from liposomes (Thongborisute et al. 2006c). The diffusion of FD4 through the mucus gel layer might be inhibited by the long chain of pectin with high molecular weight, resulting in a decrease in FD4 intensity.

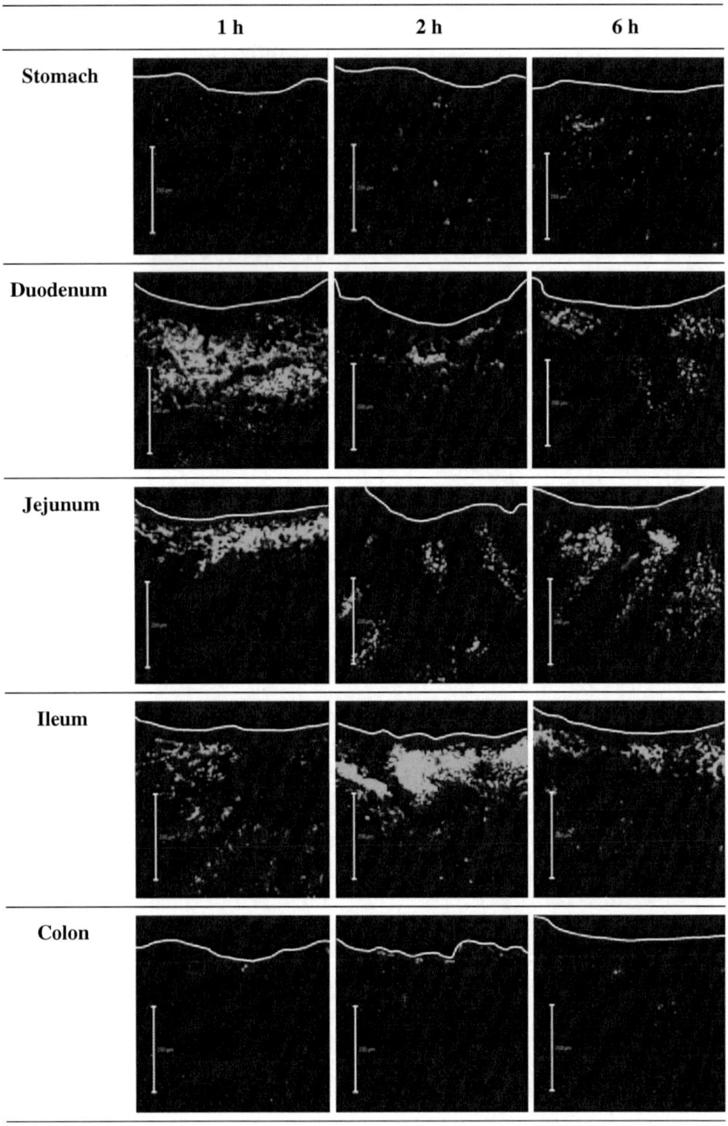

Fig. 10.10 CLSM images of FD4 intensities in the stomach, small intestine, and colon at 1, 2, and 6 h after intragastric administration of PLNs (taken from Thirawong et al. 2008b)

10.3 Absorption of Macromolecules with Liposomal Formulations

The mucosa of the small intestine is lined by a simple columnar epithelium, which evaginates into villi and invaginates into crypts, as shown schematically in Fig. 10.11. The lamina propria of the small intestine forms the core of villi

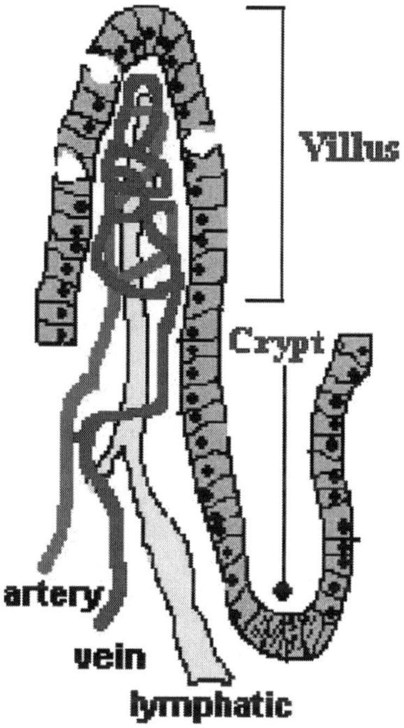

Fig. 10.11 Schematic drawing of the villus showing the blood capillaries and lymphatic vessel (source: http://www.xinminss.moe.edu.sg:1285/biology/Notes/lesson/voyagelesson.ppt)

where large lymphatic and blood vessels are located to facilitate the transport of the ingested nutrients or drugs (Schumacher and Schumacher 1999). The lamina propria of villi includes lacteals (lymphatic capillaries) which provide passage for absorbed fat (the chylomicrons) as well as lipophilic drugs and lipid formulations into the lymphatic drainage of the intestine.

According to the histology of the intestine, the drug-loaded liposomes (probably in the form of liposomes themselves, polymer-coated liposomes, or polymer–liposome complexes) could be easily absorbed into the lymphatic circulation, especially by lacteals. Compounds absorbed by the intestinal lymphatics drain via the thoracic lymph and enter the systemic circulation at the junction of the left internal jugular vein and the left subclavian vein, thereby avoiding potential first-pass metabolism. Consequently, drug transport via the intestinal lymphatics may confer delivery advantages in terms of increased bioavailability (via a reduction in presystemic metabolism) and the possibility of directing delivery to the lymphatic system (Porter and Charman 1997).

In fact, the CLSM studies were used to confirm the mucoadhesion of the dosage forms, not the absorption of the active agent into the circulation. However, the adherence of the dosage form at the mucosa would increase the chance for its absorption and then would allow the drug to take action.

Permeation of macromolecules through the GI epithelial cells after oral administration has been reported by several research groups. We have also confirmed the permeation of the hydrophilic macromolecules by using Caco-2 cells (Nagata et al. 2006). However, there are no data clearly showing the actual absorption of macromolecular drugs into the GI tract after oral administration.

The actual absorption of a model macromolecule (i.e., FD4) encapsulated in chitosan-coated liposomes through the mucus layer of the intestinal tract in rats was determined (Thongborisute et al. 2008). FD4 was used as a model drug because of its molecular weight (which is close to those of insulin and calcitonin) as well as its water solubility and hydrophilic properties. FD4 was also preferable because it is not susceptible to enzymatic degradation in the intestinal tract. In the experiment, the chitosan-coated liposomes containing FD4 at a dose of 8 mg FD4 per rat (five, 9-week-old male Wistar rats weighing 190–200 g were used after being fasted for 12 h) were intragastrically administered. The FD4 concentrations in plasma were measured at each of several time intervals with a fluorescence microplate reader at λ_{Ex} 485 nm and λ_{Em} 520 nm. The percentages of FD4 entrapment efficiency in liposomes were 22, 12, and 12% for non-coated liposomes (Non-Lips), CS-Lips, and LCS-Lips, respectively. As shown in Fig. 10.12, after the chitosan-coated liposomes containing FD4 were intragastrically administered to the rats, the FD4 concentration in plasma immediately increased, and then gradually decreased and remained constant up to 72 h. In

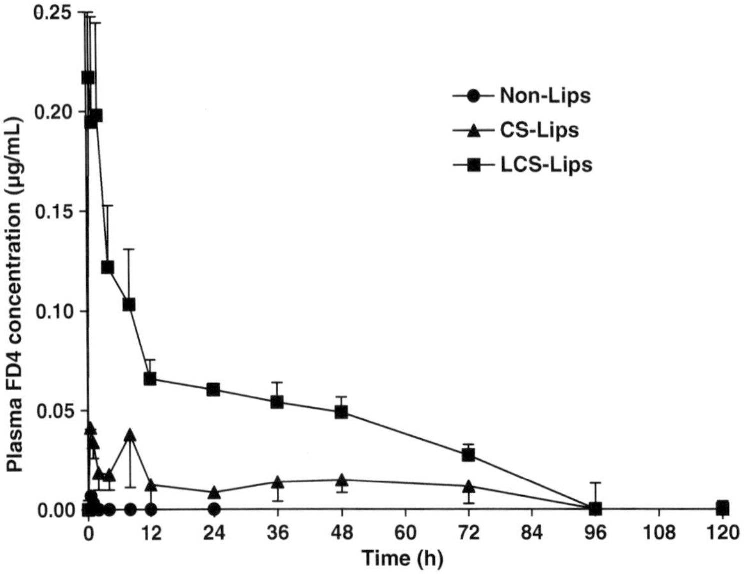

Fig. 10.12 Blood FD4 concentration profiles after oral administration of Non-, CS-, and LCS-coated liposomes encapsulating FD4 in rats ($n = 5$). CS: chitosan of molecular weight 150,000. LCS: chitosan of molecular weight 20,000

contrast, the plasma FD4 concentration after administration of Non-Lips containing FD4 was very low in the first 1 h and no FD4 could be detected in the bloodstream after 2 h. Following the FD4 intravenous injection, the FD4 was completely eliminated from the blood circulation within 1 h (data not shown). Thus, the maintenance of the FD4 concentration in plasma after administration of these coated liposomes could be attributed to continuous drug absorption over a long period. In comparing the two types of chitosan, the plasma FD4 concentration after administration of LCS-Lips was higher than that after administration of CS-Lips, which correlated well with the results of the pharmacological effects (discussed later in Section 10.4).

10.4 Pharmacological Action After Oral Administration of Liposomal Formulations

It was expected that the mucoadhesive properties of polymer-coated liposomes and polymer–liposome complexes would lead to a prolongation of the residence time in the intestinal tract for more effective drug absorption. Several pharmacological studies have been performed to determine the drug absorption and pharmacological action following this delivery modality. For example, insulin-loaded liposomes coated with chitosan have been shown to reduce the blood glucose level in animal models. The pharmacological test showed a continuous decrease of blood glucose level for 12 h. Liposomes without coating did not show such an hour-long pharmacological effect. These results confirmed the effectiveness of the mucoadhesive properties of chitosan-coated liposomes in vivo (Takeuchi et al. 1996). In this case, it was estimated that the liposomes coated with chitosan released the drug during the residence time on or within the mucus layer in the GI tract. This led to an increase in the drug concentration at the mucus layer, followed by absorption of the drug. When calcitonin was used as another model peptide drug, liposomes coated with chitosan showed a much greater pharmacological effect (decrease of calcium concentration in plasma) than non-coated liposomes (Takeuchi et al. 2001, 2003). When the size of liposomes was decreased to within the submicron range, the pharmacological effect was prolonged much further (Fig. 10.13) (Takeuchi et al. 2001, 2005a). After the administration of submicron-sized liposomes coated with chitosan to rats, their intestinal tubes were observed with CLSM. The images demonstrated that the liposomal particles penetrated into the enteric mucus layer and then into the mucus layer. These findings indicate that the pharmacological efficiency after oral administration of the liposomal systems was in good agreement with the drug absorption results in all cases.

The effectiveness of PLNs for the oral administration of calcitonin was also confirmed by a pharmacological test (Thirawong et al. 2008b). The efficiency of liposome complexes was indicated by a decrease in the plasma calcium

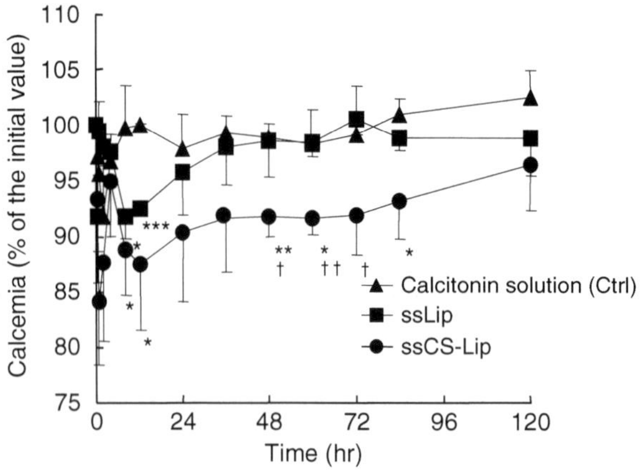

Fig. 10.13 Profiles of plasma calcium levels after intragastric administration of submicron-sized liposomes, ssLip, and ssCS-Lip containing calcitonin (taken from Takeuchi et al. 2005a) The measured mean particle sizes of ssLip and ssCS-Lip were 196.4 and 473.4 nm, respectively. The formulation of liposomes was DSPC:DCP:cholesterol = 8:2:1. The concentration of chitosan for coating was 0.3%. *$p<0.05$, **$p<0.01$, ***$p<0.001$: significantly different from the level for calcitonin solution; $^{\dagger}p<0.05$, $^{\dagger\dagger}p<0.01$: significantly different from the level for ssLip ($n=3$ in each case).

concentration. When calcitonin-loaded PLNs were administered to the rats, the calcium concentration in plasma was reduced significantly compared with that following administration of calcitonin solution. A significant difference in the plasma calcium concentration was observed up to 48 h, suggesting the effectiveness of the mucoadhesive liposomal formulation in enhancing the absorption of the peptide drug, similar to those of chitosan-coated liposomes. The area above the plasma calcium concentration versus time curve (AAC) value was the lowest after administration of calcitonin solution (Fig. 10.14), owing to the degradation of calcitonin in the crucial physiological environment of the GI tract (Thongborisute et al. 2006c). The calcitonin-loaded PLNs (0.5–1.0%) significantly decreased the plasma calcium concentration after oral administration, as the AAC values were significantly higher than the amount of calcitonin solution. At the same administration dose (i.e., 500 IU/kg rat), the AAC values of both PLNs made of 0.5 and 1.0% pectin were not significantly different. Nevertheless, the twofold increase in the percentage calcitonin dose administered to rats did not result in a significant difference in the plasma calcium concentration. It was thought that the increase in the percentage of pectin would inhibit the diffusion/penetration of calcitonin-loaded liposomes through intestinal epithelial cells due to coil entanglement. However, the calcitonin-loaded PLNs demonstrated a much stronger pharmacological effect, i.e., a significant reduction in plasma calcium concentration, compared with administration of calcitonin solution alone. This probably resulted from the ability of

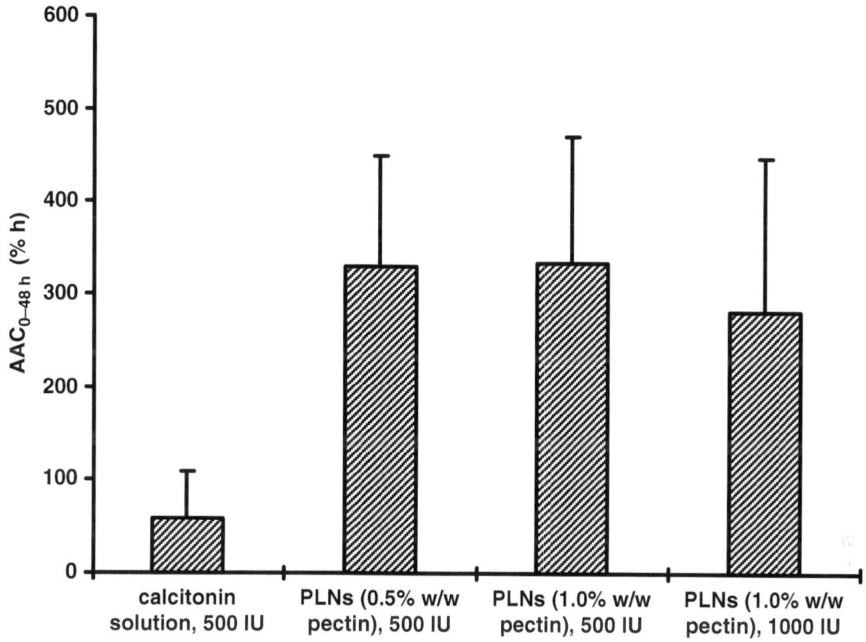

Fig. 10.14 AAC values of calcitonin solution and PLNs over 48 h after intragastric administration into rats, $n=3$ (modified from Thirawong et al. 2008b)

pectin to adhere to the mucus layer and prolong retention in the intestinal mucosa, facilitating the penetration or absorption of calcitonin-loaded liposomes through intestinal epithelial cells.

10.5 Conclusion

Liposome-based mucoadhesive dosage forms are able to adhere and penetrate through the mucus layer of the GI tissues. In this chapter, the liposome-based mucoadhesive dosage form containing mucoadhesive polymers (e.g., chitosan, pectin, carbopol, and PVA-R) was designed and evaluated. Liposomes with different compositions were prepared and coated with various polymers. The surface coating of liposomes was confirmed by monitoring the changes in zeta potential before and after surface modification. The complexation between the cationic liposomes and pectin was also examined by determination of surface charges and AFM.

The mucoadhesion of the liposome-based formulations was examined both in vitro and in vivo. Using CLSM, the penetration of polymer-coated liposomes or polymer–liposome complexes into a layer of mucus in the rats after oral administration was confirmed. The pharmacological action of the model peptide drugs (e.g., insulin and calcitonin) was observed after oral administration. In most cases, the polymer-coated liposomes and/or polymer–liposome complexes showed superior pharmacological action compared to other systems without mucoadhesive polymers. To conclude, all these data indicate that the liposome-based formulations designed herein are potentially useful candidates for oral mucoadhesive delivery systems for macromolecular drugs such as peptides or proteins through the gastrointestinal mucosa.

References

Aboubakar M, Puisieux F, Couvreur P, Vauthier C (1999) Physico-chemical characterization of insulin-loaded poly(isobutylcyanoacrylate) nanocapsules obtained by interfacial polymerization. Int J Pharm 183(1):63–66

Agnihotri SA, Mallikarjuna NN, Aminabhavi TM (2004) Recent advances on chitosan-based micro- and nanoparticles in drug delivery. J Control Rel 100(1):5–28

Akiyama Y, Nagahara N, Kashihara T, Hirai S, Toguchi H (1995) In vitro and in vivo evaluation of mucoadhesive microspheres prepared for the gastrointestinal tract using polyglycerol esters of fatty acids and a poly(acrylic acid) derivatives. Pharm Res 12:397–405

Antonietti M, Wenzel A (1998) Structure control of polyelectrolyte-lipid complexes by variation of charge density and addition of cholesterol. Colloids Surf A Physicochem Eng Asp 135(1–3):141–147

Bernkop-Schnurch A, Humenberger C, Valenta C (1998) Basic studies on bioadhesive delivery systems for peptide and protein drugs. Int J Pharm 165(2):217–225

Chae SY, Jang M-K, Nah J-W (2005) Influence of molecular weight on oral absorption of water soluble chitosans. J Control Rel 102(2):383–394

Cho EC, Lim HJ, Shim J, Park JY, Dan N, Kim J, Chang I-S (2007) Effect of polymer characteristics on structure of polymer-liposome complexes. J Colloid Interf Sci 311(1):243–252

Duchene D, Touchard F, Peppas NA (1988) Pharmaceutical and medical aspects of bioadhesive systems for drug administration. Drug Dev Ind Pharm 14:283–318

Galovic Rengel R, Barisic K, Pavelic Z, Zanic Grubisic T, Cepelak I, Filipovic-Grcic J (2002) High efficiency entrapment of superoxide dismutase into mucoadhesive chitosan-coated liposomes. Eur J Pharm Sci 15(5):441–448

George M, Abraham TE (2006) Polyionic hydrocolloids for the intestinal delivery of protein drugs: Alginate and chitosan – a review. J Control Rel 114(1):1–14

Grabovac V, Guggi D, Bernkop-Schnurch A (2005) Comparison of the mucoadhesive properties of various polymers. Adv Drug Del Rev 57:1713–1723

Guo J, Ping Q, Jiang G, Huang L, Tong Y (2003) Chitosan-coated liposomes: characterization and interaction with leuprolide. Int J Pharm 260(2):167–173

Gurny R, Meyer JM, Peppas NA (1984) Bioadhesive intraoral release systems: design, testing and analysis. Biomaterials 5:336–340

Gutierrez de Rubalcava C, Rodriguez JL, Duro R, Alvarez-Lorenzo C, Concheiro A, Seijo B (2000) Interactions between liposomes and hydroxypropylmethylcellulose. Int J Pharm 203(1–2):99–108

Henriksen I, Smistad G, Karlsen J (1994) Interactions between liposomes and chitosan. Int J Pharm 101(3):227–236

Henriksen I, Vaagen SR, Sande SA, Smistad G, Karlsen J (1997) Interactions between liposomes and chitosan II: Effect of selected parameters on aggregation and leakage. Int J Pharm 146(2):193–203

Hu LD, Tang X, Cui FD (2004) Solid lipid nanoparticles (SLNs) to improve oral bioavailability of poorly soluble drugs. J Pharm Pharmacol 56:1527–1535

Ichikawa H, Peppas NA (2003) Novel complexation hydrogels for oral peptide delivery: in-vitro evaluation of their cytocompatibility and insulin-transport enhancing effects using Caco-2 cell monolayers. J Biomed Mater Res A 67(2):609–617

Iwanaga K, Ono S, Narioka K, Morimoto K, Kakemi M, Yamashita S, Nango M, Oku N (1997) Oral delivery of insulin by using surface coating liposomes: Improvement of stability of insulin in GI tract. Int J Pharm 157(1):73–80

Katayama K, Kato Y, Onishi H, Nagai T, Mashida Y (2003) Double liposomes: hypoglycemic effects of liposomal insulin on normal rats. Drug Dev Ind Pharm 29(7):725–731

Kim A, Yun M-O, Oh Y-K, Ahn W-S, Kim C-K (1999) Pharmacodynamics of insulin in polyethylene glycol-coated liposomes. Int J Pharm 180(1):75–81

Kotze AF, de Leeuw BJ, Lue[ss]en HL, de Boer AG, Verhoef JC, Junginger HE (1997) Chitosans for enhanced delivery of therapeutic peptides across intestinal epithelia: in vitro evaluation in Caco-2 cell monolayers. Int J Pharm 159(2):243–253

Lehr CM, Bouwstra JA, Kok W, De Boer AG, Tukker JJ, Verhoef JC, Breimer DD, Junginger HE (1992) Effects of the mucoadhesive polymer polycarbophil on the intestinal absorption of a peptide drug in the rat. J Pharm Pharmacol 44(5):402–407

Lehr CM, Bouwstra JA, Tukker JJ, Junginger HE (1990) Intestinal transit of bioadhesive microspheres in an in situ loop in the rat – A comparative study with copolymers and blends based on poly acrylic (acid). J Control Rel 13:51–62

Lenaerts V, Couvreur P, Grislain L, Maincent P (1990) Nanoparticles as a gastroadhesive drug delivery system. In: Lenaerts V, Gurny R (eds) Bioadhesive Drug Delivery Systems, CRC Press, Florida, pp 93–104

Li H, Song J-H, Park J-S, Han K (2003) Polyethylene glycol-coated liposomes for oral delivery of recombinant human epidermal growth factor. Int J Pharm 258(1–2):11–19

Liu L, Fishman ML, Hicks KB, Kende M (2005) Interaction of various pectin formulations with porcine colonic tissues. Biomaterials 26:5907–5916

Longer MA, Ch'ng HS, Robinson JR (1985) Bioadhesive polymers as platforms for oral controlled drug delivery III: Oral delivery of chlorothiazide using a bioadhesive polymer. J Pharm Sci 74:406–411

Luangtana-anan M, Opanasopit P, Ngawhirunpat T, Nunthanid J, Sriamornsak P, Limmatvapirat S, Lim LY (2005) Effect of chitosan salts and molecular weight on a nanoparticulate carrier for therapeutic protein. Pharm Dev Technol 10(2):189–196

Mackay M, Phillips J, Hastewell J (1997) Peptide drug delivery: Colonic and rectal absorption. Adv Drug Del Rev 28(2):253–273

Nagata T, Tozuka Y, Takeuchi H (2006) Design and evaluation of polymer-coated liposomes for mucosal absorption of drugs. Proceedings of the 23th symposium on particulate preparation and designs 23:168–171

Pan Y, Li YJ, Zhao HY, Zheng JM, Xu H, Wei G, Hao JS, Cui FD (2002) Bioadhesive polysaccharide in protein delivery system: chitosan nanoparticles improve the intestinal absorption of insulin in vivo. Int J Pharm 249(1–2):139–147

Park K, Robinson JR (1984) Bioadhesive polymers as platforms for oral-controlled drug delivery: method to study bioadhesion. Int J Pharm 19:107–127

Patel H, Ryman BE (1981) Systemic and oral administration of liposomes. In: Knight CG (ed) Liposomes: From Physical Structure To Therapeutic Applications, Elsevier, Amsterdam, pp 409–441

Peppas NA, Buri PA (1985) Surface, interfacial and molecular aspects of polymer bioadhesion on soft tissues. J Control Rel 2:257–275

Pimienta C, Chouinard F, Labib A, Lenaerts V (1992) Effect of various poloxamer coatings on in vitro adhesion of isohexylcyanoacrylate nanospheres to rat ileal segments under liquid flow. Int J Pharm 80:1–8

Porter CJH, Charman WN (1997) Uptake of drugs into the intestinal lymphatics after oral administration. Adv Drug Del Rev 25(1):71–89

Prabaharan M, Mano JF (2004) Chitosan-based particles as controlled drug delivery systems. Drug Delivery 12(1):41–57

Rao KV, Buri P (1989) A novel in situ method to test polymers and coated microparticles for bioadhesion. Int J Pharm 52:265–270

Raviv U, Needleman DJ, Li Y, Miller HP, Wilson L, Safinya CR (2005) Cationic liposome-microtubule complexes: Pathways to the formation of two-state lipid-protein nanotubes with open or closed ends. Proc Nat Acad Sci 102:11167–11172

Sakuma S, Sudo R, Suzuki N, Kikuchi H, Akashi M, Ishida Y, Hayashi M (2002) Behavior of mucoadhesive nanoparticles having hydrophilic polymeric chains in the intestine. J Control Rel 81(3):281–290

Schmidgall J, Hensel A (2002) Bioadhesive properties of polygalacturonides against colonic epithelial membranes. Int J Biol Macromol 30:217–225

Schumacher U, Schumacher D (1999) Functional histology of epithelia relevant for drug delivery: Respiratory tract, digestive tract, eye, skin, and vagina. In: Mathiowitz E, Chickering III DE, Lehr CM (eds) Bioadhesive Drug Delivery Systems: Fundamentals, Novel Approaches and Development, Marcel Dekker, New York, pp 67–83

Seki K, Tirrell DA (1984) Interactions of synthetic polymers with cell membranes and model membrane systems. V. pH-Dependent complexation of poly(acrylic acid) derivatives with phospholipid vesicle membranes. Macromol 17:1692–1698

Sriamornsak P (1998) Investigation of pectin as a carrier for oral delivery of proteins using calcium pectinate gel beads. Int J Pharm 169(2):213–220

Sriamornsak P (1999) Effect of calcium concentration, hardening agent and drying condition on release characteristics of oral proteins from calcium pectinate gel beads. Eur J Pharm Sci 8(3):221–227

Sriamornsak P, Thirawong N, Nunthanid J, Puttipipatkhachorn S, Thongborisute J, Takeuchi H (2008) Atomic force microscopy imaging of novel self-assembling pectin-liposome nanocomplexes. Carbohydr Polym 71(2):324–329

Takeuchi H, Yamamoto H, Niwa T, Hino T, Kawashima Y (1994) Mucoadhesion of polymer-coated liposomes to rat intestine in vitro. Chem Pharm Bull 42(9):1954–1956

Takeuchi H, Yamamoto H, Niwa K, Hino T, Kawashima Y (1996) Enteral absorption of insulin in rats from mucoadhesive chitosan-coated liposomes. Pharm Res 13:896–901

Takeuchi H, Kojima H, Yamamoto H, Kawashima Y (2000) Polymer coating of liposomes with a modified polyvinyl alcohol and their systemic circulation and RES uptake in rats. J Control Rel 68(2):195–205

Takeuchi H, Yamamoto H, Kawashima Y (2001) Mucoadhesive nanoparticulate systems for peptide drug delivery. Adv Drug Del Rev 47(1):39–54

Takeuchi H, Matsui Y, Yamamoto H, Kawashima Y (2003) Mucoadhesive properties of carbopol or chitosan-coated liposomes and their effectiveness in the oral administration of calcitonin to rats. J Control Rel 86(2–3):235–242

Takeuchi H, Matsui Y, Sugihara H, Yamamoto H, Kawashima Y (2005a) Effectiveness of submicron-sized, chitosan-coated liposomes in oral administration of peptide drugs. Int J Pharm 303(1–2):160–170

Takeuchi H, Thongborisute J, Matsui Y, Sugihara H, Yamamoto H, Kawashima Y (2005b) Novel mucoadhesion tests for polymers and polymer-coated particles to design optimal mucoadhesive drug delivery systems. Adv Drug Del Rev 57(11):1583–1594

Thirawong N, Nunthanid J, Puttipipatkhachorn S, Sriamornsak P (2007) Mucoadhesive properties of various pectins on gastrointestinal mucosa: An in vitro evaluation using texture analyzer. Eur J Pharm Biopharm 67(1):132–140

Thirawong N, Kennedy RA, Sriamornsak P (2008a) Viscometric study of pectin-mucin interaction and its mucoadhesive bond strength. Carbohydr Polym 71(2):170–179

Thirawong N, Thongborisute J, Takeuchi H, Sriamornsak P (2008b) Improved intestinal absorption of calcitonin by mucoadhesive delivery of novel pectin-liposome nanocomplexes. J Control Rel 125(3):236–245

Thongborisute J, Takeuchi H, Yamamoto H, Kawashima Y (2006a) Visualization of the penetrative and mucoadhesive properties of chitosan and chitosan-coated liposomes through the rat intestine. J Liposome Res 16(2):127–141

Thongborisute J, Takeuchi H, Yamamoto H, Kawashima Y (2006b) Properties of liposomes coated with hydrophobically modified chitosan in oral liposomal drug delivery. Pharmazie 61(2):106–111

Thongborisute J, Tsuruta A, Kawabata Y, Takeuchi H (2006c) The effect of particle structure of chitosan-coated liposomes and type of chitosan on oral delivery of calcitonin. J Drug Target 14(3):147–154

Thongborisute J, Tsuruta A, Takeuchi H (2008) Correlation of drug absorption level and pharmacological efficiency of oral chitosan-coated liposomal systems. Unpublished

Torres-Lugo M, Garcia M, Record R, Peppas NA (2002) Physicochemical behavior and cytotoxic effects of p(methacrylic acid-g-ethylene glycol) nanospheres for oral delivery of proteins. J Control Rel 80(1–3):197–205

van der Merwe SM, Verhoef JC, Kotze AF, Junginger HE (2004a) N-Trimethyl chitosan chloride as absorption enhancer in oral peptide drug delivery. Development and characterization of minitablet and granule formulations. Eur J Pharm Biopharm 57(1):85–91

van der Merwe SM, Verhoef JC, Verheijden JHM, Kotze AF, Junginger HE (2004b) Trimethylated chitosan as polymeric absorption enhancer for improved peroral delivery of peptide drugs. Eur J Pharm Biopharm 58(2):225–235

Wangerek LA, Dahl HHM, Senden TJ, Carlin JB, Jans DA, Dunstan DE, Ioannou PA, Williamson R, Forrest SM (2001) Atomic force microscopy imaging of DNA-cationic liposome complexes optimised for gene transfection into neuronal cells. J Gene Med 3:72–81

Chapter 11
Strategies in Oral Immunization

Pavla Simerska, Peter Moyle, Colleen Olive, and Istvan Toth

Contents

Abstract Development of mucosal vaccine delivery system is an important area for improving public health. Oral vaccines have large implications for rural and remote populations since the access to trained medical staff to administer vaccines by injection is limited. New mucosal vaccine strategies are focused on development of non-replicating subunit vaccines, DNA, plant, and other types of recombinant vaccines. The conjugation of lipids to peptide antigens is one approach which enables the production of highly customized all-in-one self-adjuvanting vaccines. Lipid-modified peptide vaccines have been successfully investigated in humans and demonstrated to be potent and more importantly very safe.

I. Toth (✉)
School of Chemistry & Molecular Biosciences, School of Pharmacy, University of Queensland, Brisbane, QLD, Australia
e-mail: i.toth@uq.edu.au

A. Bernkop-Schnürch (ed.), *Oral Delivery of Macromolecular Drugs*,
DOI 10.1007/978-1-4419-0200-9_11,

11.1 Introduction

Immunization is one of the most effective public health interventions (Aziz et al. 2007). Traditionally vaccines play a prophylactic role, with the development of treatment vaccines [e.g. anti-cancer, malaria, and human immunodeficiency virus (HIV) vaccines] also an interest for many researchers (Jones 2007). Vaccination involves the administration of whole (live or killed) microorganisms, or microbial components, with the aim of inducing long-lasting immune responses capable of protecting the vaccinee against infection if they come into contact with micro-organisms against which they have been immunized. In many cases this has involved immunization via parenteral routes (e.g. intramuscular or subcutaneous). In general, people do not like receiving injections, thus patient compliance with both primary immunizations and subsequent boosts is reduced when this route of administration is used (Mestecky et al. 2007; Silin et al. 2007). Parenteral immunization also requires the use of trained medical professionals for vaccine administration (access to which may be problematic in developing nations or in rural communities) and poses safety risks including the risk of needle-stick injuries and the transmission of blood-borne diseases. The ability to administer vaccines via the oral route would overcome these problems, greatly simplify vaccine administration, and potentially allow for the elicitation of immune responses at mucosal surfaces (this is usually not observed following parenteral vaccine administration). Since most infections originate at mucosal surfaces, prior to becoming systemic, elicitation of mucosal immune responses would provide a first line of defence and the potential to prevent systemic infections.

Two features are required for oral vaccine development: (1) appropriate delivery systems to protect the delivered antigens from the harsh environment of the gastrointestinal tract (GIT) (e.g. low pH and the presence of many proteolytic enzymes), as well as enabling antigen uptake from the GIT, and antigen-presentation to appropriate cells of the immune system for the generation of desired immune responses, and (2) adjuvants which can stimulate the immune system to mount appropriate immune responses against the delivered antigens. Examples of delivery systems and adjuvants which have been studied for the development of orally administered vaccines (Table 11.1) are described in the following sections.

Table 11.1 Approaches for enhancement of mucosal immunity to vaccines

Goal	Approach	Example	Ref.
Reduce virulence and enhance antigen load	• Recombinant proteins • Live vectors • Subunit vaccines • DNA vaccines • Transgenic edible plants	• Attenuated *Salmonella* serovar Typhimurium expressing secreted *Yersinia pestis* F1 and V antigen • Human papillomavirus 16L1 *Lactococcus lactis* • *Lactobacillus reuteri* secreting fusion protein of *Escherichia coli*	Liu et al. (2007) Cho et al. (2007) Wu and Chung (2007)

Table 11.1 (continued)

Goal	Approach	Example	Ref.
		enterotoxin subunit protein	
Improve delivery into the mucosa	• Non-living microparticles (microspheres, liposomes) • VLP • Immune-stimulating complexes	• Bacille Calmette–Guerin encapsulated in alginate microspheres • Poly D,L-lactic-*co*-glycolic acid nanoparticles against hepatitis B • Norovirus capsid protein expressed in yeast forms virus-like particles	Ajdary et al. (2007) Gupta et al. (2007) Xia et al. (2007)
Improve mucosal interaction with antigens	• Adhesive antigens • Adjuvants	• *Helicobacter pylori* adhesin A • *E. coli* heat-labile enterotoxin	Nystrom and Svennerholm (2007) Tritto et al. (2007)
Enhancement of immune response	• Mucosal adjuvants • Combination systemic-mucosal immunization • Transcutaneous and other routes of immunization	• Tetanus toxoid with cytotoxic necrotizing factor 1 (CNF1) • Compound mucosal immune adjuvants (cMIA I and II) mixed with Newcastle-disease vaccine	Munro et al. (2007) Zhang et al. (2007)

11.2 Gastrointestinal Anatomy

An understanding of the anatomy of the gastrointestinal tract (GIT) is important for the development of oral vaccines. The surface of the GIT is lined with enterocytes, which form a physical barrier preventing the uptake of many macromolecules. In order for an effective immune response to be mounted against orally administered antigens, the antigen must be taken up and presented to the immune system (Silin et al. 2007). This process occurs in the small intestine, utilizing specialized lymphoid organs known as Peyer's patches. The luminal surface of Peyer's patches is made up of M ('microfold' or 'membrane') cells, which function to sample the intestinal contents (especially particulates), and present them to the underlying gut-associated lymphoid tissue (GALT) (Mestecky et al. 2007). The M cells, which lack apical brush border microvilli and feature a small mucous layer and glycocalyx, appear as patches of smooth cells in the small intestine (Shalaby 1995). Factors affecting M-cell uptake include particle size (nanometer-sized particles are the best) (Brayden and Baird 2001), lipophilicity (decreasing lipophilicity increases uptake) (Hillery and Florence 1996), and the conjugation of M-cell-specific ligands (e.g. tomato lectin) (Hussain et al. 1997). The Peyer's patches contain up to 90% of the bodies immunocompetent cells including large numbers of professional antigen-presenting cells (APCs; e.g. macrophages and dendritic cells) and feature a B-cell and a T-cell zone (Mestecky et al. 2007). Antigen-stimulated B- and T-cells migrate from the Peyer's patches to other mucosal sites via the lymph, facilitating the spread of mucosal immunity to all mucosal surfaces.

The uptake of antigen by M cells is believed to be an important process for the development of mucosal immunity. The ability to develop vaccines that target M cells to enhance antigen uptake would therefore increase the likelihood that an effective immune response could be elicited.

Increasing the uptake of antigens by enterocytes may represent an alternative means for oral vaccine delivery. It is, however, commonly believed that antigen uptake by enterocytes leads to the production of antigen tolerance rather than immunity (Brayden and Baird 2001).

11.3 Oral Vaccination

11.3.1 Advantages and Disadvantages

As aforementioned, oral vaccination has many advantages when compared to parenteral vaccine administration. In terms of the immune responses elicited following oral administration, a major advantage of oral vaccination (or any other form of mucosal administration) is its ability to stimulate the production of antigen-specific mucosal secretary IgA (sIgA) antibodies (McNeela and Mills 2001). sIgA antibodies have been demonstrated to be capable of protecting against pathogen colonization at mucosal sites, thus preventing systemic infection from occurring (Yokoyama and Harabuchi 2002). In addition, mucosal vaccine administration has the ability to elicit mucosal immune responses at all mucosal sites, not just at the site of administration (Shalaby 1995). Thus, the aim when producing an oral vaccine, in most cases, would be to produce a vaccine capable of eliciting both mucosal and systemic immune responses, to protect against infection arising from the mucosal surfaces and/or infection arising from other means (e.g. from skin damage).

Despite the above mentioned advantages, few orally administered vaccines exist. Several reasons are proposed for why a generic technology for oral vaccine delivery does not currently exist: The harsh conditions in the GIT (e.g. the low pH in the stomach and the presence of proteases and other enzymes) has the potential to damage antigens through processes including enzymatic proteolysis, acid hydrolysis, or acid denaturation of conformational epitopes. In addition, orally administered vaccines must be taken up and presented to appropriate cells of the immune system. This process is inefficient, necessitating the administration of large doses of antigen, and may be made worse by the presence of food or chyme in the GIT, which may entrap vaccines preventing the necessary proximity between the vaccine and the GIT epithelia for vaccine uptake (Russell-Jones 2000; Aziz et al. 2007). Oral tolerance, whereby antigens are recognized as food antigens or normal flora instead of eliciting protective immune responses, may also be observed against orally administered soluble antigens (Mestecky et al. 2007). Vaccines against different diseases may also need to stimulate different types of immune responses (e.g. different antibody isotypes or humoral versus cellular immunity). Thus, oral vaccine delivery systems must be capable of protecting antigens against degradation in the GIT, as well as enabling their uptake and presentation to appropriate cells of the immune system, while potential adjuvants must be capable of stimulating appropriate immune responses to provide protection against infection or potentially treat diseases.

11.3.2 Clinical and Pre-clinical Studies

Oral vaccines that are licensed for human administration are limited. Presently, there are few oral vaccines licensed for human use (Table 11.2), although many orally administered vaccine candidates are in development. Live attenuated oral polio (OPV) is the vaccine of choice for the prevention of poliomyelitis

Table 11.2 Internationally licensed oral/nasal vaccines

Infection and vaccine	Trade name(s)	Protection
Polio	Many	Almost complete
Live attenuated vaccine (OPV)		
Cholera	Dukoral (SBL Vaccin)	85–90% (first 6 months), 60% (first 3 years)
Cholera toxin B subunit + inactivated *V. cholerae* 01 whole cells	Orochol (Berna, SSVI)	60–100% in experimental trials, but not significant in endemic populations
CVD 103.HgR live attenuated *V. cholerae* 01 strain		
Typhoid	Vivotif (Berna, SSVI)	67% in first 3 years
Ty21a live attenuated vaccine		
Rotavirus	Rotarix	61–92%
Live attenuated monovalent human rotavirus strain	RotaTeq	Three doses, 59–77% against any strain
Attenuated pentavalent bovine–human vaccine contains human rotavirus genes from each serotype		
Influenza	FluMist (MedImmune)	60–90%
Live attenuated cold-adapted influenza virus reassortant strains adjusted to the needs of the flu season		
Bird flu	Amivax	
H5N1 influenza virus		
Enteric Redmouth Disease (ERM) in rainbow trout	AquaVac ERM Oral	Total protection strategy

and together with vaccination involving inactivated polio vaccine is a strategy that is currently designed towards achieving global polio eradication (Sutter et al. 2000; Wagner and Earn 2008). Of interest, an outbreak of poliomyelitis cases caused by wild poliovirus (WPV) type 1 (WPV1) occurred in India in 2006 (2007-MMWR Report). This resulted in the use of the monovalent oral poliovirus vaccine type 1, which has a higher efficacy against WPV1 than trivalent OPV, as a measure to reduce WPV1 cases. While the outcome has been positive, an outbreak of poliomyelitis cases caused by WPV type 3 has since been reported primarily in northern India where polio remains endemic. These findings highlight the challenges and strategic adaptations of polio eradication by oral vaccination in India.

Several strategies are being explored for the development of oral vaccines particularly for paediatric rotavirus gastroenteritis. Early clinical studies used a tetravalent human–rhesus reassortant rotavirus vaccine (RRV-TV) to cover the four serotypes – G1, G2, G3, and G4 – that predominantly cause rotavirus gastroenteritis in humans (Foster and Wagstaff 1998). In infants, the vaccine induced IgA antibodies and neutralizing IgG antibodies to RRV. Evaluation of protective efficacy has shown that RRV-TV is moderately effective in reducing the incidence of rotavirus gastroenteritis. Most importantly, RRV-TV is highly effective in protecting against severe cases of the disease. The tetravalent rhesus rotavirus vaccine was licensed in 1998, however, later reports linking vaccination to intussusception led to withdrawal of this vaccine, although an oral bovine reassortant vaccine and an attenuated human rotavirus strain are in trial (Bernstein 2000). A recent clinical study has shown that vaccination of children with the live attenuated human rotavirus vaccine RIX4414 significantly decreased the rate of rotavirus gastroenteritis even in malnourished children (Perez-Schael et al. 2007). Overall, there have been significant advances in oral rotavirus vaccine development, which is promising for the prevention of paediatric rotavirus gastroenteritis worldwide.

Killed whole cell-based oral cholera vaccines have been licensed and available for human use for the past 20 years, but have not been used for the control of endemic cholera (Longini et al. 2007). However, the logistics of mass immunization for the potential control of endemic cholera taking into account vaccine coverage rates and natural immunity are being considered (Chaignat and Monti 2007). A live attenuated cholera vaccine candidate CholeraGarde Peru-15, which expresses high levels of the cholera toxin B subunit, is currently under investigation as a bivalent cholera/enterotoxigenic *Escherichia coli* vaccine (Roland et al. 2007).

Several clinical studies have been conducted with oral influenza vaccines (Avtushenko et al. 1996; Lazzell et al. 1984). These include water in oil emulsion form of inactivated virus vaccine and an enteric-coated killed virus vaccine. Although these vaccines induced IgA responses, there were inadequate levels of virus-neutralizing IgG antibodies in the serum to fulfil regulatory requirements for vaccine immunogenicity. This warrants refinement and further development of these types of vaccines.

Several approaches for the development of oral vaccines are being investigated in pre-clinical studies. These include oral vaccination with live *Mycobacterium bovis* BCG in lipid formulation to protect against tuberculosis (Cross et al. 2007), *Saccharomyces cerevisiae* expressing recombinant *Actinobacillus pleuropneumoniae* antigens (Shin et al. 2007), adjuvanted influenza subunit vaccine (Amorij et al. 2007), *Lactococcus lactis* as a live vector or virus-like particles expressing human papillomavirus type 16 L1 (Cho et al. 2007; Thones and Muller 2007), and enteric-coated capsules containing adenoviral vectors expressing HIV antigens and peptides (Mercier et al. 2007). Recent experimental studies have investigated the feasibility of PEGylated PLGA-based nanoparticles to target antigens directly to M cells in the gut following oral administration (Garinot et al. 2007). M cells are instrumental in antigen uptake and the induction of mucosal immune responses – targeting these cells, in addition to antigen delivery within a protective nanoparticle carrier – has been shown to elicit antibody responses against a model antigen (Garinot et al. 2007) and is a promising strategy for oral vaccine delivery.

11.4 Approaches and Different Oral Vaccine Delivery Systems

11.4.1 Recombinant Vaccines and Live Vector Vaccines

Recombinant techniques (Ogra et al. 2001) for generating purified antigens in large quantities have been used for the development of several vaccines including the hepatitis B virus vaccine (oral hepatitis B vaccine based on live recombinant adenovirus) (Lubeck et al. 1989). Other examples of recombinantly produced vaccines include vaccines containing tetanus toxoid, diphtheria toxin, and acellular pertussis toxoid.

Live vector vaccines consist of attenuated bacterial or viral strains capable of expressing antigens from pathogenic microorganisms. Mucosal vaccination with live vector vaccines via the oral and intranasal routes has been demonstrated using *Lactobacillus plantarum* expressing tetanus toxin fragment C (Seegers 2002). Vaccinia virus (Sharon et al. 2007), poliovirus (Zimmerman and Spann 1999), adenovirus (Piedra et al. 1998) and chimeric HIV-poliovirus (Zhang et al. 2007) are other attractive vectors for delivery of mucosal vaccines.

11.4.2 Virus-Like Particles

Virus-like particles (VLPs) (Noad and Roy 2003) are produced from viral capsid proteins. The capsid proteins self-assemble to form particles that resemble the virus from which the capsid proteins were derived, but lack any associated viral genetic material. Virus-like particles are therefore non-infectious and represent non-replicating virus analogues. Virus-like particles may be further modified by conjugating antigens to capsid proteins such that the antigens will be displayed on the VLP surface. Alternatively, non-viral genetic material (e.g. DNA vaccines) may be encapsulated in VLPs. Oral and nasal routes of VLP administration have been investigated, with intranasal administration proving more efficient and requiring lower antigen doses (Niikura et al. 2002). Inexpensive manufacture, high immunogenicity, and the ability to utilize natural viral transmission routes for vaccine delivery are the main advantages of using VLPs. Successful vaccination has been achieved in animals and humans with VLPs that were derived from several mucosal viral pathogens including human papilloma virus (Da Silva et al. 2001), Norwalk virus (Guerrero et al. 2001) and hepatitis E virus (Li et al. 2001).

11.4.3 DNA Vaccines

DNA vaccination, which involves administration of coding DNA rather than peptide antigens, is another modern vaccination technology. There are several advantages of mucosal DNA vaccination; pure plasmid DNA is quick and easy to manufacture; DNA does not integrate into the host's chromosome; DNA vaccines are stable; they are easy to transport; and they are capable of targeting peptides for presentation by major histocompatability complex (MHC) class I molecules. Bacterial DNA has large amounts of unmethylated cytosine-phosphate-guanosine (CpG) motifs which help to increase the immune response invoked by an antigen. DNA vaccine technology has been applied to the development of oral DNA vaccines; e.g. CpG DNA tetanus toxoid vaccine administered orally in rodents demonstrated systemic humoral and cellular responses, with mucosal immunity induced at both local and distant sites (McCluskie et al. 2000). Human trials of CpG DNA have demonstrated its efficacy as an adjuvant for human use (Krieg 2002).

DNA vaccination has been attempted by intranasal administration of a plasmid expression system for herpes simplex virus (Kuklin et al. 1997), HIV (Robinson 2007), and influenza virus (Pertmer et al. 2000), with significant protection observed against mucosal challenge.

11.4.4 Plant-Based Vaccines

A plant-based vaccine combines the concepts of a subunit vaccine and the use of plant as an expression system. Several advantages including inexpensive means of expressing proteins, elimination of the risk of contamination with animal pathogens, and a heat-stable environment, indicate that the production of vaccines in plants will be successful (Aziz et al. 2007). Moreover, plants are able to carry out post-translational modifications, can be produced in large quantities, and can express multiple transgenes at one time, giving thus the potential for using one formulation to deliver several vaccines. The feasibility of using recombinant plants for the generation of vaccine antigens has been demonstrated in tobacco plants and potato tubers in which protein antigens (for example, *E. coli* heat-labile toxin B subunit, hepatitis B virus surface antigen, rotavirus, and VLPs) for a number of human pathogens were successfully expressed (Haq et al. 1995).

There are some disadvantages, however, of using plant-based vaccines. These are the need to plant genetically modified plants, which may cross-breed with wild-type plants, unknown health risks, government bans on genetically modified products/farming, and the need to buy seed from the supplier as genetically modified plants are usually produced such that they cannot reproduce – a safeguard to prevent spread in the wild (this may make this approach expensive in developing countries).

11.4.5 Microcapsules

Microspheres are solid colloidal particles with diameters in the micrometer range, into which antigenic components and/or adjuvants may be entrapped, chemically attached, or adsorbed onto the surface. Vaccines encapsulated in biodegradable polymers, which have been used in humans for drug delivery, exhibit many advantages. For example, the ability to target antigen to M cells for uptake; ability to control antigen release thus reducing the number of doses required for immunization; and protection to encapsulated antigens from harsh GIT conditions. A unique feature of the polymeric delivery system is that it can be manipulated to meet specific physical, chemical, and immunogenic requirements for a particular antigen. Polymers used in microsphere preparation include hydrophobic materials based on poly-lactide acid or poly-γ-D-glutamic acid and hydrophilic polymers prepared from polyacrylamide and polycyanoacrylates. Difficulties in comparing particle uptake data between animal species, antigen degradation, and in determining the amount of antigen present in microparticles and of that released still have to be solved for this approach to be feasible. Nevertheless, generation of a successful immune response against *Staphylococcus* enterotoxin B, pertussis filamentous hemagglutinin, simian immunodeficiency virus, influenza virus, and malaria synthetic peptide SPF66, is evidence of the functionality of microcapsules as oral vaccine delivery vehicles (Aziz et al. 2007).

11.4.6 Liposomes, Adhesins, Saponins

Antigen delivery through liposomes, hollow membrane-bound spheres, can be achieved by entrapping the molecule in the lipid membrane or inside the hollow cavity. Modified liposomes have been able to induce mucosal IgA responses compared to free antigen (Ann Clark et al. 2001; Aziz et al. 2007). Liposomes containing pertussis toxin (Guzman et al. 1993), *Streptococcus mutans* (Childers et al. 2002), or bovine serum albumin (Therien et al. 1990) as vaccine antigens have been tested in experimental models and induced effective antibody- and cell-mediated immune responses.

Adhesins, which are proteins that exhibit adhesive properties, have been demonstrated to induce mucosal immune responses and elicit serum antibodies when administered orally. Some of the examples are pili, the B subunit of *E. coli* toxin, the hemagglutinin of influenza virus, cholera toxin, meningococcal outer membrane proteins, and proteosomes hydrophobically bound to meningococcal or *Haemophilus influenzae* type B polysaccharide or *Shigella* lipopolysaccharide (Ogra et al. 2001). Saponins are triterpenoid glycosides derived from *Quillaja saponaria*, which have been used as adjuvants for many years in veterinary vaccines and recently have been used against measles (Pickering et al. 2006).

11.4.7 Mucosal Adjuvants

Adjuvants play an important role in vaccine development, but unfortunately only a few adjuvants have gained regulatory approval for human administration. The most commonly used adjuvant for human administration is alum, however, it is not suitable for eliciting cell-mediated immunity, and therefore is not an appropriate adjuvant for the development of treatment vaccines. Furthermore, alum is not suitable for mucosal vaccine administration, and in general vaccines formulated with alum do not yield mucosal antibodies following their administration (Brewer 2006). The need for new non-toxic, safe, and potent adjuvants for human use, with the capacity to elicit both cellular and mucosal immune responses, is obvious. The combination of bacterial or synthetic lipidic adjuvants and peptide antigens, to produce self-adjuvanting lipopeptide vaccines, has been tested in human clinical trials and demonstrated a high degree of safety with few or no side effects (Steller et al. 1998; Seth et al. 2000; Durier et al. 2006) and was also demonstrated to be effective for administration via mucosal routes (e.g. nasal or oral) (Mittenbuhler et al. 1997; BenMohamed et al. 2002a; Batzloff et al. 2006; Deliyannis et al. 2006; Olive et al. 2006a).

Adjuvants have been shown to dramatically increase the immunogenicity and efficacy of vaccines by increasing antigen contact time in the body, targeting antigen to the optimum location for presentation by APCs, and optimizing the environment for antigen presentation. Adjuvants thereby have the capacity to influence the types of cellular and antibody responses induced. The most common adjuvant utilized in experimental vaccine development studies is complete Freund's adjuvant (CFA), an emulsion containing heat-killed *Mycobacterium tuberculosis*. While potent as an adjuvant in mice, CFA is too toxic for use in humans (Chapel and August 1976). Other adjuvants containing microbial constituents (bacterial toxins, CpG oligodeoxynucleotides, cytokines/chemokines, live vectors, virus-like particles, monophosphoryl lipid A) or particulate systems targeting M cells (microspheres, liposomes, lipopeptides) have been studied for possible use in humans. The most-widely used and highly reported adjuvants for mucosal vaccines are cholera toxin and *E. coli* heat-labile enterotoxin. Many animal studies have utilized these adjuvants for oral and nasal vaccine delivery but unfortunately, due to severe toxicity, they are unsuitable for clinical use in humans (Aguilar and Rodriguez 2007).

11.4.8 Subunit Vaccines and Synthetic Peptides

For many years scientists have been trying to develop peptide-based pharmaceuticals for administration by parenteral routes; the oral delivery of these compounds is more difficult and remains an area of intense investigation to overcome problems, such as antigenic/enzymatic degradation in the gut that will ultimately be associated with its success in the future. Synthetic peptide-based vaccines boast many attractive features for vaccine development including the capacity to focus immune responses towards specific epitopes, the ability to be produced in a highly characterized state, their non-infectious nature, the capacity to incorporate non-natural components, and high stability in freeze-dried form, thus negating any requirement for transportation and storage under cold-chain conditions (Babiuk 1999; Wiesmüller et al. 2001; BenMohamed et al. 2002b; Purcell et al. 2007). Despite many advantages of peptide vaccines, synthetic peptides tend to be poorly immunogenic and therefore need to be administered with powerful adjuvants (BenMohamed et al. 2002c; Purcell et al. 2007). Because of the small size of peptide antigens, they tend to lack the mix of helper T-lymphocyte epitopes necessary for eliciting long-lasting immunity.

Subunit vaccines consisting of antigenic fragments from pathogenic organisms are preferable nowadays due to safety concerns. However, these fragments exhibit poor immunogenicity when administered orally, due to poor GIT absorption and GALT uptake and degradation in the GIT. Tolerance to orally delivered antigens has also been reported (Mattingly and Waksman 1978). To increase subunit vaccine immunogenicity, certain properties, which were lost during separation of the whole virulent organism into its component parts, must be replaced by an adjuvant (see Section 11.4.6), which has the ability to activate the innate immune system.

The polysaccharide capsule which surrounds bacterial species like *Haemophilus influenzae*, *Neisseria meningitidis*, *Streptococcus pneumoniae*, and *Salmonella typhi* is a potent virulence factor and has been successfully used in the preparation of vaccines (Lindberg 1999). Covalent linkage of the polysaccharide to immunogenic carrier proteins creates glycoconjugates which are T-dependent antigens and which prime for boosting either with the glycoconjugate or with the capsular polysaccharide (Lindberg 1999). Rabbits and monkeys orally immunized with diphtheria (incorporated in egg proteins) and tetanus antigens have demonstrated significant immune responses and total protection against lethal challenge (Mirchamsy et al. 1994). Mucosal immunization with filamentous hemagglutinin of *Bordetella pertussis* by the respiratory or enteric route was found to protect mice against *B. pertussis* infection (Shahin et al. 1992).

11.4.9 Lipid-Based Vaccines

The combination of bacterial or synthetic lipidic adjuvants and peptide antigens, to produce self-adjuvanting lipopeptide vaccines, has been tested in human clinical trials and have demonstrated a high degree of safety with few or no side effects (Steller et al. 1998; Seth et al. 2000; Durier et al. 2006) and was also demonstrated to be effective for administration via mucosal routes (e.g. nasal or oral) (Mittenbuhler et al. 1997; BenMohamed et al. 2002a; Batzloff et al. 2006; Deliyannis et al. 2006; Olive et al. 2006a). Lipopeptide vaccines have many advantages including the capacity to be administered via mucosal routes (orally or nasally) instead of by injection, elicitation of antigen-specific cytotoxic T-lymphocyte (CTL) responses, and mucosal immunity, as well as almost no toxicity. Several lipopeptide vaccine systems have been studied, incorporating not only single fatty acid chains, but also more complex lipids and glycolipids into the peptide vaccine candidate.

For mucosal vaccine delivery, liposomes (self-aggregating, enclosed lipid membranous vesicles) have been formulated with entrapped, surface-linked, or admixed antigens. Liposomes exhibit many advantages including particle uptake by M cells and the ability to induce CTLs by targeting antigens to endogenous processing pathways. Otherwise poorly immunogenic by themselves, lipidated peptides display increased immunogenicity and many features advantageous for mucosal vaccine delivery. Lipidated peptides have the capacity to form vesicles, facilitating M-cell uptake. Lipidation also reduces peptide degradation and so may prove valuable for oral vaccine delivery. Trials delivering lipidated peptides via parenteral, oral, and nasal routes suggest that lipidated peptides appear safe for human use, with few or no side effects reported (Gahery-Segard et al. 2000).

Interactions between lipids and receptors for structurally conserved pathogen-associated molecular patterns, most commonly the Toll-like receptors (TLRs), have been studied in detail for their involvement in adjuvant activity. TLRs appear to play an important role in the functioning of many adjuvants and the most important TLRs for lipopeptide or glycolipopeptide vaccines are TLR1, TLR2, TLR4, and TLR6 (van Duin et al. 2006).

11.4.9.1 Monophosphoryl Lipid A

Lipopolysaccharide (LPS), a major cell wall component of Gram-negative bacteria, is a powerful adjuvant capable of enhancing both innate and adaptive immune responses and its administration results in the production of antimicrobial peptides (defensins) (Ouellette et al. 2000), various cytokines, and chemokines (Th1 and Th2), leading to the recruitment and activation of immune effector cells and elicitation of adaptive immune responses (Persing et al. 2006). Administration of LPS without antigen has been shown to provide protection against tumours and various microorganisms in mice (Baldridge et al. 2004) and proved to have adjuvant activity when administered at a different site, or at a different time, to an antigen (Johnson 1964).

Despite the powerful adjuvant activity of LPS, its high toxicity, associated with the release of excessive inflammatory cytokines, has prevented the study of LPS in human clinical trials (Persing et al. 2006). In order to separate the toxicity of LPS from its adjuvant activity, monophosphoryl lipid A (MPL) (Fig. 11.1), an adjuvant consisting of a mixture of six glycolipids from *Salmonella minnesota* R595 LPS (Ulrich and Myers 1995), was developed and has been tested in clinical trials, with a high degree of safety and superior adjuvant activity observed compared to alum (Persing et al. 2006). As an example, MPL-adjuvanted hepatitis B vaccine (Fendrixä) has received marketing approval in Europe.

Mucosal vaccine administration of MPL performed via the nasal (hepatitis B (Baldridge et al. 2000), tetanus (Baldridge et al. 2000), HIV-1 (VanCott et al. 1998; Egan et al. 2004), *Streptococcus mutans* (Childers et al. 2000), and influenza) (Baldridge et al. 2000), and oral (*M. tuberculosis*) (Doherty et al. 2002) routes resulted in elicitation of antigen-specific mucosal IgA and systemic IgG antibodies, as well as increased levels of cell-mediated immunity. Due to the unique biological effects, and strong adjuvant activity of MPL, investigations have sought to produce self-adjuvanting vaccines by conjugating various antigens to MPL.

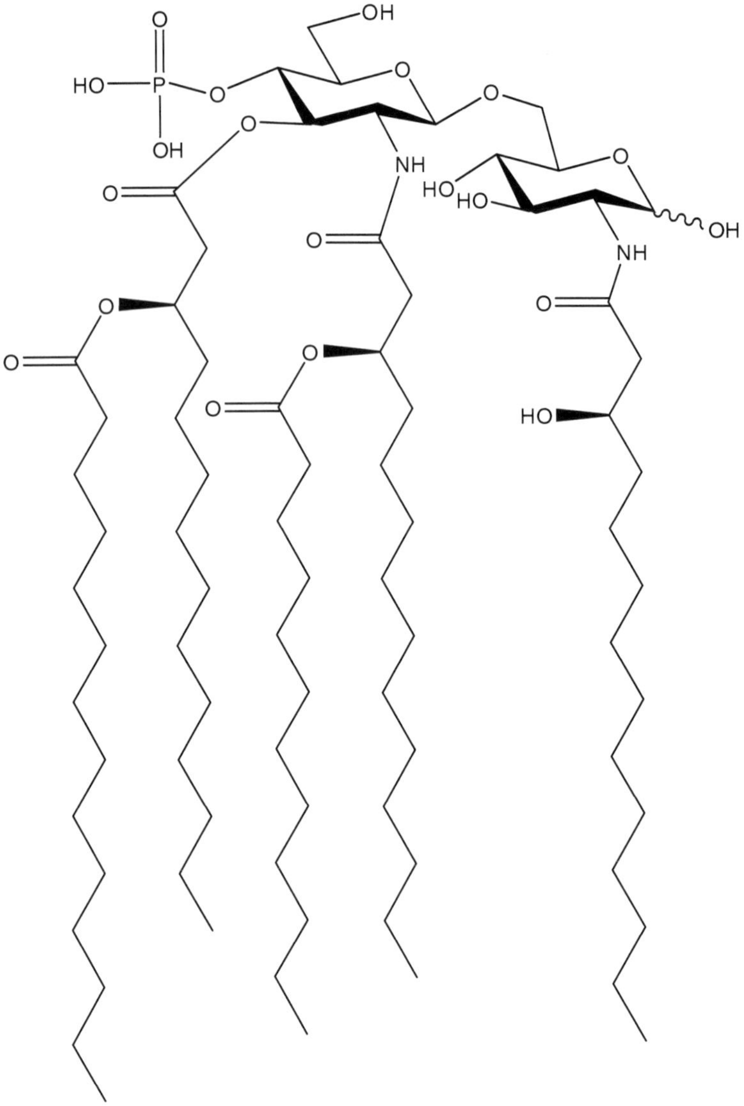

Fig. 11.1 The most common glycolipid in Corixa's MPL adjuvant

11.4.9.2 Bacterial Lipoprotein

Bacterial lipoproteins (LP) are structural components of bacterial cell walls consisting of an *S*-glycerylcysteine moiety, where the glyceryl moiety is di-O-acylated and the cysteine residue is conjugated to the amino-terminus of various polypeptides. In addition, the cysteine α-amine may be acylated with an amide-linked fatty acid to give a tri-acylated LP. The structures of some representative mono-, di- and tri-acylated LPs are illustrated in Fig. 11.2.

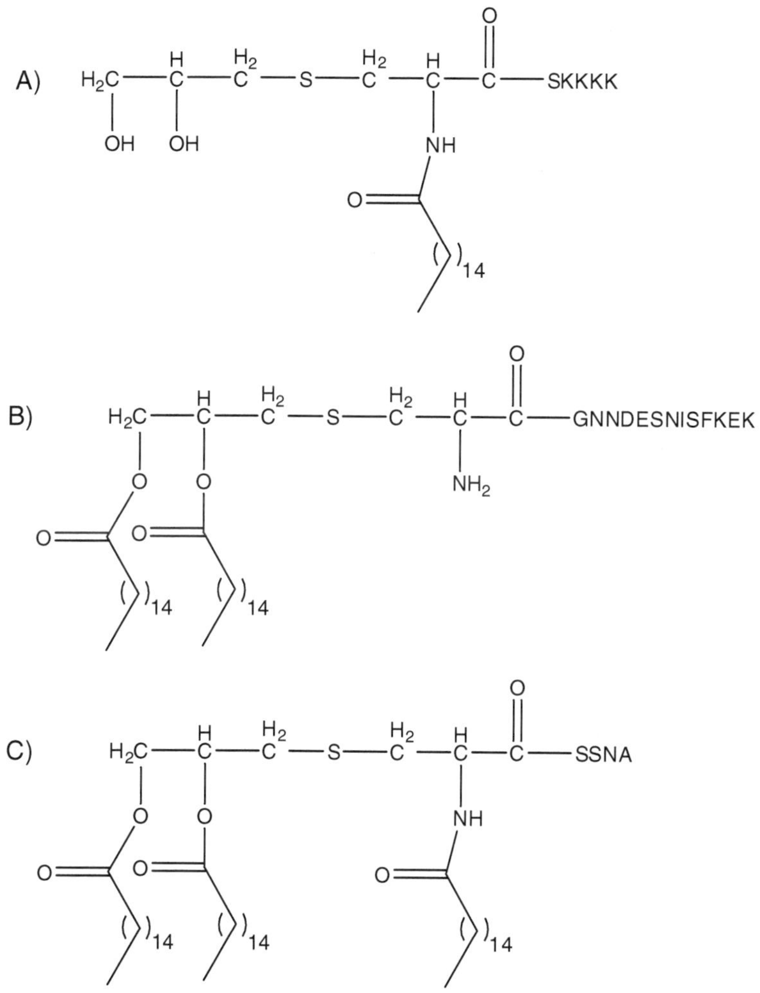

Fig. 11.2 Mono- (Ghielmetti et al. 2005), di- (Buwitt-Beckmann et al. 2005), and tri-acylated (Bessler et al. 1985) examples of lipopeptide vaccines

Antigens conjugated to or co-administered with lipopeptides have been demonstrated in some cases to be immunogenic when administered via nasal (Zeng et al. 2000; Zeng et al. 2002; Batzloff et al. 2006) or oral routes (Nardelli et al. 1994) and are capable of eliciting both cell-mediated and humoral immune responses. Recent studies have investigated the use of bacterial lipopeptides as adjuvants for vaccines against influenza virus (Jackson et al. 2004), group A streptococcus (GAS) (Batzloff et al. 2006), hepatitis C virus (Langhans et al. 2004), leishmaniasis (Hewitt and Seeberger 2001), *Listeria monocytogenes* (Jackson et al. 2004), and other disease-causing microorganisms.

11.4.9.3 Palmitoylated Peptides

Palmitoylated peptides are able to elicit both cellular and humoral immune responses, as shown, for example, using lipopeptide vaccines against human immunodeficiency virus type 1 (HIV-1) (Deprez et al. 1996; Klinguer et al. 1999; Andrieu et al. 2000; Gahéry-Ségard et al. 2000; Pialoux et al. 2001; Gahéry-Ségard et al. 2003; Lévy et al. 2005; Durier et al. 2006; Gahery et al. 2006) and *Plasmodium falciparum* malaria (BenMohamed et al. 1997, 2000, 2004). Mucosal routes of administration have also been investigated using lipopeptides conjugated to *P. falciparum*, cytomegalovirus, and Herpes simplex virus type-1 (HSV-1) peptide antigens, and it was shown that administration of these lipopeptides via the intranasal route, compared with subcutaneous administration, elicited significantly higher cell-mediated immunity and systemic IgG antibodies, as well as a Th1-biased immune response (BenMohamed et al. 2002c). In human phase I and II clinical trials of HIV-1 vaccines containing a mixture of up to six different lipopeptides derived from HIV-1 regulatory and structural proteins, strong B- and T-cell responses (including CTLs) were observed against most of the concomitantly injected lipopeptides, and these were long-term immune responses (Gahéry-Ségard et al. 2003).

11.4.9.4 The Lipid-Core Peptide System

Coupling of lipids to peptide antigens may be performed directly or via an amplifying dendrimer carrier system. A commonly investigated carrier is the multiple antigen peptide (MAP)-system, which connects peptide antigens through a branched polylysine core.

The lipid-core peptide (LCP) system (Toth et al. 1993; Moyle et al. 2003) (Fig. 11.3) is a delivery system which conjugates synthetic lipoamino acids (α-amino acids with long alkyl side chains) through a polylysine MAP system (or a carbohydrate) (McGeary et al. 2001, 2002) to multiple copies of one or several different peptide antigens. The LCP system induces similar immune responses when LCP-based vaccines are co-administered with conventional adjuvants and represents a promising system for mucosal vaccine development.

The physicochemical properties of LCP systems may be varied by changing the number of lipoamino acids, the length of the lipoamino acid alkyl side chains, the level of MAP system branching, and incorporation of different spacer molecules. The LCP systems are stable to a wide range of pH conditions in solution, as well as to peptidase enzymes, and do not need to be stored under refrigeration.

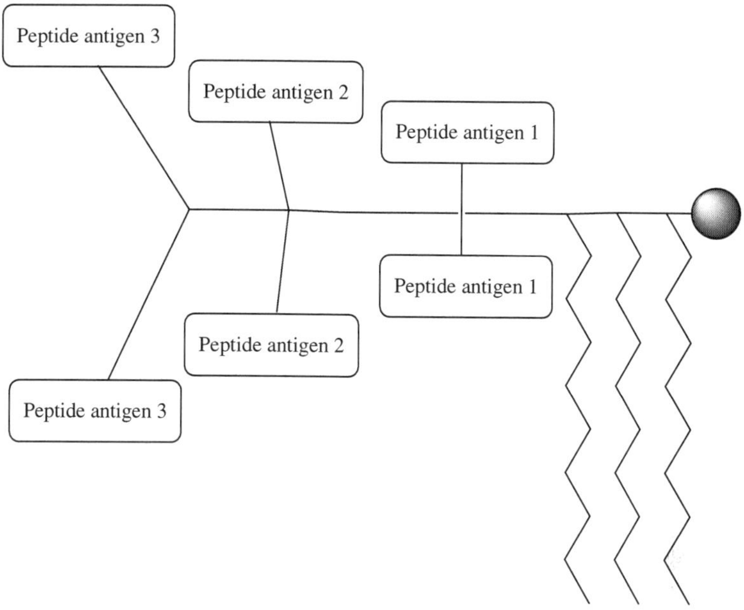

Fig. 11.3 The structure of the lipid-core peptide (LCP) system incorporating three different peptide antigens synthesized on solid phase

The LCP system has been used when developing vaccines against, for example, GAS (*Streptococcus pyogenes*) (Horváth et al. 2002a, b; Olive et al. 2002, 2003; Horváth et al. 2004; Olive et al. 2004, 2005; Moyle et al. 2006; Olive et al. 2006b), *Chlamydia trachomatis* (Zhong et al. 1993), foot-and-mouth disease virus (France et al. 1994), and human papillomavirus type-16 (Moyle et al. 2005) and in many cases, vaccines were immunogenic without the need for additional adjuvants and sometimes the titers of systemic IgG antibodies against the attached antigens were even higher when compared to administration of the antigen in CFA. Antigen-specific systemic IgG antibodies were elicited against the attached peptide antigens also in cases in which the LCP systems were administered to mice via the intranasal route (Olive et al. 2006a).

11.4.10 Carbohydrate-Based Vaccines

The advent of new, efficient, and sensitive analytical and synthetic methods in carbohydrate chemistry, together with an increasing understanding of glycobiology and glycoimmunology, has made possible the development of carbohydrate-based vaccines. The use of carbohydrates in drug delivery offers many advantages, including the potential to induce active transport and increase water solubility of drug candidates. In addition, carbohydrates can serve as carriers or aid targeting to receptors expressed on various cells. A large content of structural information of membrane carbohydrates along with their strategic location on the cell surface equip carbohydrates with a major role in cell–cell recognition processes. The ability to target drugs to particular cells or organs using selective carbohydrate receptors is one of the main features of carbohydrate-based drug delivery science. For example, macrophages, dendritic cells, and some epithelial cells express mannose receptors, a multidomain membrane-associated receptor, which can be judiciously retained for drug targeting. Moreover, the conjugation of carbohydrates to bioactive compounds is being recognized as an effective method of manipulating their physicochemical properties.

Three major discoveries are responsible for the rapid development of carbohydrate-based drug candidates (Roy 2004): in the first instance, the recent commercialization of the first semi-synthetic vaccine, which contains capsular polysaccharide, against bacterial infections caused by *Haemophilus influenzae* type B; secondly, the rapid access to sizeable amounts of complex saccharide structures through one-pot, solution-phase syntheses (for example, programmable, one-pot solution-phase synthesis technique, Optimer Pharmaceuticals) (Koeller and Wong 2000); and thirdly by utilizing automated solid-phase processes which have greatly improved the timing and yield of otherwise inefficient manual syntheses of oligosaccharides. The automated solid-phase technique has been exploited by Seeberger and co-workers at Anchora Pharmaceuticals, who developed an automated solid-phase synthetic methodology to provide access to carbohydrate epitopes that can be incorporated into potential vaccines against malaria, HIV, tuberculosis, and bacterial infections (Plante et al. 2001).

Glycosylation has also been reported to improve intestinal absorption of peptide drugs which demonstrate poor membrane penetrability. For instance, glycosylation at the *N*-terminus of tetrapeptide (Gly-Gly-Tyr-Arg) increases its resistance to degradation by peptidases; in addition, Na^+-dependent glucose transporters were shown to play an important role in the intestinal absorption of both *p*-(succinylamido)phenyl α- or β-D-glucopyranosides (Nomoto et al. 1998).

Another example is glycosylation of endomorphin-1. Endomorphin-1 is an endogenous opioid peptide with high affinity for opioid receptors in the brain that has been investigated as a potential pain relieving drug. Oral delivery would be ideal for such a drug candidate but presents problems for small peptides due to poor oral absorption, low metabolic stability, and inability to cross the blood–brain barrier. Chemical modification of short peptides with lipidic groups in combination with mono- and di-saccharides has been shown to dramatically improve peptide stability, membrane permeability, and bioavailability (Johnstone et al. 2005).

11.4.10.1 Carbohydrates as Targeting Moieties

Human serum containing anti-group A streptococcus carbohydrate (GAS CHO) antibodies were opsonic for different M protein-carrying serotypes.

Significant immunogenicity and protection against systemic or nasal challenge with live strains of GAS in mice was observed (Sabharwal et al. 2006) when subjected to subcutaneous and intranasal immunization with GAS CHO conjugated to tetanus toxoid. In parallel, analyses of serum samples and throat cultures from Mexican children revealed an inverse relationship between high serum titers of anti-GAS CHO antibodies and the presence of GAS in the throat. Moreover, no cross-reactivity of anti-GAS CHO antibodies with human tissues or cytoskeletal proteins was observed.

Malignancy is often associated with profound alterations in cell surface bound carbohydrate components of glycoconjugates. Such structural changes are due to incomplete glycosylation or novel glycosylation by tumour cells, which in turn arise from either down-regulation or up-regulation of certain glycosyl transferases (Kuberan and Linhardt 2000). Synthetic carbohydrate cancer vaccines have been shown to stimulate antibody-based immune responses in both pre-clinical and clinical settings. The antibodies have been observed to react in vitro with the corresponding natural carbohydrate antigens expressed on the surface of tumour cells and are able to mediate complement-dependent and/or antibody-dependent cell-mediated cytotoxicity. These multivalent vaccines with several different tumour-associated carbohydrate antigens have proven to be safe when administered to cancer patients (Ragupathi et al. 2002).

11.4.10.2 Carbohydrates as Carriers

Carbohydrates, as carriers, provide numerous attachment points for the conjugation of one or more peptide antigens (like the MAP system) (Fig. 11.4). By using different carbohydrates, the stereochemistry of the vaccine may be altered to provide an optimal orientation for the recognition of peptide epitopes by cells of the immune system. Carbohydrate carriers also help to reduce degradation of the attached peptide antigens. Therefore, the conjugation of peptide antigens to lipids (as an adjuvant) and carbohydrates (as a carrier) represents a highly promising strategy for the development of peptide vaccines.

One of the first examples where carbohydrates were used as carriers of peptide antigens was prophylactic GAS vaccines which were synthesized by conjugating multiple copies of a single GAS M protein derived specific peptide antigen onto the carbohydrate cores (D-glucose and D-galactose). These antigens contain peptide sequences which are highly conserved and offer the potential to prevent infections caused by up to 70% of GAS strains. Lipophilic amino acids were also conjugated to the D-glucose anomeric carbon to produce a self-adjuvanting liposaccharide vaccine. High serum IgG antibody titers against each of the incorporated peptide epitopes were detected following subcutaneous immunization of mice with the liposaccharide vaccine candidates (Simerska et al. 2008). This drug delivery system employs a glycolipid core construct as an adjuvant-carrier coupled to multiple copies of immunogenic peptides to induce antibody responses without the co-administration of additional adjuvants.

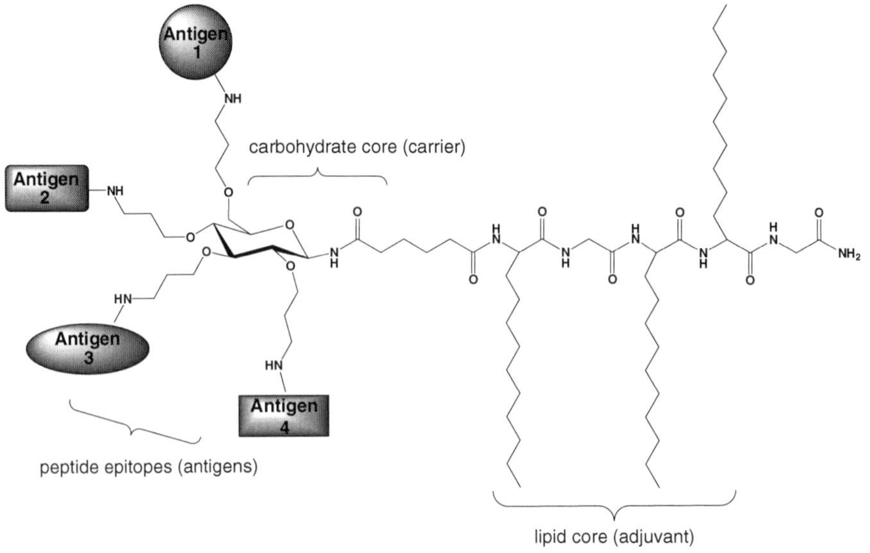

Fig. 11.4 The structure of a liposaccharide vaccine employing different peptide antigens

It is our belief that the ideal vaccines of the future will be constituted by synthetic peptide antigens, carbohydrates, and lipids. Because of the complexity of human T cells, it is also likely that several T-cell peptide epitopes would be required to create a vaccine that is able to induce a universal protective immune response. As vaccination of infants in developing countries is not a simple task, multivalent vaccines or cocktails of vaccines are highly recommended.

11.5 Conclusion

Mucosal adjuvant and vaccine delivery system development is an area of importance for improving public health. Mucosal immunization can serve in the future in increasing mucosal immune function, induction of protective immunity against infections, and induction of tolerance or modifying autoimmune disorders, allergies, and autoimmune diseases. Development of oral vaccines would have large implications for rural and remote populations where access to trained medical staff to administer vaccines by injection can be lacking.

The mucosal vaccines approved for human use include typhoid, cholera, adenovirus, Sabin oral polio, and rotavirus vaccines. New mucosal vaccine strategies are focused on development of non-replicating subunit vaccines, DNA, plant, and other types of recombinant vaccines as well as the use of mucosal adjuvants preferably inbuilt into the vaccine. The conjugation of lipids to peptide antigens is one approach which enables the production of highly customized all-in-one self-adjuvanting vaccines, which have the ability to direct the immune response towards important disease-specific antigens. Moreover, lipidated peptide vaccines have been successfully investigated in humans and demonstrated to be potentially safe.

References

Aguilar JC, Rodriguez EG (2007) Vaccine adjuvants revisited. Vaccine 25(19): 3752–3762

Ajdary S, Dobakhti F, Taghikhani M, Riazi-Rad F, Rafiei S, Rafiee-Tehrani M (2007) Oral administration of BCG encapsulated in alginate microspheres induces strong Th1 response in BALB/c mice. Vaccine 25(23): 4595–4601

Amorij JP, Westra TA, Hinrichs WLJ, Huckriede A, Frijlink HW (2007) Towards an oral influenza vaccine: Comparison between intragastric and intracolonic delivery of influenza subunit vaccine in a murine model. Vaccine 26(1): 67–76

Andrieu M, Loing E, Desoutter J-F, Connan F, Choppin J, Gras-Masse H, Hanau D, Dautry-Varsat A, Guillet J-G, Hosmalin A (2000) Endocytosis of an HIV-derived lipopeptide into human dendritic cells followed by class I-restricted $CD8^+$ T lymphocyte activation. Eur J Immunol 30(11): 3256–3265

Ann Clark M, Blair H, Liang L, Brey RN, Brayden D, Hirst BH (2001) Targeting polymerised liposome vaccine carriers to intestinal M cells. Vaccine 20(1–2): 208–217

Avtushenko SS, Sorokin EM, Zoschenkova NY, Naichin AN (1996) Clinical and immunological characteristics of the emulsion form of inactivated influenza vaccine delivered by oral immunization. J Biotechnol 44(1–3): 21–28

Aziz MA, Midha S, Waheed SM, Bhatnagar R (2007) Oral vaccines: new needs, new possibilities. Bioessays 29(6): 591–604

Babiuk LA (1999) Broadening the approaches to developing more effective vaccines. Vaccine 17(13–14): 1587–1595

Baldridge JR, McGowan P, Evans JT, Cluff C, Mossman S, Johnson D, Persing DH (2004) Taking a Toll on human disease: Toll-like receptor 4 agonists as vaccine adjuvants and monotherapeutic agents. Exp Op Biol Ther 4(7): 1129–1138

Baldridge JR, Yorgensen Y, Ward JR, Ulrich JT (2000) Monophosphoryl lipid A enhances mucosal and systemic immunity to vaccine antigens following intranasal administration. Vaccine 18(22): 2416–2425

Batzloff MR, Hartas J, Zeng W, Jackson DC, Good MF (2006) Intranasal vaccination with a lipopeptide containing a conformationally constrained conserved minimal Peptide, a universal T cell epitope, and a self-adjuvanting lipid protects mice from group a streptococcus challenge and reduces throat colonization. J Infect Dis 194(3): 325–330

BenMohamed L, Belkaid Y, Loing E, Brahimi K, Gras-Masse H, Druilhe P (2002a) Systemic immune responses induced by mucosal administration of lipopeptides without adjuvant. Eur J Immunol 32(8): 2274–2281

BenMohamed L, Gras-Masse H, Tartar A, Daubersies P, Brahimi K, Bossus M, Thomas A, Druilhe P (1997) Lipopeptide immunization without adjuvant induces potent and long-lasting B, T helper, and cytotoxic T lymphocyte responses against a malaria liver stage antigen in mice and chimpanzees. Eur J Immunol 27(5): 1242–1253

BenMohamed L, Krishnan R, Auge C, Primus JF, Diamond DJ (2002b) Intranasal administration of a synthetic lipopeptide without adjuvant induces systemic immune responses. Immunology 106(1): 113–121

BenMohamed L, Thomas A, Bossus M, Brahimi K, Wubben J, Gras-Masse H, Druilhe P (2000) High immunogenicity in chimpanzees of peptides and lipopeptides derived from four new *Plasmodium falciparum* pre-erythrocytic molecules. Vaccine 18(25): 2843–2855

BenMohamed L, Thomas A, Druilhe P (2004) Long-term multiepitopic cytotoxic-T-lymphocyte responses induced in chimpanzees by combinations of *Plasmodium falciparum* liver-stage peptides and lipopeptides. Infect Immun 72(8): 4376–4384
BenMohamed L, Wechsler SL, Nesburn AB (2002c) Lipopeptide vaccines-yesterday, today, and tomorrow. Lancet Infect Dis 2(7): 425–431
Bernstein DI (2000) Rotavirus vaccine – Current status and future prospects. Biodrugs 14(5): 275–281
Bessler WG, Cox M, Lex A, Suhr B, Wiesmuller KH, Jung G (1985) Synthetic lipopeptide analogs of bacterial lipoprotein are potent polyclonal activators for murine lymphocytes-B. J Immunol 135(3): 1900–1905
Brayden DJ, Baird AW (2001) Microparticle vaccine approaches to stimulate mucosal immunisation. Microb Infect 3(10): 867–876
Brewer JM (2006) (How) do aluminium adjuvants work? Immunol Lett 102(1): 10–15
Buwitt-Beckmann U, Heine H, Wiesmüller KH, Jung G, Brock R, Akira S, Ulmer AJ (2005) Toll-like receptor 6-independent signaling by diacylated lipopeptides. Eur J Immunol 35(1): 282–289
Chaignat CL, Monti V (2007) Use of oral cholera vaccine in complex emergencies: What next? Summary report of an expert meeting and recommendations of WHO. J Health Pop Nutr 25(2): 244–261
Chapel HM, August PJ (1976) Report of nine cases of accidental injury to man with Freund's complete adjuvant. Clin Exp Immunol 24(3): 538–541
Childers NK, Miller KL, Tong G, Llarena JC, Greenway T, Ulrich JT, Michalek SM (2000) Adjuvant activity of monophosphoryl lipid A for nasal and oral immunization with soluble or liposome-associated antigen. Infect Immun 68(10): 5509–5516
Childers NK, Tong G, Li F, Dasanayake AP, Kirk K, Michalek SM (2002) Humans immunized with Streptococcus mutans antigens by mucosal routes. J Dent Res 81(1): 48–52
Cho H-J, Shin H-J, Han I-K, Jung W-W, Kim YB, Sul D, Oh Y-K (2007) Induction of mucosal and systemic immune responses following oral immunization of mice with Lactococcus lactis expressing human papillomavirus type 16 L1. Vaccine 25(47): 8049–8057
Cho HJ, Shin HJ, Han IK, Jung WW, Kim YB, Sul D, Oh YK (2007) Induction of mucosal and systemic immune responses following oral immunization of mice with Lactococcus lactis expressing human papillomavirus type 16 L1. Vaccine 25(47): 8049–8057
Cross ML, Lambeth MR, Aldwell FE (2007) Murine cytokine responses following multiple oral immunizations using lipid-formulated mycobacterial antigens. Immunol Cell Biol 86(2): 214–217
Da Silva DM, Velders MP, Nieland JD, Schiller JT, Nickoloff BJ, Kast WM (2001) Physical interaction of human papillomavirus virus-like particles with immune cells. Int Immunol 13(5): 633–641
Deliyannis G, Kedzierska K, Lau YF, Zeng W, Turner SJ, Jackson DC, Brown LE (2006) Intranasal lipopeptide primes lung-resident memory $CD8^+$ T cells for long-term pulmonary protection against influenza. Eur J Immunol 36(3): 770–778
Deprez B, Sauzet J-P, Boutillon C, Martinon F, Tartar A, Sergheraert C, Guillet J-G, Gomard E, Gras-Masse H (1996) Comparative efficiency of simple lipopeptide constructs for in vivo induction of virus-specific CTL. Vaccine 14(5): 375–382
Doherty TM, Olsen AW, van Pinxteren L, Andersen P (2002) Oral vaccination with subunit vaccines protects animals against aerosol infection with Mycobacterium tuberculosis. Infect Immun 70(6): 3111–3121
Durier C, Launay O, Meiffrédy V, Saïdi Y, Salmon D, Lévy Y, Guillet JG, Pialoux G, Aboulker JP (2006) Clinical safety of HIV lipopeptides used as vaccines in healthy volunteers and HIV-infected adults. Aids 20(7): 1039–1049
Egan MA, Chong SY, Hagen M, Megati S, Schadeck EB, Piacente P, Ma BJ, Montefiori DC, Haynes BF, Israel ZR, Eldridge JH, Staats HF (2004) A comparative evaluation of nasal and parenteral vaccine adjuvants to elicit systemic and mucosal HIV-1 peptide-specific humoral immune responses in cynomolgus macaques. Vaccine 22(27–28): 3774–3788

Foster RH, Wagstaff AJ (1998) Tetravalent human-rhesus reassortant rotavirus vaccine – A review of its immunogenicity, tolerability and protective efficacy against paediatric rotavirus gastroenteritis. Biodrugs 9(2): 155–178

France LL, Piatti PG, Newman JFE, Toth I, Gibbons WA, Brown F (1994) Circular dichroism, molecular modeling, and serology indicate that the structural basis of antigenic variation in foot-and-mouth disease virus is α-helix formation. Proc Natl Acad Sci U S A 91(18): 8442–8446

Gahéry-Ségard H, Pialoux G, Charmeteau B, Sermet S, Poncelet H, Raux M, Tartar A, Lévy JP, Gras-Masse H, Guillet J-G (2000) Multiepitopic B- and T-cell responses induced in humans by a human immunodeficiency virus type 1 lipopeptide vaccine. J Virol 74(4): 1694–1703

Gahery-Segard H, Pialoux G, Charmeteau B, Sermet S, Poncelet H, Raux M, Tartar A, Levy JP, Gras-Masse H, Guillet JG (2000) Multiepitopic B- and T-cell responses induced in humans by a human immunodeficiency virus type 1 lipopeptide vaccine. J Virol 74(4): 1694–1703

Gahéry-Ségard H, Pialoux G, Figueiredo S, Igea C, Surenaud M, Gaston J, Gras-Masse H, Lévy JP, Guillet JG (2003) Long-term specific immune responses induced in humans by a human immunodeficiency virus type 1 lipopeptide vaccine: characterization of $CD8^+$-T-cell epitopes recognized. J Virol 77(20): 11220–11231

Gahery H, Daniel N, Charmeteau B, Ourth L, Jackson A, Andrieu M, Choppin J, Salmon D, Pialoux G, Guillet JG (2006) New $CD4^+$ and $CD8^+$ T cell responses induced in chronically HIV type-1-infected patients after immunizations with an HIV type 1 lipopeptide vaccine. AIDS Res Hum Retroviruses 22(7): 684–694

Garinot M, Fievez V, Pourcelle V, Stoffelbach F, des Rieux A, Plapied L, Theate I, Freichels H, Jerome C, Marchand-Brynaert J, Schneider YJ, Preat V (2007) PEGylated PLGA-based nanoparticles targeting M cells for oral vaccination. J Control Release 120(3): 195–204

Ghielmetti M, Reschner A, Zwicker M, Padovan E (2005) Synthetic bacterial lipopeptide analogs: structural requirements for adjuvanticity. Immunobiol 210(2–4): 211–215

Guerrero RA, Ball JM, Krater SS, Pacheco SE, Clements JD, Estes MK (2001) Recombinant Norwalk Virus-Like Particles Administered Intranasally to Mice Induce Systemic and Mucosal (Fecal and Vaginal) Immune Responses. J Virol 75(20): 9713–9722

Gupta PN, Khatri K, Goyal AK, Mishra N, Vyas SP (2007) M-cell targeted biodegradable PLGA nanoparticles for oral immunization against hepatitis B. J Drug Target 15(10): 701–713

Guzman CA, Molinari G, Fountain MW, Rohde M, Timmis KN, Walker MJ (1993) Antibody-Responses in the Serum and Respiratory-Tract of Mice Following Oral Vaccination with Liposomes Coated with Filamentous Hemagglutinin and Pertussis Toxoid. Infect Immun 61(2): 573–579

Haq TA, Mason HS, Clements JD, Arntzen CJ (1995) Oral Immunization with a Recombinant Bacterial-Antigen Produced in Transgenic Plants. Science 268(5211): 714–716

Hewitt MC, Seeberger PH (2001) Solution and solid-support synthesis of a potential leishmaniasis carbohydrate vaccine. J Org Chem 66(12): 4233–4243

Hillery AM, Florence AT (1996) The effect of adsorbed poloxamer 188 and 407 surfactants on the intestinal uptake of 60-nm polystyrene particles after oral administration in the rat. Int J Pharm 132(1–2): 123–130

Horváth A, Olive C, Karpati L, Sun HK, Good M, Toth I (2004) Toward the development of a synthetic group A streptococcal vaccine of high purity and broad protective coverage. J Med Chem 47(16): 4100–4104

Horváth A, Olive C, Wong A, Clair T, Yarwood P, Good M, Toth I (2002a) Lipoamino acid-based adjuvant carrier system: enhanced immunogenicity of group A streptococcal peptide epitopes. J Med Chem 45(6): 1387–1390

Horváth A, Olive C, Wong A, Clair T, Yarwood P, Good M, Toth I (2002b) A lipophilic adjuvant carrier system for antigenic peptides. Lett Pept Sci 8(3–5): 285–288

Hussain N, Jani PU, Florence AT (1997) Enhanced Oral Uptake of Tomato Lectin-Conjugated Nanoparticles in the Rat. Pharm Res 14(5): 613–618

Jackson DC, Lau YF, Le T, Suhrbier A, Deliyannis G, Cheers C, Smith C, Zeng W, Brown LE (2004) A totally synthetic vaccine of generic structure that targets Toll-like receptor 2 on dendritic cells and promotes antibody or cytotoxic T cell responses. Proc Natl Acad Sci U S A 101(43): 15440–15445

Johnson AG (1964). Adjuvant action of bacterial endotoxins on the primary antibody response. Bacterial endotoxins: Proc Symp Inst Microb Rutgers, State Uni Natl Sci Found. M. Landy and W. Braun. New Brunswick, NJ, Rutgers University Press: 252–262.

Johnstone KD, Dieckelmann M, Jennings MP, Toth I, Blanchfield JT (2005) Chemo-Enzymatic Synthesis of a Trisaccharide-Linked Peptide Aimed at Improved Drug-Delivery. Curr Drug Deliv 2: 215–222

Jones D (2007) Cancer vaccines on the horizon. Nat Rev Drug Discov 6(5): 333–334

Klinguer C, David D, Kouach M, Wieruszeski JM, Tartar A, Marzin D, Lévy JP, Gras-Masse H (1999) Characterization of a multi-lipopeptides mixture used as an HIV-1 vaccine candidate. Vaccine 18(3–4): 259–267

Koeller KM, Wong CH (2000) Synthesis of Complex Carbohydrates and Glycoconjugates: Enzyme-Based and Programmable One-Pot Strategies. Chem Rev 100(12): 4465–4494

Krieg AM (2002) From A to Z on CpG. Trends Immunol 23(2): 64–65

Kuberan B, Linhardt RJ (2000) Carbohydrate based vaccines. Curr Org Chem 4: 653–677

Kuklin N, Daheshia M, Karem K, Manickan E, Rouse BT (1997) Induction of mucosal immunity against herpes simplex virus by plasmid DNA immunization. J Virol 71(4): 3138–3145

Langhans B, Schweitzer S, Nischalke HD, Braunschweiger I, Sauerbruch T, Spengler U (2004) Hepatitis C virus-derived lipopeptides differentially induce epitope-specific immune responses in vitro. J Infect Dis 189(2): 248–253

Lazzell V, Waldman RH, Rose C, Khakoo R, Jacknowitz A, Howard S (1984) Immunization against influenza in humans using an oral enteric-coated killed virus-vaccine. J Biol Standard 12(3): 315–321

Lévy Y, Gahéry-Ségard H, Durier C, Lascaux AS, Goujard C, Meiffrédy V, Rouzioux C, El Habib R, Beumont-Mauviel M, Guillet J-G, Delfraissy J-F, Aboulker J-P (2005) Immunological and virological efficacy of a therapeutic immunization combined with interleukin-2 in chronically HIV-1 infected patients. Aids 19(3): 279–286

Li TC, Takeda N, Miyamura T (2001) Oral administration of hepatitis E virus-like particles induces a systemic and mucosal immune response in mice. Vaccine 19: 3476–3484

Lindberg AA (1999) Glycoprotein conjugate vaccines. Vaccine 17: S28–S36

Liu W-T, Hsu H-L, Liang C-C, Chuang C-C, Lin H-C, Liu Y-T (2007) A comparison of immunogenicity and protective immunity against experimental plague by intranasal and/or combined with oral immunization of mice with attenuated Salmonella serovar Typhimurium expressing secreted Yersinia pestis F1 and V antigen. FEMS Immunol Medic Microb 51(1): 58–69

Longini IM, Nizam A, Ali M, Yunus M, Shenvi N, Clemens JD (2007) controlling endemic cholera with oral vaccines. PLoS Med 4(11): e336

Lubeck MD, Davis AR, Chengalvala M, Natuk RJ, Morin JE, Molnar-Kimber K, Mason BB, Bhat BM, Mizutani S, Hung PP, Purcell RH (1989) Immunogenicity and Efficacy Testing in Chimpanzees of an Oral Hepatitis B Vaccine Based on Live Recombinant Adenovirus. Proc Natl Acad Sci U S A 86(17): 6763–6767

Mattingly JA, Waksman BH (1978) Immunological suppression after oral-administration of antigen .1. Specific suppressor cells formed in rat Peyers patches after oral-administration of sheep erythrocytes and their systemic migration. J Immunol 121(5): 1878–1883

McCluskie MJ, Weeratna RD, Krieg AM, Davis HL (2000) CpG DNA is an effective oral adjuvant to protein antigens in mice. Vaccine 19(7–8): 950–957

McGeary RP, Jablonkai I, Toth I (2001) Carbohydrate-based templates for synthetic vaccine and drug delivery. Tetrahedron 57(41): 8733–8742

McGeary RP, Jablonkai I, Toth I (2002) Towards synthetic vaccines built on carbohydrate cores. Lett Pept Sci 8(3–5): 273–276

McNeela EA, Mills KHG (2001) Manipulating the immune system: humoral versus cell-mediated immunity. Adv Drug Deliv Rev 51(1–3): 43–54

Mercier GT, Nehete PN, Passeri MF, Nehete BN, Weaver EA, Templeton NS, Schluns K, Buchl SS, Sastry KJ, Barry MA (2007) Oral immunization of rhesus macaques with adenoviral HIV vaccines using enteric-coated capsules. Vaccine 25(52): 8687–8701

Mestecky J, Russell MW, Elson CO (2007) Perspectives on mucosal vaccines: Is mucosal tolerance a barrier? J Immunol 179(9): 5633–5638

Xia M, Farkas T, Jiang X (2007) Norovirus capsid protein expressed in yeast forms virus-like particles and stimulates systemic and mucosal immunity in mice following an oral administration of raw yeast extracts. J Med Virol 79(1): 74–83

Mirchamsy H, Hamedi M, Fateh G, Sassani A (1994) Oral immunization against diphtheria and tetanus infections by fluid diphtheria and tetanus toxoids. Vaccine 12(13): 1167–1172

Mittenbuhler K, Baier W, van der Esche U, Heinevetter L, Wiesmuller K-H, Jung G, Weckesser J, Bessler WG, Hoffmann P (1997) Lipopeptides as efficient novel immunogens and adjuvants in parenteral and oral immunization. Curr Top Pept Prot Res 2: 125–135

Moyle PM, Barozzi N, Wimmer N, Olive C, Good M, Toth I (2005) Development of peptide vaccines against HPV-16 associated cervical cancer and group A streptococci. Biopolymers 80(4): 556–557

Moyle PM, Horvath A, Olive C, Good MF, Toth I (2003) Development of lipid-core-peptide (LCP) based vaccines for the prevention of group A streptococcal (GAS) infection. Lett Pept Sci 10(5–6): 605–613

Moyle PM, Olive C, Karpati L, Barozzi N, Ho M-F, Dyer J, Sun HK, Good M, Toth I (2006) Synthesis and immunological evaluation of M protein targeted tetra-valent and tri-valent group A streptococcal vaccine candidates based on the lipid-core peptide system. Int J Pept Res Ther 12(3): 317–326

Munro P, Flatau G, Lemichez E (2007) Intranasal immunization with tetanus toxoid and CNF1 as a new mucosal adjuvant protects BALB/c mice against lethal challenge. Vaccine 25(52): 8702–8706

Nardelli B, Haser PB, Tam JP (1994) Oral administration of an antigenic synthetic lipopeptide (MAP-P3C) evokes salivary antibodies and systemic humoral and cellular responses. Vaccine 12(14): 1335–1339

Niikura M, Takamura S, Kim G, Kawai S, Saijo M, Morikawa S, Kurane I, Li T-C, Takeda N, Yasutomi Y (2002) Chimeric recombinant Hepatitis E virus-like particles as an oral vaccine vehicle presenting foreign epitopes. Virology 293(2): 273–280

Noad R, Roy P (2003) Virus-like particles as immunogens. Trends Microbiol 11(9): 438–444

Nomoto M, Yamada K, Haga M, Hayashi M (1998) Improvement of intestinal absorption of peptide drugs by glycosylation: Transport of tetrapeptide by the sodium ion-dependent D-glucose transporter. J Pharm Sci 87(3): 326–332

Nystrom J, Svennerholm A-M (2007) Oral immunization with HpaA affords therapeutic protective immunity against H. pylori that is reflected by specific mucosal immune responses. Vaccine 25(14): 2591–2598

Ogra PL, Faden H, Welliver RC (2001) Vaccination strategies for mucosal immune responses (vol 14, p. 430, 2001). Clin Microbiol Rev 14(3): 641–641

Olive C, Batzloff M, Horváth A, Clair T, Yarwood P, Toth I, Good MF (2003) Potential of lipid core peptide technology as a novel self-adjuvanting vaccine delivery system for multiple different synthetic peptide immunogens. Infect Immun 71(5): 2373–2383

Olive C, Batzloff M, Horváth A, Clair T, Yarwood P, Toth I, Good MF (2004) Group A streptococcal vaccine delivery by immunization with a self-adjuvanting M protein-based lipid core peptide construct. Indian J Med Res 119(Suppl): 88–94

Olive C, Batzloff MR, Horváth A, Wong A, Clair T, Yarwood P, Toth I, Good MF (2002) A lipid core peptide construct containing a conserved region determinant of the group A streptococcal M protein elicits heterologous opsonic antibodies. Infect Immun 70(5): 2734–2738

Olive C, Ho M-F, Dyer J, Lincoln D, Barozzi N, Toth I, Good MF (2006b) Immunization with a tetraepitopic lipid core peptide vaccine construct induces broadly protective immune responses against group A streptococcus. J Infect Dis 193(12): 1666–1676

Olive C, Hsien K, Horvath A, Clair T, Yarwood P, Toth I, Good MF (2005) Protection against group A streptococcal infection by vaccination with self-adjuvanting lipid core M protein peptides. Vaccine 23(17–18): 2298–2303

Olive C, Sun HK, Ho M-F, Dyer J, Horváth A, Toth I, Good MF (2006a) Intranasal administration is an effective mucosal vaccine delivery route for self-adjuvanting lipid core peptides targeting the group A streptococcal M protein. J Infect Dis 194(3): 316–324

Ouellette AJ, Satchell DP, Hsieh MM, Hagen SJ, Selsted ME (2000) Characterization of luminal paneth cell alpha-defensins in mouse small intestine. Attenuated antimicrobial activities of peptides with truncated amino termini. J Biol Chem 275(43): 33969–33973

Perez-Schael I, Salinas B, Tomat M, Linhares AC, Guerrero ML, Ruiz-Palacios GM, Bouckenooghe A, Yarzabal JP (2007) Efficacy of the human rotavirus vaccine RIX4414 in malnourished children. J Infect Dis 196(4): 537–540

Persing DH, McGowan P, Evans JT, Cluff C, Mossman S, Johnson D, Baldridge JR (2006). Toll-like receptor 4 agonists as vaccine adjuvants. Immunopotentiators in modern vaccines. E. J. C. Virgil and D. T. O'Hagan. Burlington, MA, Elsevier Academic Press: 93–108.

Pertmer TM, Oran AE, Moser JM, Madorin CA, Robinson HL (2000) DNA Vaccines for Influenza Virus: Differential Effects of Maternal Antibody on Immune Responses to Hemagglutinin and Nucleoprotein. J Virol 74(17): 7787–7793

Pialoux G, Gahéry-Ségard H, Sermet S, Poncelet H, Fournier S, Gérard L, Tartar A, Gras-Masse H, Lévy JP, Guillet JG (2001) Lipopeptides induce cell-mediated anti-HIV immune responses in seronegative volunteers. AIDS 15(10): 1239–1240

Pickering RJ, Smith SD, Strugnell RA, Wesselingh SL, Webster DE (2006) Crude saponins improve the immune response to an oral plant-made measles vaccine. Vaccine 24(2): 144–150

Piedra PA, Poveda GA, Ramsey B, McCoy K, Hiatt PW (1998) Incidence and prevalence of neutralizing antibodies to the common adenoviruses in children with cystic fibrosis: Implication for gene therapy with adenovirus vectors. Pediatrics 101(6): 1013–1019

Plante OJ, Palmacci ER, Seeberger PH (2001) Automated Solid-Phase Synthesis of Oligosaccharides Automated Solid-Phase Synthesis of Oligosaccharides. Science 291(5508): 1523

Progress Toward Poliomyelitis Eradication – India, January 2006–September (2007). MMWR Morb Mortal Wkly Rep. 56(45): 1187–91

Purcell AW, McCluskey J, Rossjohn J (2007) More than one reason to rethink the use of peptides in vaccine design. Nat Rev Drug Discov 6(5): 404–414

Ragupathi G, Coltart DM, Williams LJ, Koide F, Kagan E, Allen J, Harris C, Glunz PW, Livingston PO, Danishefsky SJ (2002) On the power of chemical synthesis: Immunological evaluation of models for multiantigenic carbohydrate-based cancer vaccines. Proc Natl Acad Sci 99(21): 13699–13704

Robinson HL (2007) HIV/AIDS vaccines: 2007. Clin Pharmacol Ther 82(6): 686–693

Roland KL, Cloninger C, Kochi SK, Thomas LJ, Tinge SA, Rouskey C, Killeen KP (2007) Construction and preclinical evaluation of recombinant Peru-15 expressing high levels of the cholera toxin B subunit as a vaccine against enterotoxigenic Escherichia coli. Vaccine 25(51): 8574–8584

Roy R (2004) New trends in carbohydrate-based vaccines. Drug Discov Today: Technol 1(3): 327–336

Russell-Jones GJ (2000) Oral vaccine delivery. J Control Release 65(1–2): 49–54
Sabharwal H, Michon F, Nelson D, Dong W, Fuchs K, Manjarrez RC, Sarkar A, Uitz C, Viteri-Jackson A, Suarez RSR, Blake M, Zabriskie JB (2006) Group A Streptococcus (GAS) Carbohydrate as an Immunogen for Protection against GAS Infection. J Infect Dis 193(1): 129–135
Seegers JFML (2002) Lactobacilli as live vaccine delivery vectors: progress and prospects. Trends Biotechnol 20(12): 508–515
Seth A, Yasutomi Y, Jacoby H, Callery JC, Kaminsky SM, Koff WC, Nixon DF, Letvin NL (2000) Evaluation of a lipopeptide immunogen as a therapeutic in HIV type 1-seropositive individuals. AIDS Res Hum Retroviruses 16(4): 337–343
Shahin RD, Amsbaugh DF, Leef MF (1992) Mucosal immunization with filamentous hemagglutinin protects against Bordetella pertussis respiratory infection. Infect Immun 60(4): 1482–1488
Shalaby WSW (1995) Development of oral vaccines to stimulate mucosal and systemic immunity: barriers and novel strategies. Clin Immunol Immunopath 74(2): 127–134
Sharon M, Nir P, Lior K, David BN, Tomer I, Paula S, Reuven L, Shlomo L (2007) Tail scarification with Vaccinia virus Lister as a model for evaluation of smallpox vaccine potency in mice. Vaccine 25(45): 7743–7753
Shin SJ, Shin SW, Kang ML, Lee DY, Yang MS, Jang YS, Yoo HS (2007) Enhancement of protective immune responses by oral vaccination with Saccharomyces cerevisiae expressing recombinant Actinobacillus pleuropneumoniae ApxIA or ApxIIA in mice. J Vet Sci 8(4): 383–392
Silin DS, Lyubomska OV, Jirathitikal V, Bourinbaiar AS (2007) Oral vaccination: where we are? Exp Opin Drug Deliv 4(4): 323–340
Simerska P, Abdel-Aal ABM, Fujita Y, McGeary RP, Moyle PM, Batzloff MR, Olive C, Good M, Toth I (2008) Development of a liposaccharide-based delivery system and its application to the design of group A streptococcal vaccines. J Med Chem 51(5): 1447–1452
Steller MA, Gurski KJ, Murakami M, Daniel RW, Shah KV, Celis E, Sette A, Trimble EL, Park RC, Marincola FM (1998) Cell-mediated immunological responses in cervical and vaginal cancer patients immunized with a lipidated epitope of human papillomavirus type 16 E7. Clin Cancer Res 4(9): 2103–2109
Sutter RW, Prevots DR, Cochi SL (2000) Poliovirus vaccines – progress toward global poliomyelitis eradication and changing routine immunization recommendations in the United States. Pediatr Clin North Am 47(2): 287–308
Therien HM, Lair D, Shahum E (1990) Liposomal vaccine – influence of antigen association on the kinetics of the humoral response. Vaccine 8(6): 558–562
Thones N, Muller M (2007) Oral immunization with different assembly forms of the HPV 16 major capsid protein L1 induces neutralizing antibodies and cytotoxic T-lymphocytes. Virol 369(2): 375–388
Toth I, Danton M, Flinn N, Gibbons WA (1993) A combined adjuvant and carrier system for enhancing synthetic peptides immunogenicity utilizing lipidic amino acids. Tetrahedron Lett 34(24): 3925–3928
Tritto E, Muzzi A, Pesce I, Monaci E, Nuti S, Galli G, Wack A, Rappuoli R, Hussell T, De Gregorio E (2007) The acquired immune response to the mucosal adjuvant LTK63 imprints the mouse lung with a protective signature. J Immunol 179(8): 5346–5357
Ulrich JT, Myers KR (1995) Monophosphoryl lipid A as an adjuvant. Past experiences and new directions. Pharm Biotechnol 6: 495–524
van Duin D, Medzhitov R, Shaw AC (2006) Triggering TLR signaling in vaccination. Trends Immunol 27(1): 49–55
VanCott TC, Kaminski RW, Mascola JR, Kalyanaraman VS, Wassef NM, Alving CR, Ulrich JT, Lowell GH, Birx DL (1998) HIV-1 neutralizing antibodies in the genital and respiratory tracts of mice intranasally immunized with oligomeric gp160. J Immunol 160(4): 2000–2012

Wagner B, Earn D (2008) Circulating vaccine derived polio viruses and their impact on global polio eradication. Bull Math Biol 70(1): 253–280

Wiesmüller K-H, Fleckenstein B, Jung G (2001) Peptide vaccines and peptide libraries. Biol Chem 382(4): 571–579

Wu C-M, Chung T-C (2007) Mice protected by oral immunization with Lactobacillus reuteri secreting fusion protein of Escherichia coli enterotoxin subunit protein. FEMS Immunol & Med Microbiol 50(3): 354–365

Yokoyama Y, Harabuchi Y (2002) Intranasal immunization with lipoteichoic acid and cholera toxin evokes specific pharyngeal IgA and systemic IgG responses and inhibits streptococcal adherence to pharyngeal epithelial cells in mice. Int J Pediatr Otorhinolaryngol 63(3): 235–241

Zeng W, Ghosh S, Lau YF, Brown LE, Jackson DC (2002) Highly immunogenic and totally synthetic lipopeptides as self-adjuvanting immunocontraceptive vaccines. J Immunol 169(9): 4905–4912

Zeng W, Jackson DC, Murray J, Rose K, Brown LE (2000) Totally synthetic lipid-containing polyoxime peptide constructs are potent immunogens. Vaccine 18(11–12): 1031–1039

Zhang X, Zhang X, Yang Q (2007) Effect of compound mucosal immune adjuvant on mucosal and systemic immune responses in chicken orally vaccinated with attenuated Newcastle-disease vaccine. Vaccine 25(17): 3254–3262

Zhang Y-D, Lu X-L, Li N-F (2007) The prospective preventative HIV vaccine based on modified poliovirus. Med Hyp 68(6): 1258–1261

Zhong G, Toth I, Reid R, Brunham RC (1993) Immunogenicity evaluation of a lipidic amino acid-based synthetic peptide vaccine for Chlamydia trachomatis. J Immunol 151(7): 3728–3736

Zimmerman RK, Spann SJ (1999) Poliovirus vaccine options. Am Fam Phys 59(1): 113–118

Chapter 12
Oral Delivery of Nucleic Acid Drugs

Ronny Martien

Contents

Abstract Nucleic acid molecules have emerged as versatile tools with promising utility in a variety of biochemical, diagnostic, and therapeutic applications. A parenteral administration of a nucleic acid is inconvenient because of pain, fear, and risks being associated with this type of application. The intestinal epithelium is considered to be an attractive site for oral delivery of therapeutic genes.

The successful development of oral nucleic acid delivery systems is challenged by a variety of barriers encountered with the GI tract. The intestinal mucosa is both a physical and a biochemical barrier, separating the external environment from the internal milieu of the body.

Despite the enormous potential of gene therapy, safe and efficient delivery of nucleic acid into cells is still a dominant task in current biotechnological research. The majority of nucleic acid therapeutics are to a higher degree dependent on delivery systems for successful therapeutic intervention than conventional drugs.

Regarding safety concerns, non-viral gene delivery vehicles that have the required efficiency and safety for use in human gene therapy are being widely

---

R. Martien (✉)
Department of Pharmaceutics, Faculty of Pharmacy, Gadjah Mada University, Sekip Utara, 55281 Yogyakarta, Indonesia
e-mail: RonnyMartien@ugm.ac.id

A. Bernkop-Schnürch (ed.), *Oral Delivery of Macromolecular Drugs*,
DOI 10.1007/978-1-4419-0200-9_12,

investigated as possible alternatives. Non-viral systems show a significantly lower safety risk and can be tailored to specific therapeutic needs.

12.1 Introduction

A nucleic acid is a complex, high molecular weight biochemical macromolecule composed of nucleotide chains that convey genetic information. Nucleic acid molecules have emerged as versatile tools with promising utility in a variety of biochemical, diagnostic, and therapeutic applications (Santiago and Khachigian 2001). The past several years have witnessed the evolution of gene medicine from an experimental technology into a viable strategy for developing therapeutics for a wide range of human disorders (Patil et al. 2005). The physicochemical properties of nucleic acids with molecular weights ranging from 7 kDa for antisense oligonucleotides to over 1 MDa for plasmid DNA, and strong negative charge, however, do not favor membrane passage (Mastrobattista et al. 2007).

The parenteral administration of nucleic acid is inconvenient because of pain, fear, and risks being associated with this type of application. Injectable-to-non-invasive-conversions and in particular injectable-to-oral-conversions are consequently highly in demand (Bernkop-Schnurch et al. 2004). One of the most compelling features is its ease of access via luminal route, which would allow direct in vivo gene transfer by oral administration (Sandberg et al. 1994). The intestinal epithelium is considered to be an attractive site for oral delivery of therapeutic genes (Fig. 12.1). Furthermore, it could be an alternative target for the treatment of many metabolic and nutritional defects (Sweetser et al. 1988) and a target for gene delivery in vaccination strategies (Mestecky 1987). The concept, however, is so far strongly limited by an insufficient access to the target tissue. With a few exceptions such as colon carcinoma and sigmoid colon cancer, where a direct access to the target tissue seems to be comparatively more easily feasible via oral administration (Kai and Ochiya 2004), the gastrointestinal tract sets a variety of morphological and physiological barriers that strongly limit intestinal absorption of therapeutic nucleic acids (Borges et al. 2005).

12.2 Defining the Problems of Oral Administration

The successful development of oral nucleic acid delivery systems is challenged by a variety of barriers encountered with the GI tract. The intestinal mucosa is both a physical and a biochemical barrier, separating the external environment from the internal milieu of the body. The physical barrier arises from cell membrane and the intercellular junctions between cells – designated tight junctions. The biochemical barrier is based on acidic conditions in the stomach and

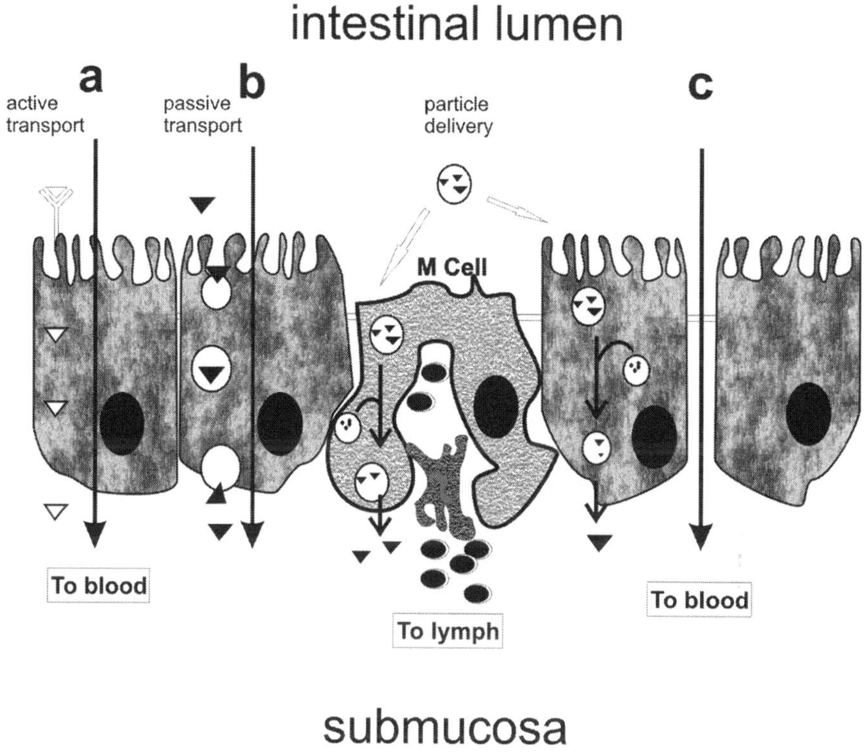

Fig. 12.1 Schematic representation of the three transepithelial intestinal pathways: (**a**) transcellular active transport, (**b**) transcellular passive transport, (**c**) paracellular transport (Fasano 1998)

on various metabolic enzymes such as nucleases present in gastrointestinal juice degrading orally administered nucleic acids. Recently, Loretz et al. (2006) determined the nuclease activity in the small intestinal juice of pigs to be 0.02 Kunits units/ml. To date, however, a more detailed knowledge about this enzymatic barrier for nucleic acids is to our notice not available.

The second challenge is the passage through the mucus gel layer. GI-epithelia are covered by a mucus gel layer, which is 80–200 μm in thickness. The viscosity of mucus affects the diffusion barrier. The presence of sulfate and sialic acid moieties in mucus causes a negative net charge. Hence, the diffusion barrier is even more pronounced for cationic nucleic acid/polymer complexes, which undergo ionical immobilization on the mucus.

In case of local gastrointestinal treatments nucleic acid drugs shall be taken up by enterocytes, usually through endocytosis. The route of uptake determines, subsequently, nucleic acid trafficking and lifetime in the cell. Endocytosis is a multistep process involving binding, internalization, formation of endosomes, fusion with lysosomes, and lysis. The low pH and enzymes within endosomes and lysosomes usually lead to degradation of entrapped nucleic acids and associated complexes. Finally, nucleic acid that has survived both endocytotic processing and cytoplasmic nucleases must dissociate from the condensed complexes either before or after entering the nucleus. Entry is thought to occur through nuclear pores which are 10 nm in diameter or during cell division. Once inside the nucleus, in case of orally administered therapeutic genes, the transfection efficiency is mostly dependent on the composition of the gene expression system. The low efficiency of nucleic acid delivery from outside the cell to inside the nucleus is a natural consequence of this multistep process. As a result, the number of nucleic acid molecules decreases at each step of the journey to the nucleus. Therefore, identifying and overcoming each hurdle along the nucleic acid entry pathways can improve nucleic acid delivery, and hence overall transfection efficiency, dramatically.

In case of systemic delivery of nucleic acid drugs via the oral route only comparatively small therapeutic agents such as oligonucleotides seem to reach the systemic circulation in significant quantities via the paracellular route of uptake.

12.3 Type of Nucleic Acid Drugs

DNA-based therapeutics include plasmids containing transgenes for gene therapy, oligonucleotides for antisense and anti-gene applications (Crooke 1998), ribozymes, DNAzymes, and aptamers. RNA-based therapeutics are mainly represented by small interfering RNAs (siRNAs) (Stull and Szoka 1995; Patil et al. 2005). A deficient gene can either be replaced or the effect of an unwanted gene can be blocked by the introduction of a counteracting one (Merdan et al. 2002). Furthermore, antisense strategy offers the potential to down-regulate selectively the expression of specific genes mainly on translational level predominantly by sequence-specific interaction with messengers. Current and planned clinical trials with antisense oligonucleotides and ribozymes are shown in Table 12.1. Despite the enormous potential of gene therapy, safe and efficient delivery of nucleic acid into cells is still a dominant task in current biotechnological research (Rubanyi 2001). The majority of nucleic acid therapeutics is to a higher degree dependent on delivery systems for successful therapeutic intervention than conventional drugs. The transfection efficiency of currently used delivery systems is still too low for many clinical applications. Therefore, new and more potent delivery systems are needed that combine a high degree of transfection with acceptable toxicity profiles (Mastrobattista et al. 2007).

12.3.1 Plasmids

Plasmids are high molecular weight, double-stranded DNA constructs containing transgenes, which encode specific proteins. On the molecular level plasmid

Table 12.1 DNA-based therapeutics currently in advanced stages of clinical development (phase 3 and beyond)* (Patil et al. 2005)

Drug candidate – company	Type of DNA-based therapeutic	Development status	Molecular basis of action	Disease indication
Gendicine – SiBiono Genetech	Plasmid	Chinese-FDA approved	Adenovirus encoding the tumor suppressor *p53* gene	Head and neck squamous cell carcinoma
Advexin – Introgen Therapeutics	Plasmid	Phase 3 alone or in combination with cisplatin and 5-fluorouracil	Adenovirus encoding the tumor suppressor *p53* gene	Refractory head and neck squamous cell carcinoma
Vitravene (fomivirsen sodium) – Isis Pharmaceuticals	Antisense oligonucleotide	FDA approved	Inhibitor of immediate early region 2 (IE2) of human cytomegalovirus	Cytomegalovirus retinitis in AIDS patients
Affinitak – Isis Pharmaceuticals	Antisense oligonucleotide	Phase 3 in combination with carboplatin and paclitaxel	Inhibitor of protein kinase C-alpha (PKC-alpha) expression	Stage IIIb or Stage IV non-small cell lung cancer
Alicaforsen – Isis Pharmaceuticals	Antisense oligonucleotide	Phase 3	Inhibitor of intracellular adhesion molecule-1 (ICAM-1)	Crohn's disease
Macugen (pegaptanib sodium) – Eyetech and Pfizer	Aptamer	Phase 3 with or without photodynamic therapy	Inhibitor of vascular endothelial growth factor (VEGF)	Age-related macular Degeneration
Genasense (oblimersen sodium) – Aventis and Genta	Antisense oligonucleotide	Late-stage phase 3 in combination with dexamethasone	Inhibitor of B-cell leukemia/lymphoma 2 (Bcl-2) protein	Malignant melanoma
Genasense (oblimersen sodium) – Aventis and Genta	Antisense oligonucleotide	Phase 3 in combination with fludarabine and cyclophosphamide	Inhibitor of B-cell leukemia/lymphoma 2 (Bcl-2) protein	Chronic lymphocytic leukemia
Genasense (oblimersen sodium) – Aventis and Genta	Antisense oligonucleotide	Phase 3 in combination with dacarbazine	Inhibitor of B-cell leukemia/lymphoma 2 (Bcl-2) protein	Multiple myeloma

*All development statuses are filed with the United States Food and Drug Administration (USFDA) unless otherwise indicated.

DNA can be considered as pro-drug that upon cellular internalization employs the DNA transcription and translation apparatus in the cell to biosynthesize the therapeutic entity, the protein (Uherek and Wels 2000). Gene therapy involves the use of plasmid DNA to introduce transgenes into cells that inherently lack the ability to produce the protein that the transgene is programmed to generate. Plasmids can be used to correct genetic errors that produce functionally incompetent copies of a given protein. In addition to disease treatment, plasmids can be used as DNA vaccines for genetic immunization (Johnston et al. 2002). DNA vaccines function through induction of immune response by introducing genes encoding antigens for specific pathogens. Especially, such DNA vaccines are of interest in terms of oral delivery.

12.3.2 Aptamers

Aptamers from the latin aptus (meaning to fit) are single-stranded or double-stranded nucleic acids. These are selected and amplified oligonucleotides that have been isolated from random pools of synthetic oligonucleotides according to their ability to bind with high affinity to biological target molecules (Ellington and Szostak 1990). Aptamers as therapeutics would most likely bind proteins involved in the regulation and expression of genes dependent upon activity of protein. First of all, sugar modifications of nucleoside triphosphates are necessary to render the resulting aptamers resistant to nucleases found in serum. Changing the 2′-OH groups of ribose to 2′-F or 2′-NH2 groups yields aptamers, which are long-lived in blood. Only a few instances of oligonucleotide aptamers displaying biological effects have been reported. Double-stranded aptamers acting on B cells, phosphorothioates acting on T cells, and phosphodiester aptamers acting on thrombin for anticoagulation are a few examples for them (Riordan and Martin 1991; Ess et al. 1994).

12.3.3 Antisense Oligonucleotides

Antisense therapy is designed to prevent or at least lower expression of a specific gene. An oligonucleotide that has a sequence that is complementary to the mRNA of the target gene is introduced into the cell. It will bind to the mRNA of the target gene and block translation of the message into protein. It may bind to DNA in the nucleus, blocking transcription, or to the transcript during its processing and transport from the nucleus to the cytoplasm: all these interactions would reduce expression of gene (Zamecnik and Stephenson 1978). A minimum length for antisense oligonucleotides in order to get specific binding is 11 bases, but most being tested are in the 15–25 base range. The most commonly used antisense oligonucleotides (ASONs) are phosphorothioates and methyl phosphonates. These derivatives are generally more stable toward an enzymatic attack in the GI tract. In addition, methyl phosphonates are uncharged and therefore more lipophilic than native DNA or RNA and may consequently penetrate cells better (Elbashir et al. 2001; Stein 2001).

12.3.4 Ribozymes

Catalytic RNAs, or ribozymes, are RNAs, which catalytically cleave covalent bonds in a target RNA. The catalytic site is the result of the conformation adopted by the RNA–RNA complex in the presence of divalent cations. Shortly thereafter, Altman and colleagues discovered the active role of the RNA component of RNase P in the process of tRNA maturation. This was the first characterization of a true RNA enzyme that catalyzes the reaction of a free substrate, i.e., possesses catalytic activity in *trans* (Guerrier et al. 1983). A variety of ribozymes, catalyzing intramolecular splicing or cleavage reactions, have subsequently been found in lower eukaryotes, viruses, and some bacteria.

12.3.5 RNA Interference

RNA interference is initiated by long double-stranded RNA molecules, which are processed into 21–23-nucleotide-long RNAs by the Dicer enzyme. This RNase III protein is thought to act as a dimer that cleaves both strands of dsRNAs and leaves two nucleotide 3′-overhanging ends. These small interfering RNAs (siRNAs) are then incorporated into the RNA-induced silencing complex, a protein–RNA complex, and guide a nuclease, which degrades the target RNA (Fire et al. 1998). The 21-nucleotide-long siRNA duplexes with 3′-overhangs can specifically suppress gene expression in mammalian cells (Yu JY et al. 2002).

12.4 Strategies

In case of gene therapy, two different approaches, namely viral and non-viral vectors, have been utilized for the delivery of nucleic acids (Mansouri et al. 2006). A vector can be described as a system fulfilling several functions, including (a) enabling delivery of genes into the target cells and their nucleus, (b) providing protection from gene degradation, and (c) ensuring gene transcription in the cell. The administration of gene therapy vectors requires not only to be targeted and safe, but also to protect from degradation, sequestration, or immune attack. Moreover, it has to be inexpensive and easy to produce in large amounts and at high concentrations (Gardlik et al. 2005). According to statistics covering all clinical trials worldwide in gene delivery, viral vectors are still the clear number one, being used in more than 70% of all protocols because of their potentially high efficiencies (Fraunhofer et al. 2004). Problems associated with viral vectors, however, are their potential oncogenicity due to insertional mutagenesis and the limited size of DNA that can be carried. Furthermore, their isolation from biological sources and their processing are very expensive. Since a patient died from viral gene therapy treatment nobody can still deny the safety concerns connected to these delivery systems (Hacein-Bey-Abina et al. 2003; Raper et al. 2003).

Regarding safety concerns, non-viral gene delivery vehicles that have the required efficiency and safety for use in human gene therapy are being widely investigated as possible alternatives (Tomlinson and Rolland 1996). Non-viral systems show a significantly lower safety risk and can be tailored to specific therapeutic needs. They are capable of carrying large DNA molecules and can be easily and inexpensively produced in large quantities (Ma and Diamond 2001). Cationic polymers and cationic lipids are by far the most widely used vector in non-viral gene and oligonucleotide delivery.

At least to some extent they can provide protection for nucleic acids toward both extracellular and intracellular degradation during the long journey to the cell nucleus (El Ouahabi, Thiry et al. 1997) as illustrated in Fig. 12.2.

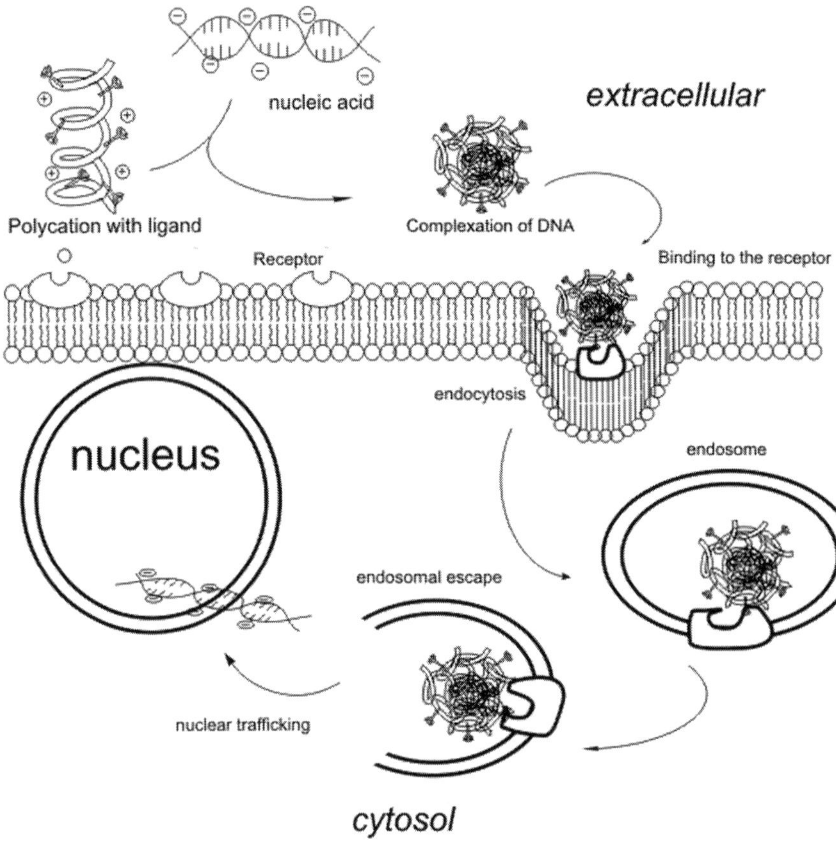

Fig. 12.2 Overview of the processes required for cationic complex-mediated gene delivery

Katayose and Kataoka (Kataoka 1998) have investigated a block copolymer, PEG–PLL (poly(ethylene glycol)–poly(L-lysine)), and have shown that copolymer–DNA complexes are highly resistant to DNase I attack. The stability of complex formation, and therefore the rate of DNA "unpackaging," must influence efficiency of gene expression. Although DNA release from vector complexes is often neglected, recent work by Schaffer, Lauffenburger, and colleagues (Katayose and Kataoka 1997; Schaffer et al. 2000) demonstrates the importance of this step. Maximal gene expression occurs at intermediate stability, because stable complexes restrict DNA transcription and unstable complexes permit rapid DNA degradation.

The group of Klibanov approached the problem of identifying suitable vectors for plasmid DNA delivery using a high-throughput synthesis coupled to combinatorial chemistry approach. Their study is based on the cationic polymer poly(ethylene imine) (PEI). Experimental observations of their group and others indicate that PEI molecular weight is positively correlated with degree of transfection but also with severity of toxicity. Their results show that superior PEI-derivatives could be identified as compared to the presently used "golden standard" 22-kD PEI both with respect to degree of transfection as well as toxicity both in vitro and in vivo (Lampela et al. 2002; Thomas et al. 2005).

Apart from that, there are also other strategies for non-viral delivery including particles bombardment and ultrasound transfection available (Newman et al. 2001) which are, however, not of practical relevance for oral gene delivery. Heading to the final goal of gene therapy polymer-mediated complexation seems to be the most promising strategy, owing to their higher stability compared to lipid- or protein-based systems (Audouy et al. 2000). Non-viral techniques of gene transfer represent a simple and, more importantly, safer alternative to viral vectors. Thanks to their relatively simple quantitative production and their low host immunogenicity, non-viral vectors are attractive tools in gene therapy (Lundstrom and Boulikas T 2003) and likely also for oral gene therapy.

Strategies to overcome the mucus barrier focus mainly on the use of mucolytic agents such as dithiothreitol (DTT) and *N*-acetyl cysteine (NAC). So far, however, results were rather disappointing in improving accessibility to the intestinal crypts for somatic gene therapy (Sandberg et. al. 1994).

The administration of oligonucleotides (ODNs) is currently limited to parenteral routes, mainly intravenous and subcutaneous, because of poor oral bioavailability. Reasons for this poor oral bioavailability are on the one hand their rapid degradation in the GI tract by nucleases and on the other hand their limited intestinal permeability due to unfavorable physicochemical properties (hydrophilicity, large molecular mass and high negative charge density). Only one class of nucleic acids, aptamers, can act extracellularly, which circumvents the need for cell membrane translocation. Conversely, all other classes need to interact with intracellular targets to be active. The problem is most prominent for plasmid DNA, which has the largest size of all proposed nucleic acid therapeutics and also needs to arrive inside the cell nucleus to be effective. Nuclear localization would in principle require passage through the nuclear pore for which the DNA molecule is too large (van der Aa et al. 2006). These qualities at least partly explain why the marketed drugs are an aptamer and an antisense oligonucleotide. In addition, nucleic acids are rapidly cleared from the body, either via glomerular filtration by the kidneys and excretion into the urine or by (scavenger) receptor uptake and intracellular degradation. Therefore, local injection at the site of the pathology is the preferred administration route for the clinically applied oligonucleotides (Mastrobattista et al. 2007).

Nevertheless, the difficult biopharmaceutical characteristics of nucleic acids put a lot of demands on the delivery systems that should compensate for these qualities by increasing stability against the action of nucleases, reducing excretion and uptake by non-target tissues, and promoting target tissue interaction, target cell association, membrane translocation, and correct intracellular trafficking (Mastrobattista et al. 2006).

Strategies to eliminate the problem of degradation by nucleases are based on chemical modification such as phosphorothioates (Kurreck et al. 2002) or methoxyethyl phosphorothioates (Agrawal et al. 1995). Furthermore, the co-administration of permeation enhancers such as medium chain fatty acid turned out to be a promising strategy to increase the gastrointestinal uptake (Raoof et al. 2002).

The literature references sodium caprate (C10) being used as a permeation enhancer in both preclinical (Ishizawa et al. 1987) and clinical studies of Class III compounds (Lindmark et al. 1997). Permeation data from in situ rodent studies calculate tight junctions opening after C10 dosing which is consistent with the cross-sectional diameter of ASOs (Ma et al. 1992; Tsutsumi et al. 2003). The work of Raoof et al. demonstrated in the pig and dog, that the use of permeation enhancers, notably C10, represents an attractive strategy to enhance the oral delivery of ASO molecules (Raoof et al. 2002, 2004). In Fig. 12.3 the pharmacokinetic of an orally administered ASO utilizing sodium caprate as permeation enhancer is illustrated.

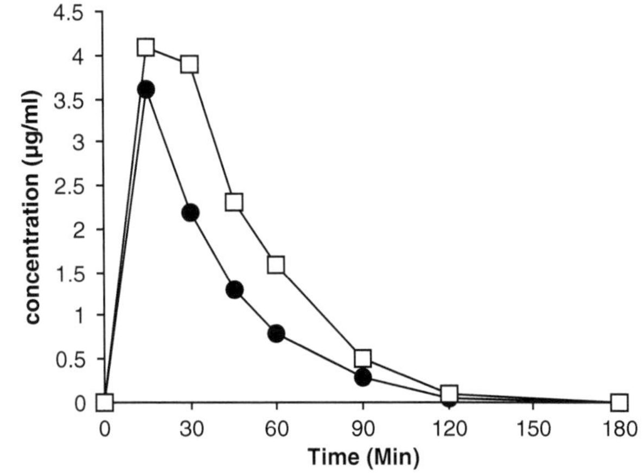

Fig. 12.3 Mean plasma concentration–time profiles following intrajejunal administration of 10 mg/kg of an ASO with 25 (●) and 50 (□) mg/kg of sodium caprate in pigs (mean $n = 8$). Adopted from Raoof et al. (2002)

Another promising strategy in oral nucleic acid delivery is based on thiolated polymers (Bernkop-Schnurch et al. 2001). Thiolated polymers – designated thiomers – are hydrophilic polymers such as polycarbophil (PCP) derivatized with thiol groups. Thiomers show enzyme inhibitory activity toward metalloenzymes like nucleases (Martien et al. 2007) which are abundant in the intestine. They exhibit furthermore the capability of opening tight junctions, which are mainly responsible for a limited paracellular uptake of hydrophilic macromolecules such as nucleic acid drugs, in a very efficient and reversible manner (Bernkop-Schnurch et al. 2003; Martien R 2008). Based on a simple oxidation process disulfide bonds are formed within the thiomer itself, resulting in comparatively higher stability of delivery systems being based on it. Thiomers are relatively low toxic (Guggi et al. 2004). These properties make thiomers a promising tool in order to improve the stability of nucleic acids in the GI tract and in order to improve their uptake.

12.5 Conclusion

The majority of nucleic acid therapeutics are to a higher degree dependent on delivery systems for successful therapeutic intervention than conventional drugs. The transfection efficiency of currently used delivery systems is still too low for many clinical applications. Therefore, new delivery agents are needed that combine a high degree of transfection with acceptable toxicity profiles. Multifunctional polymers such as PEI and thiomers improve stability of nucleic acids against the harsh environment of the gastrointestinal tract, such as enzymatic degradation, salt concentration, and low pH conditions. In addition, auxiliary agents displaying permeation-enhancing properties seem to be essential. Such delivery agents should be well characterized regarding their physicochemical and structural characteristics and should ideally be adapted to the nucleic acid payload to optimally tailor oral delivery needs.

References

Agrawal, S., X. Zhang, Z. Lu, H. Zhao, J. M. Tamburin, J. Yan, H. Cai, R. B. Diasio, I. Habus and Z. Jiang (1995). Absorption, tissue distribution and in vivo stability in rats of a hybrid antisense oligonucleotide following oral administration. Biochem Pharmacol **50**(4): 571–6.

Audouy, S., G. Molema, L. de Leij and D. Hoekstra (2000). Serum as a modulator of lipoplex-mediated gene transfection: dependence of amphiphile, cell type and complex stability. J Gene Med **2**(6): 465–76.

Bernkop-Schnurch, A., C. E. Kast and D. Guggi (2003). Permeation enhancing polymers in oral delivery of hydrophilic macromolecules: thiomer/GSH systems. J Control Release **93**(2): 95–103.

Bernkop-Schnurch, A., A. H. Krauland, V. M. Leitner and T. Palmberger (2004). Thiomers: potential excipients for non-invasive peptide delivery systems. Eur J Pharm Biopharm **58**(2): 253–263.

Bernkop-Schnurch, A., H. Zarti and G. F. Walker (2001). Thiolation of polycarbophil enhances its inhibition of intestinal brush border membrane bound aminopeptidase N. J Pharma Sci **90**(11): 1907–1914.

Borges, O., G. Borchard, J. C. Verhoef, A. de Sousa and H. E. Junginger (2005). Preparation of coated nanoparticles for a new mucosal vaccine delivery system. Int J Pharm **299**(1–2): 155–166.

Crooke, S. T. (1998). An overview of progress in antisense therapeutics. Antisense Nucleic Acid Drug Dev **8**(2): 115–22.

El Ouahabi, A., M. Thiry, V. Pector, R. Fuks, J. M. Ruysschaert and M. Vandenbranden (1997). The role of endosome destabilizing activity in the gene transfer process mediated by cationic lipids. FEBS Lett **414**(2): 187–92.

Elbashir, S.M., J. Harborth, W. Lendeckel, A. Yalcin, K. Weber and T. Tuschl (2001). Duplexes of 21-nucleotide RNAs mediate RNA inerference in cultured mammalian cells. Nature **411**: 494–498.

Ellington, A.D. and J. Szostak (1990). In vitro selection of RNA molecules that bind specific ligands. Nature **346**: 818–822.

Ess, K.C., J.J. Hutton and B.J. Aronow (1994). Double-stranded phosphorothioate oligonucleotide modulation of gene expression. Ann New York Acad Sci **716**: 321–332.

Fasano, A. (1998). Novel approaches for oral delivery of macromolecules. J Pharm Sci **87**(11): 1351–6.

Fire, A., S. Q. Xu, M. K. Montgomery, S. A. Kostas, S. E. Driver and C. C. Mello (1998). Potent and specific genetic interference by double stranded RNA in Caenorhabditis elegans. Nature **391**: 806–811.

Fraunhofer, W., G. Winter and C. Coester (2004). Asymmetrical flow field-flow fractionation and multiangle light scattering for analysis of gelatin nanoparticle drug carrier systems. Anal Chem **76**(7): 1909–20.

Gardlik, R., R. Palffy, J. Hodosy, J. Lukacs, J. Turna and P. Celec (2005). Vectors and delivery systems in gene therapy. Med Sci Monit **11**(4): RA110–21.

Guerrier, T. C., K. Gardiner, T. Marsh, N. Pace and S. Altman (1983). The RNA moiety of ribonuclease P is the catalytic subunit of the enzyme. Cell **35**: 849–857.

Guggi, D., N. Langoth, M. H. Hoffer, M. Wirth and A. Bernkop-Schnurch (2004). Comparative evaluation of cytotoxicity of a glucosamine-TBA conjugate and a chitosan-TBA conjugate. Int J Pharm **278**(2): 353–360.

Hacein-Bey-Abina, S., C. von Kalle, M. Schmidt, F. Le Deist, N. Wulffraat, E. McIntyre, I. Radford, J. L. Villeval, C. C. Fraser, M. Cavazzana-Calvo and A. Fischer (2003). A serious adverse event after successful gene therapy for X-linked severe combined immunodeficiency. N Engl J Med **348**(3): 255–6.

Ishizawa, T., M. Hayashi and S. Awazu (1987). Enhancement of jejunal and colonic absorption of fosfomycin by promoters in the rat. J Pharm Pharmacol **39**: 892–895.

Johnston, S. A., A. M. Talaat and M. J. McGuire (2002). Genetic immunization: what's in a name? Arch Med Res **33**(4): 325–9.

Kai, E. and T. Ochiya (2004). A method for oral DNA delivery with N-acetylated chitosan. Pharmaceutical Res **21**(5): 838–843.

Kataoka, K. (1998). [Intracellular gene delivery by polymer micelle vectors]. Nippon Rinsho **56**(3): 718–23.

Katayose, S. and K. Kataoka (1997). Water-soluble polyion complex associates of DNA and poly(ethylene glycol)-poly(L-lysine) block copolymer. Bioconjug Chem **8**(5): 702–7.

Kurreck, J., E. Wyszko, C. Gillen and V. A. Erdmann (2002). Design of antisense oligonucleotides stabilized by locked nucleic acids. Nucleic Acids Res **30**(9): 1911–8.

Lampela, P., J. Raisanen, P. T. Mannisto, S. Yla-Herttuala and A. Raasmaja (2002). The use of low-molecular-weight PEIs as gene carriers in the monkey fibroblastoma and rabbit smooth muscle cell cultures. J Gene Med **4**: 205–214.

Lindmark, T., J. Soderholm, G. Olaison, G. Alvan, G. Ocklind and P. Artursson (1997). Mechanism of absorption enhancement in humans after rectal administration of ampicillin in suppositories containing sodium caprate. Pharm Res **14**: 930–935.

Loretz, B., F. Foger, M. Werle and A. Bernkop-Schnurch (2006). Oral gene delivery: Strategies to improve stability of pDNA towards intestinal digestion. J Drug Target **14**(5): 311–9.

Lundstrom, K. and T. Boulikas (2003). Viral and non-viral vectors in gene therapy: technology development and clinical trials. Technol Cancer Res Treat **2**: 471–86.

Ma, H. and S. L. Diamond (2001). Nonviral gene therapy and its delivery systems. Curr Pharm Biotechnol **2**(1): 1–17.

Ma, T., D. Hollander, D. Bhalla, H. Nguyen and P. Krugliak (1992). IEC-18, a nontransformed small intestinal cell line for studying epithelial permeability. J Lab Clin Med **120**: 329–341.

Mansouri, S., Y. Cuie, F. Winnik, Q. Shi, P. Lavigne, M. Benderdour, E. Beaumont and J. C. Fernandes (2006). Characterization of folate-chitosan-DNA nanoparticles for gene therapy. Biomaterials **27**(9): 2060–2065.

Martien, R., B. Loretz, A.M. Sandbichler and Bernkop Schnürch A (2008). Thiolated chitosan nanoparticles: transfection study in the Caco-2 differentiated cell culture. Nanotechnology **19**: 045101 (9 pp).

Martien, R., B. Loretz, M. Thaler, S. Majzoob and A. Bernkop-Schnurch (2007). Chitosan-thioglycolic acid conjugate: an alternative carrier for oral nonviral gene delivery? J Biomed Mater Res A **82**(1): 1–9.

Mastrobattista, E., W. E. Hennink and R. M. Schiffelers (2007). Delivery of nucleic acids. Pharm Res **24**(8): 1561–3.

Mastrobattista, E., M. A. van der Aa, W. E. Hennink and D. J. Crommelin (2006). Artificial viruses: a nanotechnological approach to gene delivery. Nat Rev Drug Discov **5**(2): 115–21.

Merdan, T., J. Kopecek and T. Kissel (2002). Prospects for cationic polymers in gene and oligonucleotide therapy against cancer. Adv Drug Deliv Rev **54**(5): 715–58.

Mestecky, J. (1987). The common mucosal immune-system and current strategies for induction of immune-responses in external secretions. J Clin Immunol **7**(4): 265–276.

Newman, C. M., A. Lawrie, A. F. Brisken and D. C. Cumberland (2001). Ultrasound gene therapy: on the road from concept to reality. Echocardiography **18**(4): 339–47.

Patil, S. D., D. G. Rhodes and D. J. Burgess (2005). DNA-based therapeutics and DNA delivery systems: a comprehensive review. Aaps J **7**(1): E61–77.

Raoof, A. A., P. Chiu, Z. Ramtoola, I. K. Cumming, C. Teng, S. P. Weinbach, G. E. Hardee, A. A. Levin and R. S. Geary (2004). Oral bioavailability and multiple dose tolerability of an antisense oligonucleotide tablet formulated with sodium caprate. J Pharm Sci **93**(6): 1431–9.

Raoof, A. A., Z. Ramtoola, B. McKenna, R. Z. Yu, G. Hardee and R. S. Geary (2002). Effect of sodium caprate on the intestinal absorption of two modified antisense oligonucleotides in pigs. Eur J Pharm Sci **17**(3): 131–8.

Raper, S. E., N. Chirmule, F. S. Lee, N. A. Wivel, A. Bagg, G. P. Gao, J. M. Wilson and M. L. Batshaw (2003). Fatal systemic inflammatory response syndrome in a ornithine transcarbamylase deficient patient following adenoviral gene transfer. Mol Genet Metab **80**(1–2): 148–58.

Riordan, M.L and J. C. Martin (1991). Oligonucleotide based therapeutics. Nature **350**: 442–443.

Rubanyi, G. M. (2001). The future of human gene therapy. Mol Aspects Med **22**(3): 113–42.

Sandberg, J. W., C. Lau, M. Jacomino, M. Finegold and S. J. Henning (1994). Improving access to intestinal stem-cells as a step toward intestinal gene-transfer. Human Gene Therapy **5**(3): 323–329.

Santiago, F. S. and L. M. Khachigian (2001). Nucleic acid based strategies as potential therapeutic tools: mechanistic considerations and implications to restenosis. J Mol Med **79**(12): 695–706.

Schaffer, D. V., N. A. Fidelman, N. Dan and D. A. Lauffenburger (2000). Vector unpacking as a potential barrier for receptor-mediated polyplex gene delivery. Biotechnol Bioeng **67**(5): 598–606.

Stein, C. A. (2001). The experimental use of antisense oligonucleotide: a guide for the perplexed. J Clin Invest **108**: 641–644.

Stull, R. A. and F. C. Szoka, Jr. (1995). Antigene, ribozyme and aptamer nucleic acid drugs: progress and prospects. Pharm Res **12**(4): 465–83.

Sweetser, D. A., S. M. Hauft, P. C. Hoppe, E. H. Birkenmeier and J. I. Gordon (1988). Transgenic mice containing intestinal fatty acid-binding protein human growth-hormone fusion genes exhibit correct regional and cell-specific expression of the reporter gene in their small-intestine. Proc Natl Acad Sci USA **85**(24): 9611–9615.

Thomas, M., Q. Ge, J. J. Lu, J. Chen and A. M. Klibanov (2005). Crosslinked small polyethylenimines: while still nontoxic, deliver DNA efficiently to mammalian cells in vitro and in vivo. Pharm. Res **22**: 373–380.

Tomlinson, E. and A. P. Rolland (1996). Controllable gene therapy – pharmaceutics of non-viral gene delivery systems. J Controlled Release **39**(2–3): 357–372.

Tsutsumi, K., S. Li, A. Ghanem, N. Ho and H. WI. (2003). A systematic examination of the in vitro using chamber and the in situ single-pass perfusion model systems in rat ileum permeation of model solutes. J Pharm Sci **92**: 344–359.

Uherek, C. and W. Wels (2000). DNA-carrier proteins for targeted gene delivery. Adv Drug Deliv Rev **44**(2–3): 153–66.

van der Aa, M. A., E. Mastrobattista, R. S. Oosting, W. E. Hennink, G. A. Koning and D. J. Crommelin (2006). The nuclear pore complex: the gateway to successful nonviral gene delivery. Pharm Res **23**(3): 447–59.

Yu, J.Y., S.L. DeRuiter and D.L. Turnner (2002). RNA interference by expression of short interferencing and hair pin RNAs in mammalian cells. Proc Natl Acad Sci **99**: 6047–6052.

Zamecnik, P. C. and M. L. Stephenson (1978). Inhibition of Rous sarcoma virus replication and cell transformation by a specific oligodeoxynucleotide. Proc Natl Acad Sci **75**: 280–284.

Index

A. Bernkop-Schnürch (ed.), *Oral Delivery of Macromolecular Drugs*,
DOI 10.1007/978-1-4419-0200-9, © Springer Science + Business Media, LLC 2009